ELECTRICAL TRANSFORMERS AND ROTATING MACHINES

SECOND EDITION

Stephen L. Herman

THOMSON

DELMAR LEARNING

Australia • Canada • Mexico • Singapore • Spain • United Kingdom • United States

THOMSON

DELMAR LEARNING

Electrical Transformers & Rotating Machines, Second Edition

Stephen L. Herman

Vice President, Technology and Trades SBU:
Alar Elken

Editorial Director:
Sandy Clark

Senior Acquisitions Editor:
Stephen Helba

Development Editor:
Sharon Chambliss

Marketing Director:
David Garza

Channel Manager:
Dennis Williams

Marketing Coordinator:
Stacey Wiktorek

Production Director:
Mary Ellen Black

Senior Production Manager:
Larry Main

Art/Design Coordinator:
Francis Hogan

Senior Editorial Assistant:
Dawn Daugherty

For permission to use material from the text or product, contact us by
Tel.   (800) 730-2214
Fax   (800) 730-2215
www.thomsonrights.com

Library of Congress Cataloging-in-Publication Data:

Herman, Stephen L.
  Electrical transformers and rotating machines / by Stephen L. Herman.—2nd ed.
     p. cm.
  ISBN 1-4018-9942-0
  1. Electric transformers. 2. Electric generators. 3. Electric motors.
I. Title.
  TK2551.H4745 2005
  621.31'4—dc22

2005013648

NOTICE TO THE READER

Contents

Preface

Transformers and Rotating Machines, Second Edition combines theory and practical applications for those desiring employment in the industrial electrical field. This text assumes the student has knowledge of basic electrical theory. The text begins with a study of magnetism and magnetic induction and progresses through single-phase isolation transformers, current transformers, and autotransformers. A unit on three-phase power refreshes the student's knowledge of basic three-phase connections and calculations before proceeding into three-phase transformers. All the basic types of three-phase transformers are covered, such as delta-wye, delta-delta, wye-delta, wye-wye, and open-delta. Special transformer connections such as the Scott, T, and zig-zag are also presented. Examples of adding single-phase loads to three-phase transformers are included.

Transformers and Rotating Machines also provides information on direct current generators and motors. The basic types of DC machines (series, shunt, and compound) are discussed. The text also provides information on brushless motors, printed circuit motors, and permanent magnet motors.

Alternating current machines covered in this text include alternators, three-phase motors, and single-phase motors. The operating characteristics of squirrel cage, consequent pole, wound rotor, and synchronous motors are explained. Diagrams and explanations provide students with thorough coverage of both wye and delta high- and low-voltage connections for three-phase motors.

Single-phase alternating-current motors include split phase, repulsion, universal, and shaded pole. The operating characteristics of each type of motor are discussed.

Transformers and Rotating Machines includes a set of hands-on laboratory experiments for single-phase transformers, three-phase transformers, three-phase motors, and single-phase motors. All the transformer experiments require the use of common equipment such as 0.5 kVA control transformers, 100 watt lamps, voltmeters, ohmmeters, and ammeters.

NEW FOR THE SECOND EDITION

Since motors and transformers are magnetic devices, three units have been added on basic magnetism and magnetic induction. The units covering the installation of transformers and motors have been updated to reflect the changes in the 2005 *National Electric Code (NEC)*®.

ACKNOWLEDGMENTS

The author and Thomson Delmar Learning would like to thank the following reviewers for the comments and suggestions they offered during the development of this project. Our gratitude is extended to:

Robert Coy
Kentucky Technical College
Elizabethtown, KY

Greg Fletcher
Kennebec Valley Technical College
Fairfield, ME

Paul Greenwood
Bellville Area College
Granit City, IL

Douglas Teague
Mississippi Gulf Coast Community College

UNIT 1

Magnetism

Objectives

After studying this unit, you should be able to

- Discuss the properties of permanent magnets.
- Discuss the difference between the axis poles of the earth and the magnetic poles of the earth.
- Discuss the operation of electromagnets.
- Determine the polarity of an electromagnet when the direction of the current is known.
- Discuss the different systems used to measure magnetism.
- Define terms used to describe magnetism and magnetic quantities.

Magnetism is one of the most important phenomena in the study of electricity. It is the force used to produce most of the electrical power in the world. The force of magnetism has been known for over 2000 years. It was first discovered by the Greeks when they noticed that a certain type of stone was attracted to iron. This stone was first found in Magnesia in Asia Minor and was named magnetite. In the Dark Ages, the strange powers of the magnet were believed to be caused by evil spirits or the devil.

THE EARTH IS A MAGNET

The first compass was invented when it was noticed that a piece of magnetite, a type of stone that is attracted to iron, placed on a piece of wood floating in water always aligned itself north and south *(Figure 1-1)*. Because they are always able to align themselves north and south, natural magnets became known as "leading stones" or **lodestones**. The reason that the lodestone aligned itself north and south is because the earth itself contains magnetic poles. *Figure 1-2* illustrates the position of the true North and South poles, or the axis, of the earth and the position of the magnetic poles. Notice

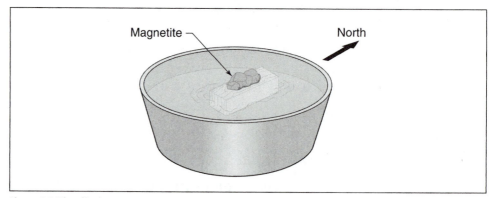

Figure 1-1 The first compass.

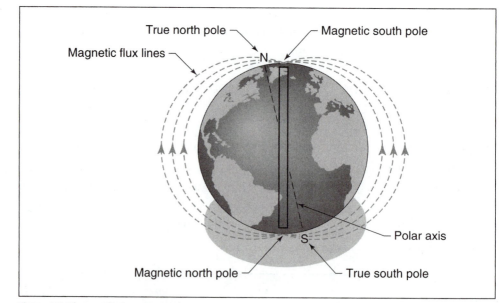

Figure 1-2 The earth is a magnet.

that what is considered as *magnetic* north is not located at the true North Pole of the earth. This is the reason that navigators must distinguish between true north and magnetic north. The angular difference between the two is known as the angle of declination. Although the illustration shows the magnetic lines of force to be only on each side of the earth, the lines actually surround the entire earth like a magnetic shell.

Also notice that the magnetic north pole is located near the southern polar axis and the magnetic south pole is located near the northern polar axis. The reason that the *geographic* poles (axes) are called north and south is because the north pole of a compass needle points in the direction of the north geographic pole. Since unlike magnetic poles attract, the north magnetic pole of the compass needle is attracted to the south magnetic pole of the earth.

PERMANENT MAGNETS

Permanent magnets are magnets that do not require any power or force to maintain their field. They are an excellent example of one of the basic laws of magnetism, which states that **Energy is required to create a magnetic field, but no energy is required to maintain a magnetic field**. Man-made permanent magnets are much stronger and can retain their magnetism longer than natural magnets.

THE ELECTRON THEORY OF MAGNETISM

There are actually only three substances that form natural magnets: iron, nickel, and cobalt. Why these materials form magnets has been the subject of complex scientific investigations, resulting in an explanation of magnetism based on **electron spin patterns**. It is believed that each electron spins on its axis as it orbits around the nucleus of the atom. This spinning motion causes each electron to become a tiny permanent magnet. Although all electrons spin, they do not all spin in the same direction. In most atoms, electrons that spin in opposite directions tend to form pairs *(Figure 1-3)*. Since the electron pairs spin in opposite directions, their magnetic effects cancel each other out as far as having any effect on distant objects. In a similar manner two horseshoe magnets connected together would be strongly attracted to each other, but would have little effect on surrounding objects *(Figure 1-4)*.

An atom of iron contains twenty-six electrons. Of these twenty-six, twenty-two are paired and spin in opposite directions, canceling each other's magnetic effect. In the next-to-the-outermost shell, however, four electrons are not paired and spin in the same direction. These four electrons account for the magnetic properties of iron. At a temperature of 1420°F, or 771.1°C, the electron spin patterns rearrange themselves and iron loses its magnetic properties.

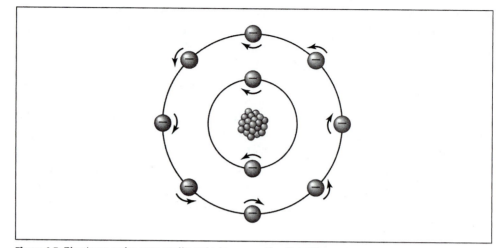

Figure 1-3 Electron pairs generally spin in opposite directions.

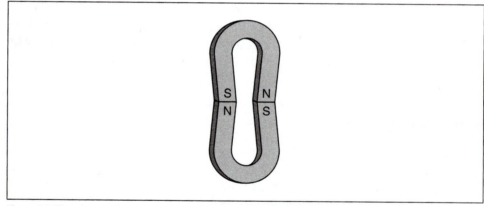

Figure 1-4 Two horseshoe magnets attract each other.

When the atoms of most materials combine to form molecules, they arrange themselves in a manner that produces a total of eight valence electrons. The electrons form a spin pattern that cancels the magnetic field of the material. When the atoms of iron, nickel, and cobalt combine, however, the magnetic field is not canceled. Their electrons combine so that they share valence electrons in such a way that their spin patterns are in the same direction, causing their magnetic fields to add instead of cancel. The additive effect forms regions in the molecular structure of the metal called **magnetic domains** or **magnetic molecules**. These magnetic domains act as small permanent magnets.

A piece of nonmagnetized metal has its molecules in a state of disarray as shown in *Figure 1-5*. When the metal is magnetized, its molecules

Figure 1-5 The molecules are disarrayed in a piece of nonmagnetized metal.

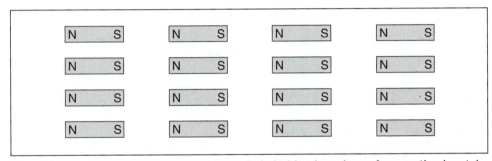

Figure 1-6 The molecules are aligned in an orderly fashion in a piece of magnetized metal.

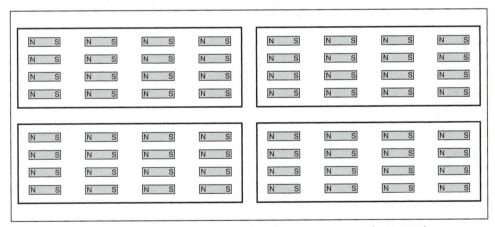

Figure 1-7 When a magnet is cut apart, each piece becomes a separate magnet.

align themselves in an orderly pattern as shown in *Figure 1-6*. In theory, each molecule of a magnetic material is itself a small magnet. If a permanent magnet were cut into pieces, each piece would be a separate magnet (*Figure 1-7*).

MAGNETIC MATERIALS

Magnetic materials can be divided into three basic classifications, as follows:

Ferromagnetic materials are metals that are easily magnetized. Examples of these materials are iron, nickel, cobalt, and manganese.

Paramagnetic materials are metals that can be magnetized, but not as easily as ferromagnetic materials. Some examples of paramagnetic materials are platinum, titanium, and chromium.

Diamagnetic materials are either metal or nonmetal materials that cannot be magnetized. The magnetic lines of force tend to go around them instead of through them. Some examples of these materials are copper, brass, and antimony.

Some of the best materials for the production of permanent magnets are alloys. One of the best permanent magnet materials is Alnico 5, which is made from a combination of aluminum, nickel, cobalt, copper, and iron. Another type of permanent magnet material is made from a combination of barium ferrite and strontium ferrite. Ferrites can have an advantage in some situations because they are insulators and not conductors. They have a resistance of approximately 1,000,000 Ω per centimeter. These two materials can be powdered. The powder is heated to the melting point and then rolled and heat-treated. This treatment changes the grain structure and magnetic properties of the material. The new type of material has a property more like stone than metal and is known as a ceramic magnet. Ceramic magnets can be powdered and mixed with rubber, plastic, or liquids. Ceramic magnetic materials mixed with liquids can be used to make magnetic ink, which is used on checks. Another frequently used magnetic material is iron oxide, which is used to make magnetic recording tape and computer diskettes.

MAGNETIC LINES OF FORCE

flux

Magnetic lines of force are called **flux**. The symbol used to represent flux is the Greek letter phi (Φ). Flux lines can be seen by placing a piece of cardboard on a magnet and sprinkling iron filings on the cardboard. The filings will align themselves in a pattern similar to the one shown in *Figure 1-8*. The pattern produced by the iron filings forms a two-dimensional figure, but the flux lines actually surround the entire magnet (*Figure 1-9*). Magnetic **lines of flux** repel each other and never cross. Although magnetic lines of flux do not flow, it is assumed they are in a direction north to south.

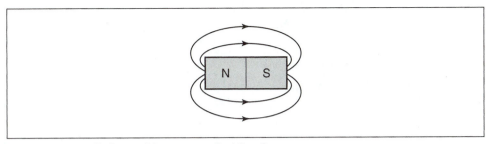

Figure 1-8 Magnetic lines of force are called flux lines.

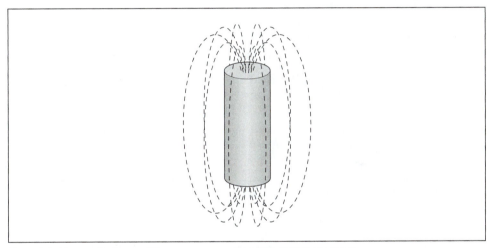

Figure 1-9 Magnetic lines of force surrounding the entire magnet.

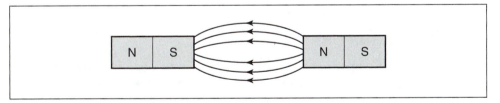

Figure 1-10 Opposite magnetic poles attract each other.

A basic law of magnetism states that **unlike poles attract and like poles repel**. *Figure 1-10* illustrates what happens when a piece of cardboard is placed over two magnets with their north and south poles facing each other and iron filings are sprinkled on the cardboard. The filings form a pattern showing that the magnetic lines of flux are attracted to each other. *Figure 1-11* illustrates the pattern formed by the iron filings when the cardboard is placed over two magnets with like poles facing each other. The filings show that the magnetic lines of flux repel each other.

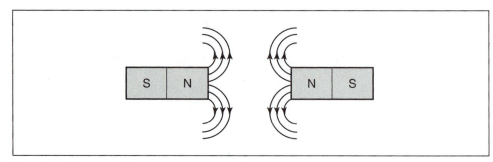

Figure 1-11 Like magnetic poles repel each other.

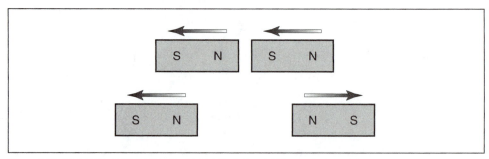

Figure 1-12 Opposite poles of a magnet attract, and like poles repel.

If the opposite poles of two magnets are brought close to each other, they will be attracted to each other as shown in *Figure 1-12*. If like poles of the two magnets are brought together, they will repel each other.

ELECTROMAGNETICS

A basic law of physics states that **whenever an electric current flows through a conductor, a magnetic field is formed around the conductor**. **Electromagnets** depend on electric current flow to produce a magnetic field. They are generally designed to produce a magnetic field only as long as the current is flowing; they do not retain their magnetism when current flow stops. Electromagnets operate on the principle that current flowing through a conductor produces a magnetic field around the conductor *(Figure 1-13)*. If the conductor is wound into a coil as shown in *Figure 1-14*, the magnetic lines of flux add to produce a stronger magnetic field. A coil with ten turns of wire will produce a magnetic field that is ten times as strong as the magnetic field around a single conductor.

Another factor that affects the strength of an electromagnetic field is the amount of current flowing through the wire. An increase in current flow will cause an increase in magnetic field strength. The two factors that determine

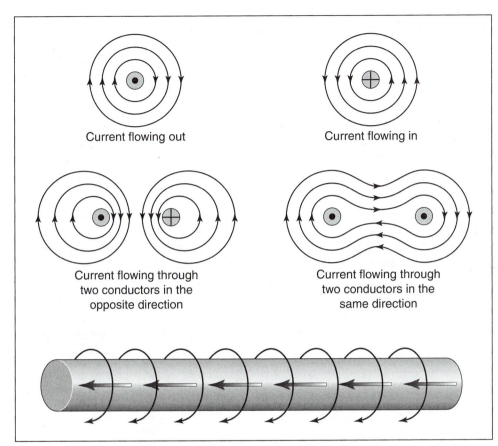

Figure 1-13 Current flowing through a conductor produces a magnetic field around the conductor.

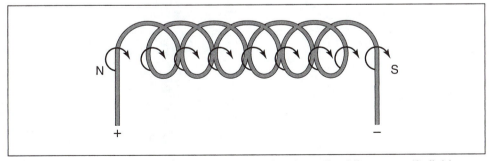

Figure 1-14 Winding the wire into a coil increases the strength of the magnetic field.

the number of flux lines produced by an electromagnet are the number of turns of wire and the amount of current flow through the wire. The strength of an electromagnet is proportional to its **ampere-turns**. Ampere-turns are determined by multiplying the number of turns of wire by the current flow.

ampere-turns

Core Material

Coils can be wound around any type of material to form an electromagnet. The base material is called the core material. When a coil is wound around a nonmagnetic material such as wood or plastic, it is known as an *air-core* magnet. When a coil is wound around a magnetic material such as iron or soft steel, it is known as an *iron-core* magnet. The addition of magnetic material to the center of the coil can greatly increase the strength of the magnet. If the core material causes the magnetic field to become fifty times stronger, the **permeability** core material has a permeability of 50 *(Figure 1-15)*. **Permeability** is a measure of a material's willingness to become magnetized. The number of flux lines produced is proportional to the ampere-turns. The magnetic core material provides an easy path for the flow of magnetic lines in much the same way a conductor provides an easy path for the flow of electrons. This increased permeability permits the flux lines to be concentrated in a smaller area, which increases the number of flux lines per square inch or per square centimeter. In a similar manner, a person using a garden hose with an adjustable nozzle attached can adjust the nozzle to spray the water in a fine mist that covers a large area or in a concentrated stream that covers a small area.

reluctance   Another common magnetic measurement is reluctance. **Reluctance** is resistance to magnetism. A material such as soft iron or steel has a high

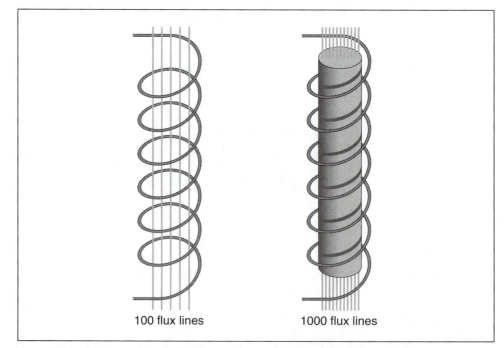

100 flux lines   1000 flux lines

Figure 1-15 An iron core increases the number of flux lines per-square-inch.

permeability and low reluctance because it is easily magnetized. A material such as copper has a low permeability and high reluctance.

If the current flow in an electromagnet is continually increased, the magnet will eventually reach a point at which its strength will increase only slightly with an increase in current. When this condition occurs, the magnetic material is at a point of saturation. **Saturation** occurs when all the molecules of the magnetic material are lined up. Saturation is similar to pouring 5 gallons of water into a 5-gallon bucket. Once the bucket is full, it simply cannot hold any more water. If it became necessary to construct a stronger magnet, a larger piece of core material would be required.

saturation

When the current flow through the coil of a magnet is stopped, there may be some magnetism left in the core material. The amount of magnetism left in a material after the magnetizing force has stopped is called **residual magnetism**. If the residual magnetism of a piece of core material is hard to remove, the material has a high coercive force. **Coercive force** is a measure of a material's ability to retain magnetism. A high coercive force is desirable in materials that are intended to be used as permanent magnets. A low coercive force is generally desirable for materials intended to be used as electromagnets. Coercive force is measured by determining the amount of current flow through the coil in the direction opposite to that required to remove the residual magnetism. Another term that is used to describe a material's ability to retain magnetism is **retentivity**.

residual magnetism

retentivity

MAGNETIC MEASUREMENT

The terms used to measure the strength of a magnetic field are determined by the system that is being used. There are three different systems used to measure magnetism: the English system, the CGS system, and the MKS system.

The English System

In the English system of measure, magnetic strength is measured in a term called flux density. **Flux density** is measured in lines per square inch. The Greek letter phi (Φ) is used to measure flux. The letter B is used to represent flux density. The following formula is used to determine flux density.

flux density

$$B \text{ (flux density)} = \frac{\Phi \text{ (flux lines)}}{A \text{ (area)}}$$

In the English system, the term used to describe the total force producing a magnetic field, or flux, is **magnetomotive force (mmf)**. Magnetomotive force can be computed using the formula:

$$mmf = \Phi \times rel \text{ (reluctance)}$$

The formula shown below can be used to determine the strength of the magnet.

$$\text{Pull (in pounds)} = \frac{B \times A}{72{,}000{,}000}$$

where B = flux density in lines per square inch
A = area of the magnet

The CGS System

In the CGS (centimeter-gram-second) system of measurement, one magnetic line of force is known as a **maxwell**. A **gauss** represents a magnetic force of one maxwell per square centimeter. In the English system, magnetomotive force is measured in ampere-turns. In the CGS system, **gilberts** are used to represent the same measurement. Since the main difference between these two systems of measurement is that one uses English units of measure and the other uses metric units of measure, a conversion factor can be used to help convert one set of units to the other.

maxwell
gauss
gilbert

$$\text{1 gilbert} = 1.256 \text{ ampere-turns}$$

The MKS System

The MKS (meter-kilogram-second) system uses metric units of measure also. In this system, the main unit of magnetic measurement is the **dyne**. The dyne is a very weak amount of force. One dyne is equal to 1/27,800 of an ounce, or it requires 27,800 dynes to equal a force of one ounce. In the MKS system, a standard called the **unit magnetic pole** is used. In *Figure 1-16*, two magnets are separated by a distance of 1 cm. These magnets repel each other with a force of 1 dyne. When two magnets separated by a distance of 1 cm exert a force on each other of 1 dyne, they are considered to be a unit magnetic pole. Magnetic force can then be determined using the formula

$$\text{Force (in dynes)} = \frac{M_1 \times M_2}{D}$$

where M_1 = strength of first magnet, in unit magnetic poles
M_2 = strength of second magnet, in unit magnetic poles
D = distance between the poles, in centimeters

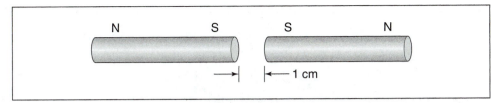

Figure 1-16 A unit magnetic pole produces a force of 1 dyne.

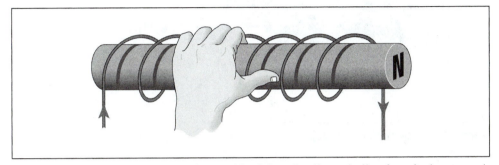

Figure 1-17 The "left-hand rule" can be used to determine the polarity of an electromagnet.

MAGNETIC POLARITY

The polarity of an electromagnet can be determined using the **left-hand rule**. When the fingers of the left hand are placed around the windings in the direction of electron current flow, the thumb will point to the north magnetic pole *(Figure 1-17)*. If the direction of current flow is reversed, the polarity of the magnetic field will reverse also.

DEMAGNETIZING

When an object is to be **demagnetized**, its molecules must be disarranged as they are in a nonmagnetized material. This can be done by placing the object in the field of a strong electromagnet connected to an alternating current (AC) line. Since the magnet is connected to AC current, the polarity of the magnetic field reverses each time the current changes direction. The molecules of the object to be demagnetized are, therefore, aligned first in one direction and then in the other. If the object is pulled away from the AC magnetic field, the effect of the field becomes weaker as the object is moved farther away *(Figure 1-18)*. The weakening of the magnetic field causes the molecules of the object to be left in a state of disarray. The ease or difficulty with which an object can be demagnetized depends on the strength of the AC magnetic field and the coercive force of the object.

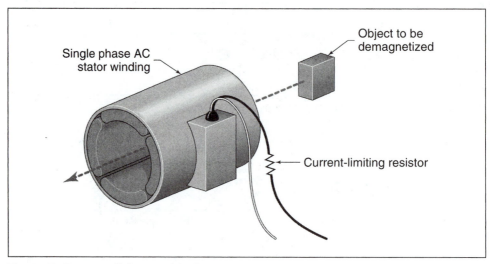

Figure 1-18 Demagnetizing an object.

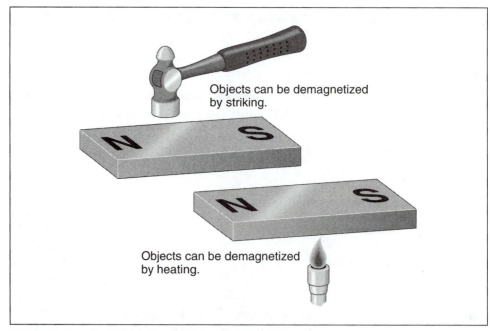

Figure 1-19 Other methods for demagnetizing objects.

An object can be demagnetized in two other ways *(Figure 1-19)*. If a magnetized object is struck, the vibration will often cause the molecules to rearrange themselves in a disordered fashion. It may be necessary to strike the object several times. Heating also will demagnetize an object. When the temperature becomes high enough, the molecules will rearrange themselves in a disordered fashion.

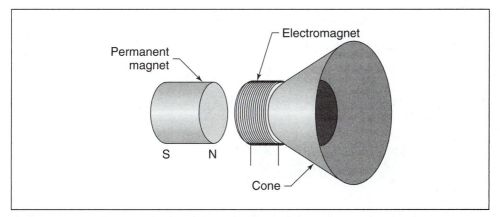

Figure 1-20 A speaker uses both an electromagnet and a permanent magnet.

MAGNETIC DEVICES

A list of devices that operate on magnetism would be very long indeed. Some of the more common devices are electromagnets, measuring instruments, inductors, transformers, and motors.

The Speaker

The speaker is a common device that operates on the principle of magnetism *(Figure 1-20)*. The speaker produces sound by moving a cone; the movement causes a displacement of air. The tone is determined by how fast the cone vibrates. Low or bass sounds are produced by vibrations in the range of 20 cycles per second. High sounds are produced when the speaker vibrates in the range of 20,000 cycles per second.

The speaker uses two separate magnets. One is a permanent magnet, and the other is an electromagnet. The permanent magnet is held stationary, and the electromagnet is attached to the speaker cone. When current flows through the coil of the electromagnet, a magnetic field is produced. The polarity of the field is determined by the direction of current flow. When the electromagnet has a north polarity, it is repelled away from the permanent magnet, causing the speaker cone to move outward and displace air. When the current flow reverses through the coil, the electromagnet has a south polarity and is attracted to the permanent magnet. The speaker cone then moves inward and again displaces air. The number of times per second that the current through the coil reverses determines the tone of the speaker.

Summary

1. Early natural magnets were known as lodestones.

2. The earth has a north and a south magnetic pole.

3. The magnetic poles of the earth and the axis poles are not the same.

4. Like poles of a magnet repel each other, and unlike poles attract each other.

5. Some materials have the ability to become better magnets than others.

6. Three basic types of magnetic material are:
   A. Ferromagnetic
   B. Paramagnetic
   C. Diamagnetic

7. When current flows through a wire, a magnetic field is created around the wire.

8. The direction of current flow through the wire determines the polarity of the magnetic field.

9. The strength of an electromagnet is determined by the ampere-turns.

10. The type of core material used in an electromagnet can increase its strength.

11. Three different systems are used to measure magnetic values:
    A. The English system
    B. The CGS system
    C. The MKS system

12. An object can be demagnetized by placing it in an AC magnetic field and pulling it away, by striking, and by heating.

Review Questions

1. Is the north magnetic pole of the earth a north polarity or a south polarity?

2. What were early natural magnets known as?

3. The south pole of one magnet is brought close to the south pole of another magnet. Will the magnets repel or attract each other?

4. How can the polarity of an electromagnet be determined if the direction of current flow is known?

5. Define the following terms:
   Flux density
   Permeability
   Reluctance
   Saturation
   Coercive force
   Residual magnetism

6. A force of 1 ounce is equal to how many dynes?

UNIT 2

Magnetic Induction

Objectives

After studying this unit, you should be able to:

- Discuss magnetic induction.
- List factors that determine the amount and polarity of an induced voltage.
- Discuss Lenz's law.
- Discuss an exponential curve.
- List devices used to help prevent inductive voltage spikes.

Magnetic induction is one of the most important concepts in the electrical field. It is the basic operating principle underlying alternators, transformers, and most alternating-current motors. It is imperative that anyone desiring to work in the electrical field have an understanding of the principles involved.

MAGNETIC INDUCTION

In Unit 1, it was stated that one of the basic laws of electricity is that whenever current flows through a conductor, a magnetic field is created around the conductor *(Figure 2-1)*. The direction of the current flow determines the polarity of the magnetic field, and the amount of current determines the strength of the magnetic field.

17

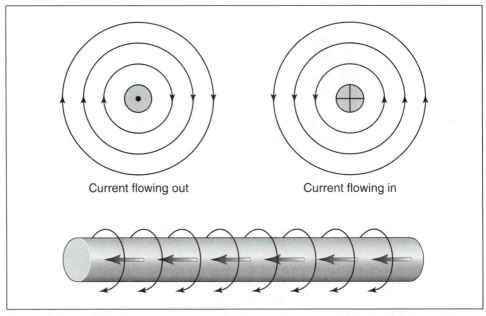

Current flowing out   Current flowing in

Figure 2-1 Current flowing through a conductor produces a magnetic field around the conductor.

That basic law in reverse is the principle of **magnetic induction**, which states that **whenever a conductor cuts through magnetic lines of flux, a voltage is induced into the conductor**. The conductor in *Figure 2-2* is connected to a zero-center microammeter, creating a complete circuit. When the conductor is moved downward through the magnetic lines of flux, the induced voltage will cause electrons to flow in the direction indicated by the arrows. This flow of electrons causes the pointer of the meter to be deflected from the center-zero position.

If the conductor is moved upward, the polarity of induced voltage will be reversed and the current will flow in the opposite direction *(Figure 2-3)*. The pointer will be deflected in the opposite direction.

The polarity of the induced voltage can also be changed by reversing the polarity of the magnetic field *(Figure 2-4)*. In this example, the conductor is again moved downward through the lines of flux, but the polarity of the magnetic field has been reversed. Therefore the polarity of the induced voltage will be the opposite of that in *Figure 2-2*, and the pointer of the meter will be deflected in the opposite direction. It can be concluded that **the polarity of the induced voltage is determined by the polarity of the magnetic field in relation to the direction of movement**.

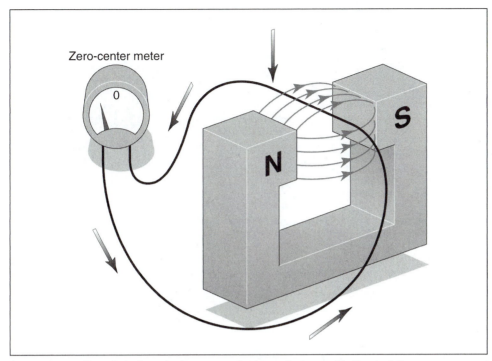

Figure 2-2 A voltage is induced when a conductor cuts magnetic lines of flux.

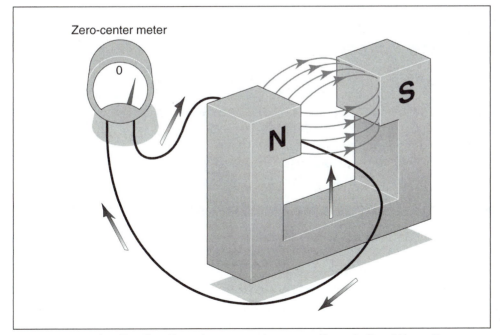

Figure 2-3 Reversing the direction of movement reverses the polarity of the voltage.

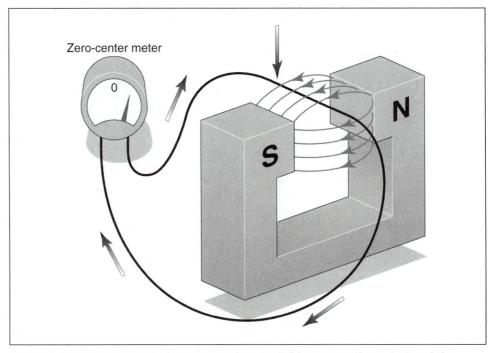

Figure 2-4 Reversing the polarity of the magnetic field reverses the polarity of the voltage.

MOVING MAGNETIC FIELDS

The important factors concerning magnetic induction are a conductor, a magnetic field, and movement. In practice, it is often desirable to move the magnet instead of the conductor. Most alternating current generators or alternators operate on this principle. In *Figure 2-5*, a coil of wire is held stationary while a magnet is moved through the coil. As the magnet is moved, the lines of flux cut through the windings of the coil and induce a voltage into them.

DETERMINING THE AMOUNT OF INDUCED VOLTAGE

Three factors determine the amount of voltage that will be induced in a conductor:
1. **the number of turns of wire**,
2. **the strength of the magnetic field** (flux density), and
3. **the speed of the cutting action**.

In order to induce 1 V in a conductor, the conductor must cut 100,000,000 lines of magnetic flux in 1 s. In magnetic measurement, 100,000,000 lines of

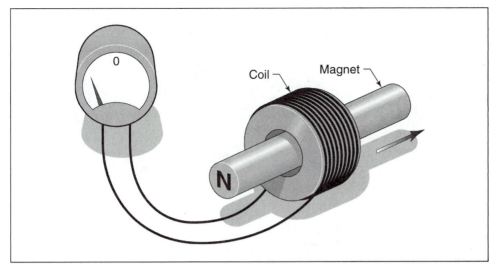

Figure 2-5 Voltage is induced by a moving magnetic field.

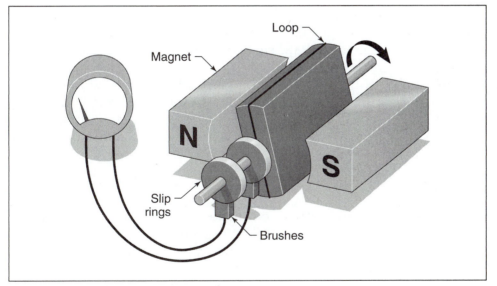

Figure 2-6 A single-loop generator.

flux are equal to one **weber (Wb)**. Therefore, if a conductor cuts magnetic **weber (Wb)** lines of flux at a rate of 1 Wb/s, a voltage of 1 V will be induced. A simple one-loop generator is shown in *Figure 2-6*. The loop is attached to a rod that is free to rotate. This assembly is suspended between the poles of two stationary magnets. If the loop is turned, the conductor cuts through magnetic lines of flux and a voltage is induced into the conductor.

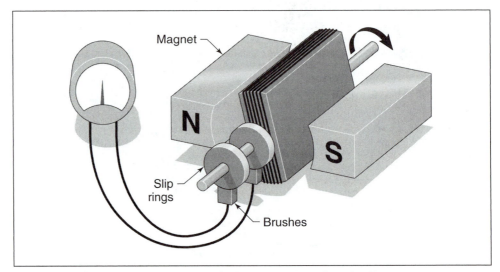

Figure 2-7 Increasing the number of turns increases the induced voltage.

If the speed of rotation is increased, the conductor cuts more lines of flux per second, and the amount of induced voltage increases. If the speed of rotation remains constant and the strength of the magnetic field is increased, there will be more lines of flux per square inch. When there are more lines of flux, the number of lines cut per second increases and the induced voltage increases. If more turns of wire are added to the loop *(Figure 2-7)*, more flux lines are cut per second and the amount of induced voltage increases again. Adding more turns has the effect of connecting single conductors in series, and the amount of induced voltage in each conductor adds.

LENZ'S LAW

When a voltage is induced in a coil and there is a complete circuit, current will flow through the coil *(Figure 2-8)*. When current flows through the coil, a magnetic field is created around the coil. This magnetic field develops a polarity opposite that of the moving magnet. The magnetic field developed by the induced current acts to attract the moving magnet and pull it back inside the coil.

If the direction of motion is reversed, the polarity of the induced current is reversed, and the magnetic field created by the induced current again opposes the motion of the magnet. This principle was first noticed by Heinrich Lenz many years ago and is summarized in **Lenz's law**, which states that **an induced voltage or current opposes the motion that causes it**. From this basic principle, other laws concerning inductors have been developed. One is that **inductors always oppose a change of current**. The coil in *Figure 2-9,*

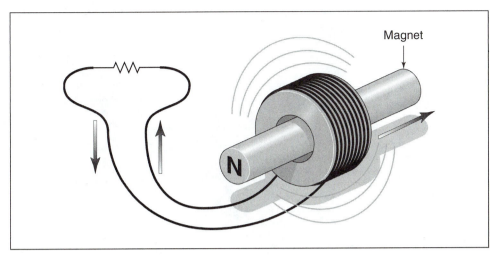

Figure 2-8 An induced current produces a magnetic field around the coil.

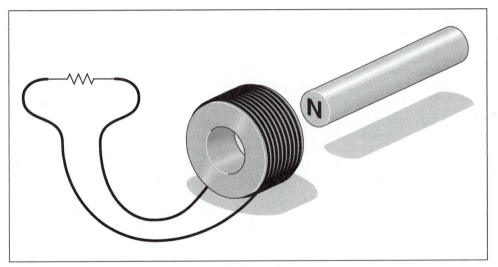

Figure 2-9 There is no current flow through the coils.

for example, has no induced voltage and therefore no induced current. If the magnet is moved toward the coil, however, magnetic lines of flux will begin to cut the conductors of the coil, and a current will be induced in the coil. The induced current causes magnetic lines of flux to expand outward around the coil *(Figure 2-10)*. As this expanding magnetic field cuts through the conductors of the coil, a voltage is induced in the coil. The polarity of the voltage is such that it opposes the induced current caused by the moving magnet.

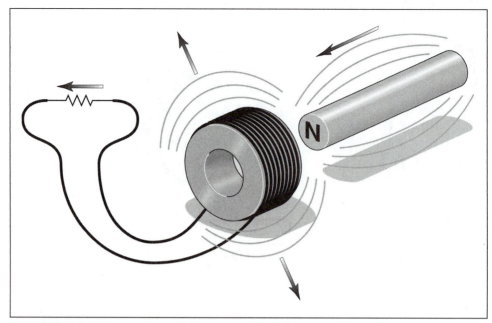

Figure 2-10 Induced current produces a magnetic field around the coil.

If the magnet is moved away, the magnetic field around the coil will collapse and induce a voltage in the coil *(Figure 2-11)*. Since the direction of movement of the collapsing field has been reversed, the induced voltage will be opposite in polarity, forcing the current to flow in the same direction.

RISE TIME OF CURRENT IN AN INDUCTOR

When a resistive load is suddenly connected to a source of direct current *(Figure 2-12)*, the current will instantly rise to its maximum value. The resistor shown in *Figure 2-12* has a value of 10 Ω and is connected to a 20-V source. When the switch is closed the current will instantly rise to a value of 2 A (20 V/10 Ω = 2 A).

If the resistor is replaced with an inductor that has a wire resistance of 10 Ω and the switch is closed, the current cannot instantly rise to its maximum value of 2 A *(Figure 2-13)*. As current begins to flow through an inductor, the expanding magnetic field cuts through the conductors, inducing a voltage into them. In accord with Lenz's law, the induced voltage is opposite in polarity to the applied voltage. The induced voltage, therefore, acts as a resistance to hinder the flow of current through the inductor *(Figure 2-14)*.

The induced voltage is proportional to the rate of change of current (speed of the cutting action). When the switch is first closed, current flow

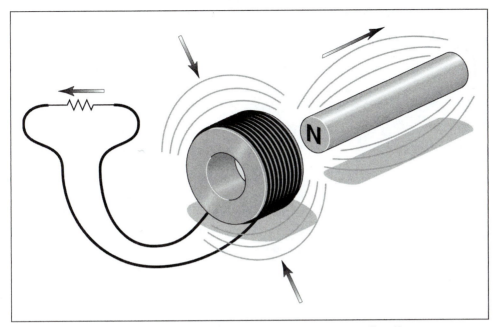

Figure 2-11 The induced voltage forces current to flow in the same direction.

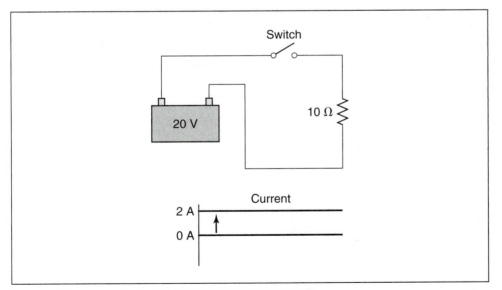

Figure 2-12 The current rises instantly in a resistive circuit.

through the coil tries to rise instantly. This extremely fast rate of current change induces maximum voltage in the coil. As the current flow approaches its maximum Ohm's law value, 2 A in this example, the rate of change becomes less and the amount of induced voltage decreases.

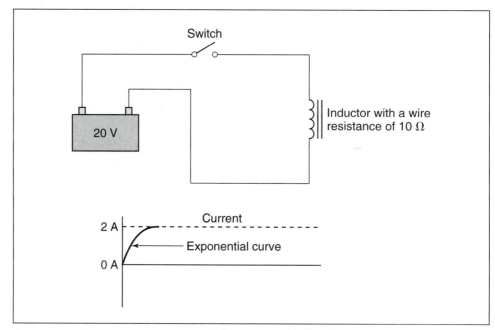

Figure 2-13 Current rises through an indicator at an exponential rate.

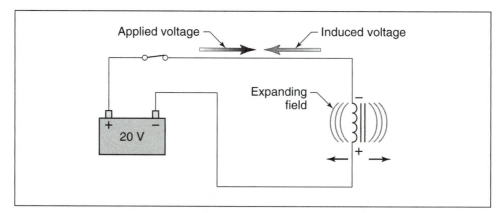

Figure 2-14 The induced voltage is opposite in polarity to the applied voltage.

THE EXPONENTIAL CURVE

The **exponential curve** describes a rate of certain occurrences. The curve is divided into five time constants. Each time constant is equal to 63.2% of some value. An exponential curve is shown in *Figure 2-15*. In this example, current must rise from zero to a value of 1.5 A in 100 ms at an exponential

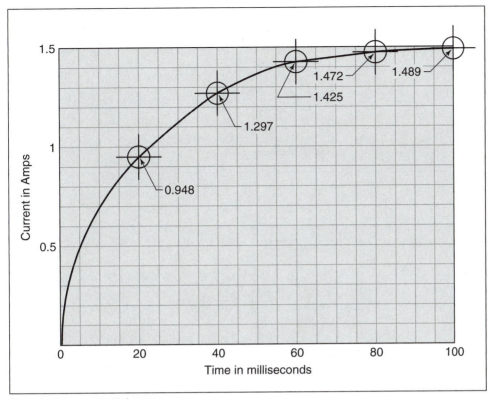

Figure 2-15 Exponential curve.

rate. Since the current requires a total of 100 ms to rise to its full value, each time constant is 20 ms (100 ms/5 time constants = 20 ms per time constant). During the first time constant, the current will rise from 0 to 63.2% of its total value, or 0.948 A (1.5 x 0.632 = 0.948). During the second time constant the current will rise to a value of 1.297 A, and during the third time constant the current will reach a total value of 1.425 A.

Because the current increases at a rate of 63.2% during each time constant, it is theoretically impossible to reach the total value of 1.5 A. After five time constants, however, the current has reached approximately 99.3% of the maximum value and for all practical purposes is considered to be complete.

The exponential curve can often be found in nature. If clothes are hung on a line to dry, they will dry at an exponential rate. Another example of the exponential curve can be seen in *Figure 2-16*. In this example, a bucket has been filled to a certain mark with water. A hole has been cut at the bottom of the bucket and a stopper placed in the hole. When the stopper is removed from the bucket, water will flow out at an exponential rate. Assume, for example, it takes 5 min for the water to flow out of the bucket. Exponential

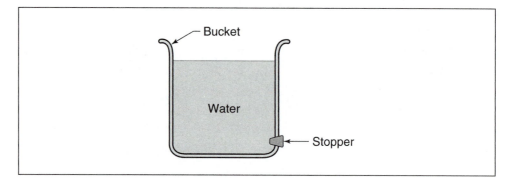

Figure 2-16 Exponential curves can be found in nature.

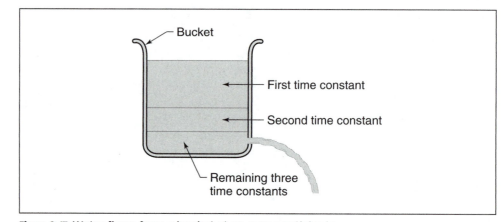

Figure 2-17 Water flows from a bucket at an exponential rate.

curves are always divided into five time constants, so in this case each time constant has a value of 1 min. In *Figure 2-17*, if the stopper is removed and water is permitted to drain from the bucket for a period of 1 min before the stopper is replaced, during that first time constant 63.2% of the water in the bucket will drain out. If the stopper is again removed for a period of 1 min, 63.2% of the water remaining in the bucket will drain out. Each time the stopper is removed for a period of one time constant, the bucket will lose 63.2% of its remaining water.

INDUCTANCE

henry (H)

Inductance is measured in units called the **henry (H)** and is represented by the letter *L*. **A coil has an inductance of one henry when a current change of one ampere per second results in an induced voltage of one volt.**

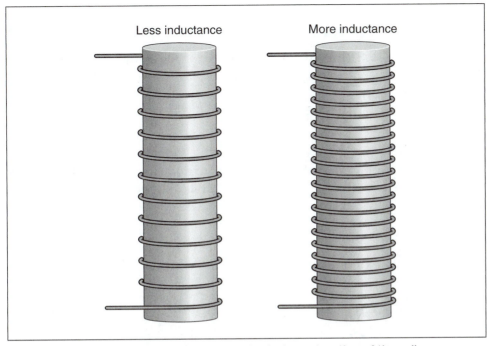

Less inductance   More inductance

Figure 2-18 Inductance is determined by the physical construction of the coil.

The amount of inductance a coil will have is determined by its physical properties and construction. A coil wound on a nonmagnetic core material such as wood or plastic is referred to as an **air-core** inductor. If the coil is wound on a core made of magnetic material such as silicon steel or soft iron it is referred to as an **iron-core** inductor. Iron-core inductors produce more inductance with fewer turns than air-core inductors because of the good magnetic path provided by the core material. Iron-core inductors cannot be used for high-frequency applications, however, because of **eddy current** loss and **hysteresis loss** in the core material.

Another factor that determines inductance is how far the windings are separated from each other. If the turns of wire are far apart they will have less inductance than turns wound closer together (*Figure 2-18*).

eddy current

hysteresis loss

R-L Time Constants

The time necessary for current in an inductor to reach its full Ohm's law value, called the **R-L time constant**, can be computed using the formula

$$T = \frac{L}{R}$$

where

$$T = \text{time in seconds}$$
$$L = \text{inductance in henrys}$$
$$R = \text{resistance in ohms}$$

This formula computes the time of one time constant.

Example 1

A coil has an inductance of 1.5 H and a wire resistance of 6 Ω. If the coil is connected to a battery of 3 V, how long will it take the current to reach its full Ohm's law value of 0.5 A (3 V/6 Ω = 0.5 A)?

Solution

To find the time of one time constant, use the formula

$$T = \frac{L}{R}$$
$$T = \frac{1.5}{6}$$
$$T = 0.25 \text{ s}$$

The time for one time constant is 0.25 s. Since five time constants are required for the current to reach its full value of 0.5 A, 0.25 s will be multiplied by 5.

$$0.25 \times 5 = 1.25 \text{ s}$$

INDUCED VOLTAGE SPIKES

A **voltage spike** occurs when the current flow through an inductor stops, and the current decreases at an exponential rate also *(Figure 2-19)*. As long as a complete circuit exists when the power is interrupted, there is little or no problem. In the circuit shown in *Figure 2-20*, a resistor and inductor are connected in parallel. When the switch is closed, the battery will supply current to both. When the switch is opened, the magnetic field surrounding the inductor will collapse and induce a voltage into the inductor. The induced voltage will attempt to keep current flowing in the same direction. Recall that inductors oppose a change of current. The amount of current flow and the time necessary for the flow to stop will be determined by the resistor and the properties of the inductor. The amount of

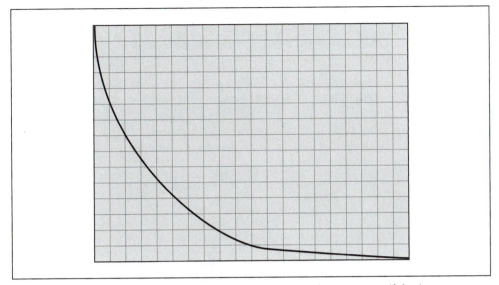

Figure 2-19 Current flow through an inductor decreases at an exponential rate.

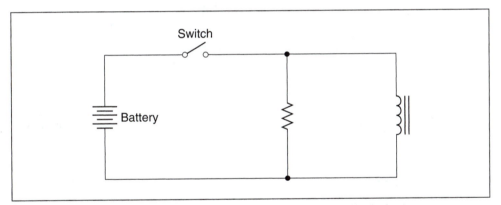

Figure 2-20 The resistor helps prevent voltage spikes caused by the inductor.

voltage produced by the collapsing magnetic field is determined by the maximum current in the circuit and the total resistance in the circuit. In the circuit shown in *Figure 2-20*, assume that the inductor has a wire resistance of 6 Ω, and the resistor has a resistance of 100 Ω. Also assume that when the switch is closed a current of 2 A will flow through the inductor.

When the switch is opened, a series circuit exists composed of the resistor and inductor *(Figure 2-21)*. The maximum voltage developed in this circuit would be 212 V (2 A x 106 Ω = 212 V). If the circuit resistance were increased, the induced voltage would become greater. If the circuit resistance were decreased, the induced voltage would become less.

Another device often used to prevent induced voltage spikes when the current flow through an inductor is stopped is the diode *(Figure 2-22)*. The diode is an electronic component that operates as an electrical check valve. The diode will permit current to flow through it in only one direction. The diode is connected in parallel with the inductor in such a manner that when voltage is applied to the circuit, the diode is reverse-biased and acts as an open switch. When the diode is reverse-biased no current will flow through it.

When the switch is opened, the induced voltage produced by the collapsing magnetic field will be opposite in polarity to the applied voltage. The

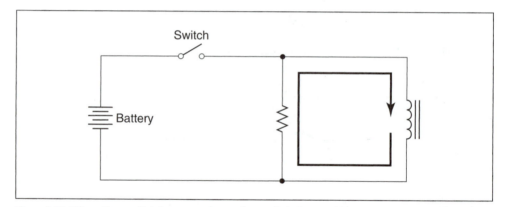

Figure 2-21 When the switch is opened, a series path is formed by the resistor and inductor.

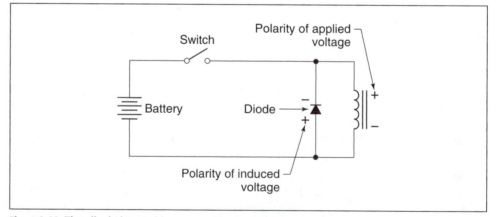

Figure 2-22 The diode is used to prevent induced voltage spikes.

diode then becomes forward-biased and acts as a closed switch. Current can now flow through the diode and complete a circuit back to the inductor. A silicon diode has a forward voltage drop of approximately 0.7 V regardless of the current flowing through it. Since the diode is connected in parallel with the inductor, and voltage drops of devices connected in parallel must be the same, the induced voltage is limited to approximately 0.7 V. The diode can be used to eliminate inductive voltage spikes in direct-current circuits only; it cannot be used for this purpose in alternating-current circuits.

A device that can be used for spike suppression in either direct- or alternating-current circuits is the **metal oxide varistor (MOV)**. The MOV is a bidirectional device, which means that it will conduct current in either direction, and can therefore be used in alternating-current circuits. The metal oxide varistor is an extremely fast-acting solid-state component that will exhibit a change of resistance when the voltage reaches a certain point. Assume that the MOV shown in *Figure 2-23* has a voltage rating of 140 V, and that the voltage applied to the circuit is 120 V. When the switch is closed and current flows through the circuit, a magnetic field will be established around the inductor *(Figure 2-24)*. As long as the voltage applied to the MOV is less than 140 V, it will exhibit an extremely high resistance, in the range of several hundred thousand ohms.

When the switch is opened, current flow through the coil suddenly stops, and the magnetic field collapses. This sudden collapse of the magnetic field will cause an extremely high voltage to be induced in the coil. When this induced voltage reaches 140 V, however, the MOV will suddenly change from a high resistance to a low resistance, preventing the voltage from becoming greater than 140 V *(Figure 2-25)*.

Metal oxide varistors are extremely fast-acting. They can typically change resistance values in less than 20 ns (nanoseconds). They are often found

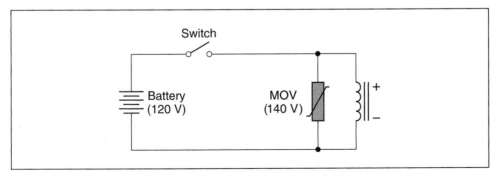

Figure 2-23 Metal oxide varistor used to suppress a voltage spike.

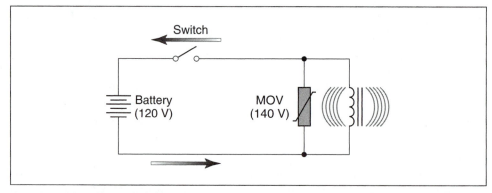

Figure 2-24 When the switch is closed, a magnetic field is established around the inductor.

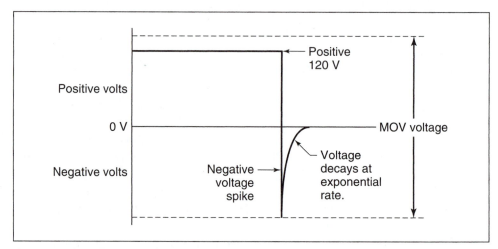

Figure 2-25 The MOV prevents the spike from becoming too high.

connected across the coils of relays and motor starters in control systems to prevent voltage spikes from being induced back into the line. They are also found in the surge protectors used to protect many home appliances such as televisions, stereos, and computers.

If nothing is connected in the circuit with the inductor when the switch opens, the induced voltage can become extremely high. In this instance, the resistance of the circuit is the air gap of the switch contacts, which is practically infinite. The inductor will attempt to produce any voltage necessary to prevent a change of current. Inductive voltage spikes can reach thousands of volts. This is the principle of operation of many high-voltage devices such as the ignition systems of many automobiles.

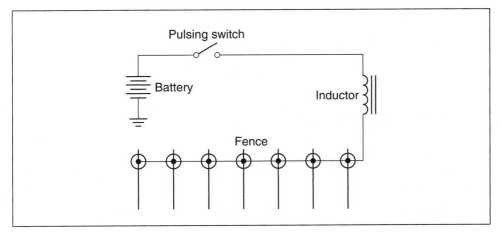

Figure 2-26 An inductor is used to produce a high voltage for an electric fence.

Another device that uses the collapsing magnetic field of an inductor to produce a high voltage is the electric-fence charger, shown in *Figure 2-26.* The switch is constructed in such a manner that it will pulse on and off. When the switch closes, current flows through the inductor, and a magnetic field is produced around the inductor. When the switch opens, the magnetic field collapses and induces a high voltage across the inductor. If anything or anyone standing on the ground touches the fence, a circuit is completed through the object or person and the ground. The coil is generally constructed of many turns of very small wire. This construction provides the coil with a high resistance and limits current flow when the field collapses.

Summary

1. When current flows through a conductor, a magnetic field is created around the conductor.

2. When a conductor is cut by a magnetic field, a voltage is induced in the conductor.

3. The polarity of the induced voltage is determined by the polarity of the magnetic field in relation to the direction of motion.

4. Three factors that determine the amount of induced voltage are:
   a. The number of turns of wire,
   b. The strength of the magnetic field, and
   c. The speed of the cutting action.

5. One volt is induced in a conductor when magnetic lines of flux are cut at a rate of one weber per second.

6. Induced voltage is always opposite in polarity to the applied voltage.

7. Inductors oppose a change of current.

8. Current rises in an inductor at an exponential rate.

9. An exponential curve is divided into five time constants.

10. Each time constant is equal to 63.2% of some value.

11. Inductance is measured in units called henrys (H).

12. A coil has an inductance of 1 H when a current change of 1 A per second results in an induced voltage of 1 V.

13. Air-core inductors are inductors wound on cores of nonmagnetic material.

14. Iron-core inductors are wound on cores of magnetic material.

15. The amount of inductance an inductor will have is determined by the number of turns of wire and the physical construction of the coil.

16. Inductors can produce extremely high voltages when the current flowing through them is stopped.

17. Two devices used to help prevent large spike voltages are the resistor and diode.

Review Questions

1. What determines the polarity of magnetism when current flows through a conductor?

2. What determines the strength of the magnetic field when current flows through a conductor?

3. Name three factors that determine the amount of induced voltage in a coil.

4. How many lines of magnetic flux must be cut in 1 s to induce a voltage of 1 V?

5. What is the effect on induced voltage of adding more turns of wire to a coil?

6. Into how many time constants is an exponential curve divided?

7. Each time constant of an exponential curve is equal to what percentage of the whole?

8. An inductor has an inductance of 0.025 H and a wire resistance of 3 Ω. How long will it take the current to reach its full Ohm's law value?

9. Refer to the circuit shown in *Figure 2-20*. Assume that the inductor has a wire resistance of 0.2 Ω and the resistor has a value of 250 Ω. If a current of 3 A is flowing through the inductor, what will be the maximum induced voltage when the switch is opened?

10. What electronic component is often used to prevent large voltage spikes from being produced when the current flow through an inductor is suddenly terminated?

UNIT 3

Inductance in Alternating-Current Circuits

Objectives

After studying this unit, you should be able to:

- Discuss the properties of inductance in an alternating current circuit.
- Discuss inductive reactance.
- Compute values of inductive reactance and inductance.
- Discuss the relationship of voltage and current in a pure inductive circuit.
- Be able to compute values for inductors connected in series or parallel.
- Discuss reactive power (VARs).
- Determine the Q of a coil.

This unit discusses the effects of inductance on alternating-current circuits. The unit explains how current is limited in an inductive circuit as well as the effect inductance has on the relationship of voltage and current.

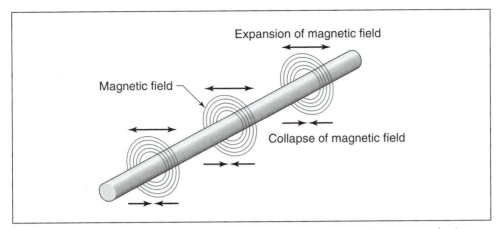

Figure 3-1 A continually changing magnetic field induces a voltage into any conductor.

INDUCTANCE

Inductance (L) is one of the primary types of loads in alternating-current circuits. Some amount of inductance is present in all alternating-current circuits because of the continually changing magnetic field *(Figure 3-1)*. The amount of inductance of a single conductor is extremely small, and in most instances it is not considered in circuit calculations. Circuits are generally considered to contain inductance when any type of load that contains a coil is used. For circuits that contain a coil, inductance *is* considered in circuit calculations. Loads such as motors, transformers, lighting ballasts, and chokes all contain coils of wire.

In Unit 2, it was discussed that whenever current flows through a coil of wire a magnetic field is created around the wire *(Figure 3-2)*. If the amount of current decreases, the magnetic field will collapse *(Figure 3-3)*. Recall from Unit 2 several facts concerning inductance:

1. When magnetic lines of flux cut through a coil, a voltage is induced in the coil.
2. An induced voltage is always opposite in polarity to the applied voltage.
3. The amount of induced voltage is proportional to the rate of change of current.
4. An inductor opposes a change of current.

The inductors in *Figure 3-2* and *Figure 3-3* are connected to an AC voltage. Therefore, the magnetic field continually increases, decreases, and reverses polarity. Since the magnetic field continually changes magnitude and direction, a voltage is continually being induced in the coil. This **induced voltage**

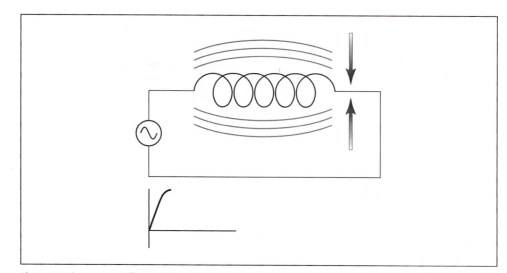

Figure 3-2 As current flows through a coil, a magnetic field is created around the coil.

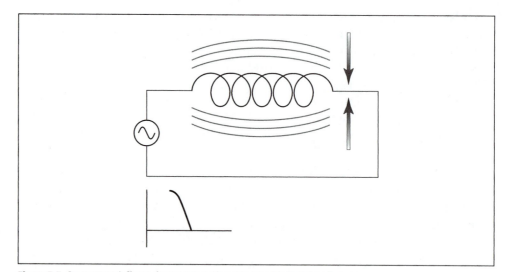

Figure 3-3 As current flow decreases, the magnetic field collapses.

is 180° out of phase with the applied voltage and is always in opposition to the applied voltage *(Figure 3-4)*. Since the induced voltage is always in opposition to the applied voltage, the applied voltage must overcome the induced voltage before current can flow through the circuit. For example, assume an inductor is connected to a 120-V AC line. Now assume that the inductor has an induced voltage of 116 V. Since an equal amount of applied

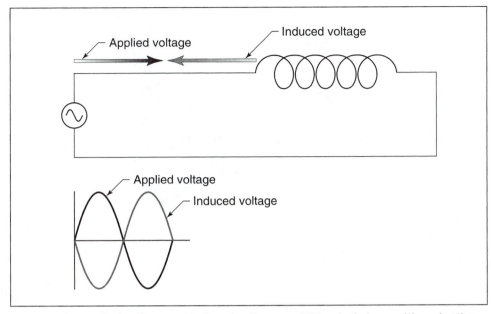

Figure 3-4 The applied voltage and induced voltage are 180° out of phase with each other.

voltage must be used to overcome the induced voltage, there will be only 4 V to push current through the wire resistance of the coil (120 – 116 = 4).

Computing the Induced Voltage

The amount of induced voltage in an inductor can be computed if the resistance of the wire in the coil and the amount of circuit current are known. For example, assume that an ohmmeter is used to measure the actual amount of resistance in a coil, and the coil is found to contain 6 Ω of wire resistance *(Figure 3-5)*. Now assume that the coil is connected to a 120-V AC circuit and an ammeter measures a current flow of 0.8 A *(Figure 3-6)*. Ohm's law can now be used to determine the amount of voltage necessary to push 0.8 A of current through 6 Ω of resistance.

$$E = I \times R$$
$$E = 0.8 \times 6$$
$$E = 4.8 \text{ V}$$

Since only 4.8 V is needed to push the current through the wire resistance of the inductor, the remainder of the 120 V is used to overcome the coil's induced voltage of 115.2 V (120 – 4.8 = 115.2).

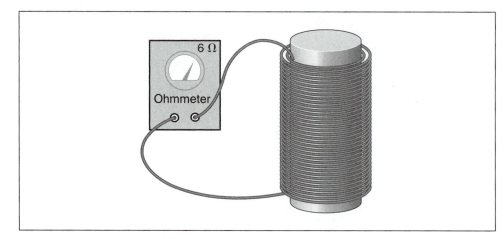

Figure 3-5 Measuring the resistance of a coil.

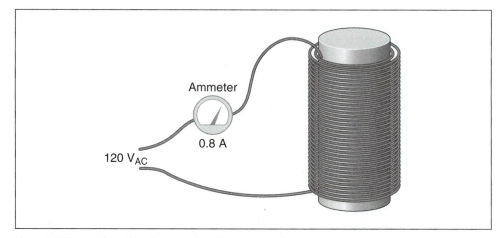

Figure 3-6 Measuring circuit current with an ammeter.

INDUCTIVE REACTANCE

Notice that the induced voltage is able to limit the flow of current through the circuit in a manner similar to resistance. This induced voltage is *not* resistance, but it can limit the flow of current just as resistance does. This current-limiting property of the inductor is called **reactance** and is symbolized by the letter *X*. This reactance is caused by inductance, so it is called **inductive reactance** and is symbolized by X_L, pronounced "X sub L." Inductive reactance is measured in ohms just as resistance is and can be computed when the values of inductance and frequency are known. The

reactance

**inductive
reactance (X_L)**

following formula can be used to find inductive reactance:

$$X_L = 2\pi fL$$

where

X_L = inductive reactance

2 = a constant

π = 3.1416

f = frequency in hertz (Hz)

L = inductance in henrys (H)

Inductive reactance is an induced voltage and is, therefore, proportional to the three factors that determine induced voltage:

1. the **number** of turns of wire,
2. the **strength** of the magnetic field, and
3. the **speed** of the cutting action (relative motion between the inductor and the magnetic lines of flux).

The number of turns of wire and strength of the magnetic field are determined by the physical construction of the inductor. Factors such as the size of wire used, the number of turns, how close the turns are to each other, and the type of core material determine the amount of inductance (in henrys, H) of the coil *(Figure 3-7)*. The speed of the cutting action is proportional to the

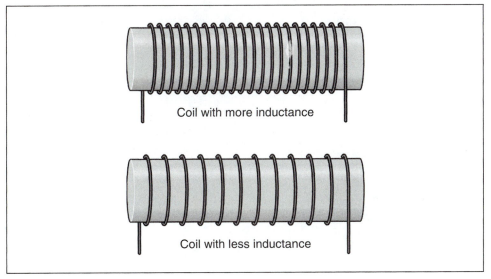

Coil with more inductance

Coil with less inductance

Figure 3-7 Coils with turns close together produce more inductance than coils with turns far apart.

frequency (Hz). An increase of frequency will cause the magnetic lines of flux to cut the conductors at a faster rate, and thus will produce a higher induced voltage or more inductive reactance.

Example 1

The inductor shown in *Figure 3-8* has an inductance of 0.8 H and is connected to a 120-V, 60-Hz line. How much current will flow in this circuit if the wire resistance of the inductor is negligible?

Solution

The first step is to determine the amount of inductive reactance of the inductor.

$$X_L = 2\pi fL$$

$$X_L = 2 \times 3.1416 \times 60 \times 0.8$$

$$X_L = 301.6 \ \Omega$$

Since inductive reactance is the current-limiting property of this circuit, it can be substituted for the value of R in an Ohm's law formula.

$$I = \frac{E}{X_L}$$

$$I = \frac{120}{301.6}$$

$$I = 0.398 \ A$$

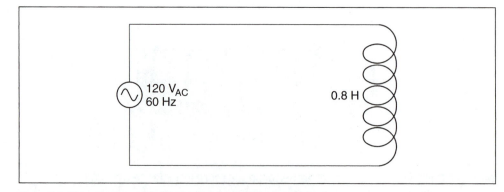

Figure 3-8 Circuit current is limited by inductive reactance.

If the amount of inductive reactance is known, the inductance of the coil can be determined using the formula

$$L = \frac{X_L}{2\pi f}$$

Example 2

Assume an inductor with a negligible resistance is connected to a 36-V, 400-Hz line. If the circuit has a current flow of 0.2 A, what is the inductance of the inductor?

Solution

The first step is to determine the inductive reactance of the circuit.

$$X_L = \frac{E}{I}$$

$$X_L = \frac{36}{0.2}$$

$$X_L = 180 \; \Omega$$

Now that the inductive reactance of the inductor is known, the inductance can be determined.

$$L = \frac{X_L}{2\pi f}$$

$$L = \frac{180}{2 \times 3.1416 \times 400}$$

$$L = 0.0716 \; H$$

Example 3

An inductor with negligible resistance is connected to a 480-V, 60-Hz line. An ammeter indicates a current flow of 24 A. How much current will flow in this circuit if the frequency is increased to 400 Hz?

Solution

The first step in solving this problem is to determine the amount of inductance of the coil. Since the resistance of the wire used to make the inductor is negligible, the current is limited by inductive reactance. The inductive reactance can be found by substituting X_L for R in an Ohm's law formula.

$$X_L = \frac{E}{I}$$

$$X_L = \frac{480}{24}$$

$$X_L = 20 \ \Omega$$

Now that the inductive reactance is known, the inductance of the coil can be found using the formula

$$L = \frac{X_L}{2\pi f}$$

NOTE: When using a frequency of 60 Hz, 2 x π x 60 = 377. Since 60 Hz is the major frequency used throughout the United States and Canada, 377 should be memorized for use when necessary.

$$L = \frac{20}{377}$$

$$L = 0.053 \ H$$

Since the inductance of the coil is determined by its physical construction, it will not change when connected to a different frequency. Now that the inductance of the coil is known, the inductive reactance at 400 Hz can be computed.

$$X_L = 2\pi fL$$

$$X_L = 2 \times 3.1416 \times 400 \times 0.053$$

$$X_L = 133.2 \ \Omega$$

The amount of current flow can now be found by substituting the value of inductive reactance for resistance in an Ohm's law formula.

$$I = \frac{E}{X_L}$$

$$I = \frac{480}{133.2}$$

$$I = 3.6 \ A$$

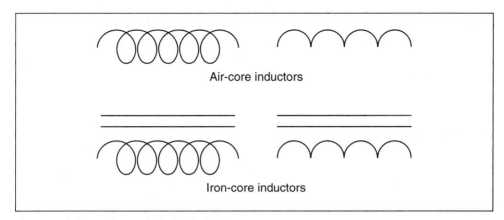

Figure 3-9 Schematic symbols for inductors.

SCHEMATIC SYMBOLS

The schematic symbol used to represent an inductor depicts a coil of wire. Several symbols for inductors are shown in *Figure 3-9*. The symbols shown with the two parallel lines represent iron-core inductors, and the symbols without the parallel lines represent air-core inductors.

INDUCTORS CONNECTED IN SERIES

When inductors are connected in series, the total inductance of the circuit (L_T) equals the sum of the inductances of all the inductors.

$$L_T = L_1 + L_2 + L_3$$

The total inductive reactance (X_{LT}) of inductors connected in series equals the sum of the inductive reactances for all the inductors.

$$X_{LT} = X_{L1} + X_{L2} + X_{L3}$$

Example 4

Three inductors are connected in series. Inductor 1 has an inductance of 0.6 H, inductor 2 has an inductance of 0.4 H, and inductor 3 has an inductance of 0.5 H. What is the total inductance of the circuit?

Solution

$$L_T = 0.6 + 0.4 + 0.5$$
$$L_T = 1.5 \text{ H}$$

Example 5

Three inductors are connected in series. Inductor 1 has an inductive reactance of 180 Ω, inductor 2 has an inductive reactance of 240 Ω, and inductor 3 has an inductive reactance of 320 Ω. What is the total inductive reactance of the circuit?

Solution

$$X_{LT} = 180\ \Omega + 240\ \Omega + 320\ \Omega$$

$$X_{LT} = 740\ \Omega$$

INDUCTORS CONNECTED IN PARALLEL

When inductors are connected in parallel, the total inductance can be found in a similar manner to finding the total resistance of a parallel circuit. The reciprocal of the total inductance is equal to the sum of the reciprocals of all the inductors.

$$\frac{1}{L_T} = \frac{1}{L_1} + \frac{1}{L_2} + \frac{1}{L_3}$$

or

$$L_T = \frac{1}{\dfrac{1}{L_1} + \dfrac{1}{L_2} + \dfrac{1}{L_3}}$$

Another formula that can be used to find the total inductance of parallel inductors is the product over sum formula.

$$L_T = \frac{L_1 \times L_2}{L_1 + L_2}$$

If the values of all the inductors are the same, total inductance can be found by dividing the inductance of one inductor by the total number of inductors.

$$L_T = \frac{L}{N}$$

Similar formulas can be used to find the total inductive reactance of inductors connected in parallel.

$$\frac{1}{X_{LT}} = \frac{1}{X_{L1}} + \frac{1}{X_{L2}} + \frac{1}{X_{L3}}$$

or

$$X_{LT} = \cfrac{1}{\cfrac{1}{X_{L1}} + \cfrac{1}{X_{L2}} + \cfrac{1}{X_{L3}}}$$

or

$$X_{LT} = \cfrac{X_{L1} \times X_{L2}}{X_{L1} + X_{L2}}$$

or

$$X_{LT} = \cfrac{X_L}{N}$$

Example 6

Three inductors are connected in parallel. Inductor 1 has an inductance of 2.5 H, inductor 2 has an inductance of 1.8 H, and inductor 3 has an inductance of 1.2 H. What is the total inductance of this circuit?

Solution

$$L_T = \cfrac{1}{\cfrac{1}{2.5} + \cfrac{1}{1.8} + \cfrac{1}{1.2}}$$

$$L_T = \cfrac{1}{1.789}$$

$$L_T = 0.559 \text{ H}$$

VOLTAGE AND CURRENT RELATIONSHIPS IN AN INDUCTIVE CIRCUIT

When current flows through a pure resistive circuit, the current and voltage are in phase with each other. **In a pure inductive circuit the current lags the voltage by 90°.** At first this may seem to be an impossible condition until the relationship of applied voltage and induced voltage is considered. How the current and applied voltage can become 90° out of phase with each other can best be explained by comparing the relationship of the current and induced voltage *(Figure 3-10).* Recall that the induced voltage is proportional to the rate of change of the current (speed of cutting action). At the beginning of the wave form, the current is shown at its

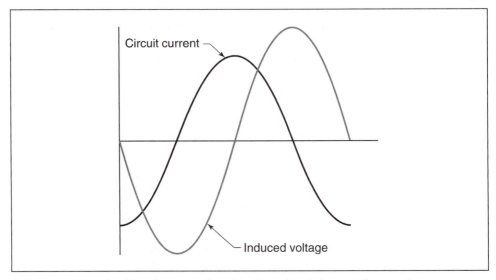

Figure 3-10 Induced voltage is proportional to the rate of change of current.

maximum value in the negative direction. At this time, the current is not changing, so induced voltage is zero. As the current begins to decrease in value, the magnetic field produced by the flow of current decreases or collapses and begins to induce a voltage into the coil as it cuts through the conductors *(Figure 3-3)*.

The greatest rate of current change occurs when the current passes from negative, through zero, and begins to increase in the positive direction *(Figure 3-11)*. Since the current is changing at the greatest rate, the induced voltage is maximum. As current approaches its peak value in the positive direction, the rate of change decreases, causing a decrease in the induced voltage. The induced voltage will again be zero when the current reaches its peak value and the magnetic field stops expanding.

It can be seen that the current flowing through the inductor is leading the induced voltage by 90°. Since the induced voltage is 180° out of phase with the applied voltage, the current will lag the applied voltage by 90° *(Figure 3-12)*.

POWER IN AN INDUCTIVE CIRCUIT

In a pure resistive circuit, the true power, or wattage, is equal to the product of the voltage and current. In a pure inductive circuit, however, no true power is produced. Recall that voltage and current must both be either positive or negative before true power can be produced. Since the voltage and

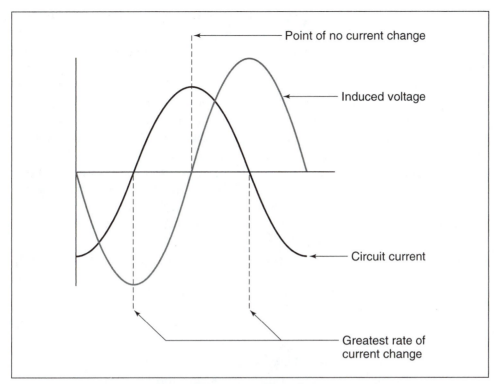

Figure 3-11 No voltage is induced when the current does not change.

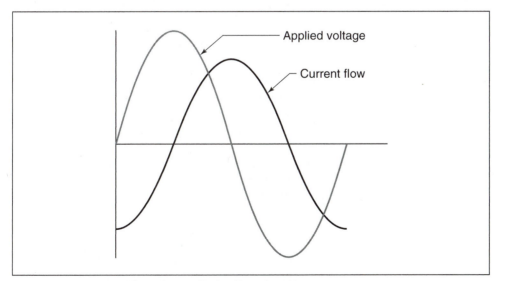

Figure 3-12 The current lags the applied voltage by 90°.

current are 90° out of phase with each other in a pure inductive circuit, the current and voltage will be at different polarities 50% of the time and at the same polarity 50% of the time. During the period of time that the current and voltage have the same polarity, power is being given to the circuit in the form of creating a magnetic field. When the current and voltage are opposite in polarity, power is being given back to the circuit as the magnetic field collapses and induces a voltage back into the circuit. Since power is stored in the form of a magnetic field and then given back, no power is used by the inductor. Any power used in an inductor is caused by losses. I^2R losses are caused by the resistance of the wire used to construct the conductor. Eddy current losses are caused by currents induced into the core material of the inductor, and Hysteresis losses are caused by the molecular friction of magnetic domains continually changing polarity each time the current flowing through the inductor changes direction.

The current and voltage waveforms in *Figure 3-13* have been divided into four sections: A, B, C, and D. During the first time period, indicated by A, the current is negative and the voltage is positive. During this period, energy is being given to the circuit as the magnetic field collapses. During the second time period, section B, both the voltage and current are positive. Power is being used to produce the magnetic field. In the third time period, C, the current is positive and the voltage is negative. Power is again being given back to the circuit as the field collapses. During the fourth time period, D,

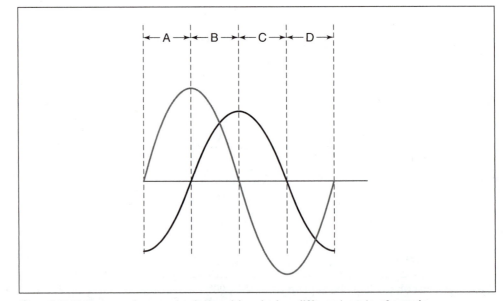

Figure 3-13 Voltage and current relationships during different parts of a cycle.

both the voltage and current are negative. Power is again being used to produce the magnetic field. If the amount of power used to produce the magnetic field is subtracted from the power given back, the result will be zero.

REACTIVE POWER

Although essentially no true power is being used, except by previously mentioned losses, an electrical measurement called **VARs** is used to measure the **reactive power** in a pure inductive circuit. **VARs** is an abbreviation for volt-amps-reactive. VARs can be computed in the same way as watts except that inductive values are substituted for resistive values in the formulas. VARs is equal to the amount of current flowing through an inductive circuit times the voltage applied to the inductive part of the circuit. Several formulas for computing VARs are:

$$VARs = E_L \times I_L$$

$$VARs = \frac{E_L^2}{X_L}$$

$$VARs = I_L^2 \times X_L$$

where

E_L = voltage applied to an inductor

I_L = current flow through an inductor

X_L = inductive reactance

Q OF AN INDUCTOR

So far in this unit, it has been generally assumed that an inductor has no resistance and that inductive reactance is the only current-limiting factor. In reality, that is not true. Since inductors are actually coils of wire they all contain some amount of internal resistance. Inductors actually appear to be a coil connected in series with some amount of resistance *(Figure 3-14)*. The amount of resistance compared with the inductive reactance determines the Q of the coil. The letter **Q** stands for **quality**. Inductors that have a higher ratio of inductive reactance to resistance are considered to be inductors of higher quality. An inductor constructed with a large wire will have a low wire resistance and, therefore, a higher Q *(Figure 3-15)*. Inductors constructed with many turns of small wire have a much higher resistance,

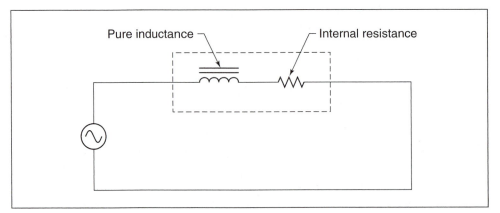

Figure 3-14 Inductors contain internal resistance.

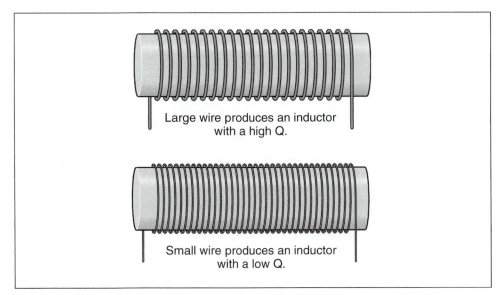

Figure 3-15 The Q of an inductor is a ratio of inductive reactance as compared to resistance. The letter "Q" stands for quality.

and, therefore, a lower Q. To determine the Q of an inductor, divide the inductive reactance by the resistance.

$$Q = \frac{X_L}{R}$$

Although inductors have some amount of resistance, inductors that have a Q of 10 or greater are generally considered to be pure inductors. Once the ratio of inductive reactance becomes 10 times as great as resistance, the amount of resistance is considered negligible. For example, assume an inductor has an

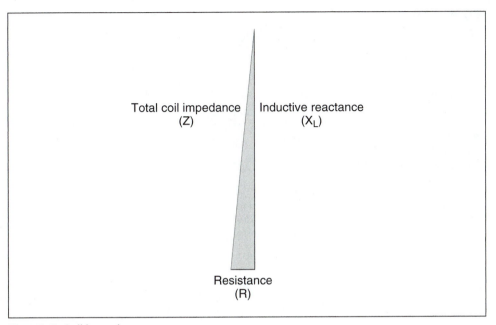

Figure 3-16 Coil impedance.

inductive reactance of 100 Ω and a wire resistance of 10 Ω. The inductive reactive component in the circuit is 90° out of phase with the resistive component. This relationship produces a right triangle *(Figure 3-16)*. The total current-limiting effect of the inductor is a combination of the inductive reactance and resistance. This total current-limiting effect is called **impedance** and is symbolized by the letter **Z**. The impedance of the circuit is represented by the hypotenuse of the right triangle formed by the inductive reactance and the resistance. To compute the value of impedance for the coil, the inductive reactance and resistance must be added. Since these two components form the legs of a right triangle and the impedance forms the hypotenuse, the value of impedance can be computed using the Pythagorean theorem: the sum of the squares of the sides of a right triangle is equal to the square of the hypotenuse.

impedance

$$Z = \sqrt{R^2 + X_L^2}$$

$$Z = \sqrt{10^2 + 100^2}$$

$$Z = \sqrt{10,100}$$

$$Z = 100.5 \ \Omega$$

Notice that the value of total impedance for the inductor is only 0.5 Ω greater than the value of inductive reactance.

Summary

1. Induced voltage is proportional to the rate of change of current.

2. Induced voltage is always opposite in polarity to the applied voltage.

3. Inductive reactance is a countervoltage that limits the flow of current, as does resistance.

4. Inductive reactance is measured in ohms.

5. Inductive reactance is proportional to the inductance of the coil and the frequency of the line.

6. Inductive reactance is symbolized by X_L.

7. Inductance is measured in henrys (H) and is symbolized by the letter L.

8. When inductors are connected in series the total inductance is equal to the sum of all the inductors.

9. When inductors are connected in parallel the reciprocal of the total inductance is equal to the sum of the reciprocals of all the inductors.

10. The current lags the applied voltage by 90° in a pure inductive circuit.

11. All inductors contain some amount of resistance.

12. The Q of an inductor is the ratio of the inductive reactance to the resistance.

13. Inductors with a Q of 10 are generally considered to be "pure" inductors.

14. Pure inductive circuits consume no true power or watts.

15. Reactive power is measured in VARs.

16. VARs is an abbreviation for volt-amps-reactive.

Review Questions

1. How many degrees are the current and voltage out of phase with each other in a pure resistive circuit?

2. How many degrees are the current and voltage out of phase with each other in a pure inductive circuit?

3. To what is inductive reactance proportional?

4. Four inductors, each having an inductance of 0.6 H, are connected in series. What is the total inductance of the circuit?

5. Three inductors are connected in parallel. Inductor 1 has an inductance of 0.06 H; inductor 2 has an inductance of 0.05 H; and inductor 3 has an inductance of 0.1 H. What is the total inductance of this circuit?

6. If the three inductors in question 5 were connected in series, what would be the inductive reactance of the circuit? Assume the inductors are connected to a 60-Hz line.

7. An inductor is connected to a 240-V, 1000-Hz line. The circuit current is 0.6 A. What is the inductance of the inductor?

8. An inductor with an inductance of 3.6 H is connected to a 480-V, 60-Hz line. How much current will flow in this circuit?

9. If the frequency in question 8 is reduced to 50 Hz, how much current will flow in the circuit?

10. An inductor has an inductive reactance of 250 Ω when connected to a 60-Hz line. What will be the inductive reactance if the inductor is connected to a 400-Hz line?

Problems

Inductive Circuits

Fill in all the missing values. Refer to the formulas given below.

$$X_L = 2\pi fL$$

$$L = \frac{X_L}{2\pi f}$$

$$f = \frac{X_L}{2\pi L}$$

Inductance (H)	Frequency (Hz)	Induct. Rct. (Ω)
1.2	60	
0.085		213.628
	1000	4712.389
0.65	600	
3.6		678.584
	25	411.459
0.5	60	
0.85		6408.849
	20	201.062
0.45	400	
4.8		2412.743
	1000	40.841

UNIT **4**

Single-Phase Isolation Transformers

Objectives

After studying this unit, you should be able to:

- Discuss the different types of transformers.
- Calculate values of voltage, current, and turns for single-phase transformers using formulas.
- Calculate values of voltage, current, and turns for single-phase transformers using the turns ratio.
- Connect a transformer and test the voltage output of different windings.
- Discuss polarity markings on a schematic diagram.
- Test a transformer to determine the proper polarity marks.

Transformers are one of the most common devices found in the electrical field. They range in size from less than one cubic inch to requiring rail cars to move them after they have been broken into sections. Their rating can range from mVA (milli-volt-amps) to GVA (giga-volt-amps).

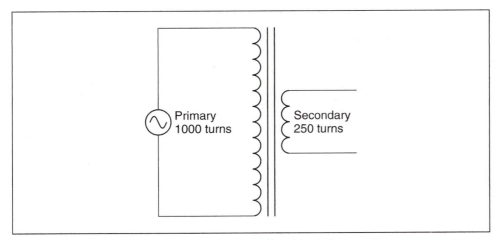

Figure 4-1 All values of a transformer are proportional to its turns ratio.

A **transformer** is a magnetically operated machine that can change values of voltage, current, and impedance without a change of frequency. Transformers are the most efficient machines known. Their efficiencies commonly range from 90% to 99% at full load. Transformers can be divided into several classifications, which include:

    1. Isolation

    2. Auto

    3. Current

A basic law concerning transformers is that *all values of a transformer are proportional to its turns ratio.* This does not mean that the exact number of turns of wire on each winding must be known to determine different values of voltage and current for a transformer. What must be known is the *ratio* of turns. For example, assume a transformer has two windings. One winding, the primary, has 1000 turns of wire, and the other, the secondary, has 250 turns of wire *(Figure 4-1)*. The **turns ratio** of this transformer is 4 to 1, or 4:1 (1000/250 = 4). This indicates that there are four turns of wire on the primary for every one turn of wire on the secondary.

transformer

turns ratio

TRANSFORMER FORMULAS

There are different formulas that can be used to find the values of voltage and current for a transformer. The following is a list of standard formulas, where:

$$N_P = \text{number of turns in the primary}$$

$$N_S = \text{number of turns in the secondary}$$

E_P = voltage of the primary

E_S = voltage of the secondary

I_P = current in the primary

I_S = current in the secondary

$$\frac{E_P}{E_S} = \frac{N_P}{N_S}$$

$$\frac{E_P}{E_S} = \frac{I_S}{I_P}$$

$$\frac{N_P}{N_S} = \frac{I_S}{I_P}$$

or

$$E_P \times N_S = E_S \times N_P$$

$$E_P \times I_P = E_S \times I_S$$

$$N_P \times I_P = N_S \times I_S$$

primary winding

secondary winding

  The **primary winding** of a transformer is the power-input winding. It is the winding that is connected to the incoming power supply. The **secondary winding** is the load winding, or output winding. It is the side of the transformer that is connected to the driven load *(Figure 4-2)*. Any winding of a transformer can be used as a primary or secondary wiring provided its voltage or current rating is not exceded. Transformers can also be operated at a lower voltage than their rating indicates, but they cannot be connected to a higher voltage. Assume the transformer shown in *Figure 4-2*, for example, has a primary voltage rating of 480 volts and the secondary has a voltage rating of 240 volts. Now assume that the primary winding is connected to a 120-volt source. No damage would occur to the transformer but the secondary winding would produce only 20 volts.

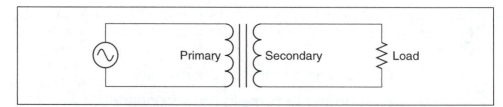

Figure 4-2 Isolation transformer.

ISOLATION TRANSFORMERS

The transformers shown in *Figure 4-1* and *Figure 4-2* are **isolation transformers**. This means that the secondary winding is physically and electrically isolated from the primary winding. There is no electrical connection between the primary and secondary winding. This transformer is magnetically coupled, not electrically coupled. This "line isolation" is often a very desirable characteristic. Since there is no electrical connection between the load and power supply, the transformer becomes a filter between the two. The isolation transformer will greatly reduce any voltage spikes that originate on the supply side before they are transferred to the load side. Some isolation transformers are built with a turns ratio of 1:1. A transformer of this type will have the same input and output voltage and is used for the purpose of isolation only.

The reason that the isolation transformer can greatly reduce any voltage spikes before they reach the secondary is because of the rise time of current through an inductor. Recall that the current in an inductor rises at an exponential rate *(Figure 4-3)*. As the current increases in value, the expanding magnetic field cuts through the conductors of the coil and induces a voltage that is opposed to the applied voltage. The amount of induced voltage is proportional to the rate of change of current. This simply means that the faster current attempts to increase, the greater the opposition to that increase will be. Spike voltages and currents are generally of very short duration, which means that they increase in value very rapidly *(Figure 4-4)*. This rapid change of value causes the opposition to the change to increase just as rapidly. By the time the spike has been transferred to the secondary

isolation transformers

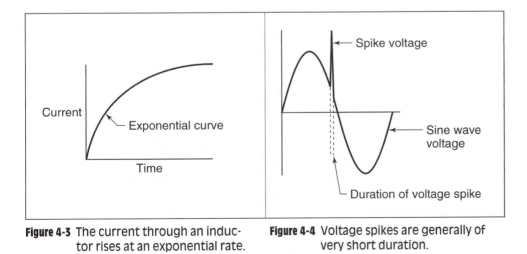

Figure 4-3 The current through an inductor rises at an exponential rate.

Figure 4-4 Voltage spikes are generally of very short duration.

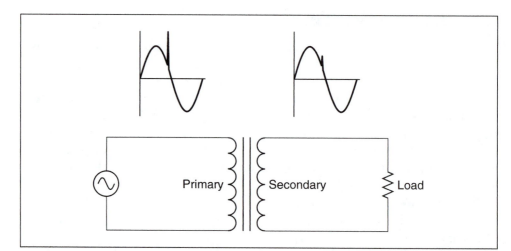

Figure 4-5 The isolation transformer greatly reduces the voltage spike.

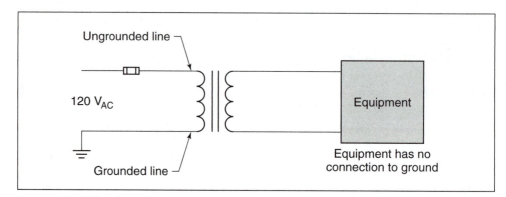

Figure 4-6 Isolation transformer used to remove a piece of electrical equipment from ground.

winding of the transformer it has been eliminated or greatly reduced *(Figure 4-5)*.

Another purpose of isolation transformers is to disconnect some piece of electrical equipment from ground. It is sometimes desirable that a piece of electrical equipment not be connected directly to ground. This is often done as a safety precaution to eliminate the hazard of an accidental contact between a person at ground potential and the ungrounded conductor. If the equipment case should come into contact with the ungrounded conductor, the isolation transformer would prevent a circuit from being completed to ground by someone touching the equipment case. Many alternating-current circuits have one side connected to ground. A familiar example of this is the common 120-volt circuit with a grounded neutral conductor *(Figure 4-6)*.

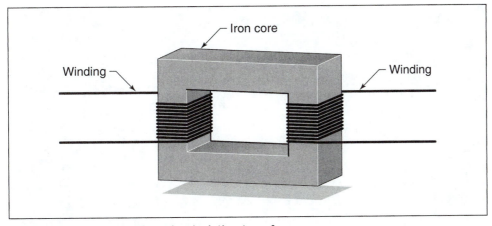

Figure 4-7 Basic construction of an isolation transformer.

Transformer Construction

The basic construction of an isolation transformer is shown in *Figure 4-7*. A metal core is used to provide good magnetic coupling between the two windings. The core is generally made of laminations stacked together. Laminating the core helps reduce power losses due to eddy current induction. This illustration shows the basic design of electrically separated winding.

Transformer Core Types

There are several different types of cores used in the construction of transformers. Most cores are made from thin steel punchings laminated to form a solid metal core. The core for a 600 MVA (mega-volt-amp) three-phase transformer is shown in *Figure 4-8*. Laminated cores are preferred because a thin layer of oxide forms on the surface of each lamination, which acts as an insulator to reduce the formation of eddy currents inside the core material. The amount of core material needed for a particular transformer is determined by the power rating of the transformer. The amount of core material must be sufficient to prevent saturation at full load. The type and shape of the core generally determines the amount of magnetic coupling between the windings, and to some extent, the efficiency of the transformer.

The transformer illustrated in *Figure 4-9* is known as a *core*-type transformer. The windings are placed around each end of the core material. The metal core provides a good magnetic path between the two windings.

The *shell*-type transformer is constructed in a manner similar to the core-type, except that the shell-type has a metal core piece through the

Figure 4-8 Core of a 600-MVA three-phase transformer (courtesy of Reliant Energy). (Note: Houston Lighting and Power is now Reliant Energy.)

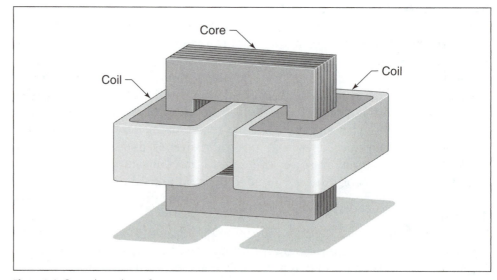

Figure 4-9 Core-type transformer.

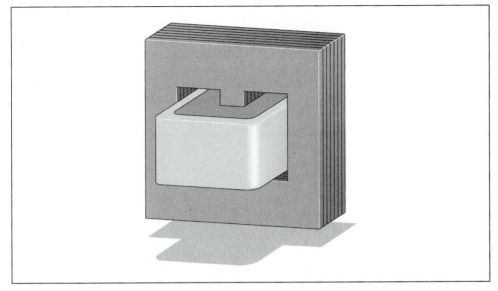

Figure 4-10 Shell-type transformer.

middle of the window *(Figure 4-10)*. The primary and secondary windings are wound around the center core piece, with the low-voltage winding being closest to the metal core. The transformer is surrounded by the core, an arrangement that provides excellent magnetic coupling. When the transformer is in operation, all the magnetic flux must pass through the center core piece. It then divides through the two outer core pieces. Shell-type cores are sometimes referred to as E-I cores because the steel punchings used to construct the core are in the shape of an E and an I *(Figure 4-11)*.

The H-type core shown in *Figure 4-12* is similar to the shell-type in that it has an iron core through its center around which the primary and secondary windings are wound. The H core, however, surrounds the windings on four sides instead of two. This extra metal helps reduce stray leakage flux and improves the efficiency of the transformer. The H-type core is often found on high-voltage distribution transformers.

The *tape wound* or *toroid* core *(Figure 4-13)* is constructed by winding tightly one long continuous silicon-steel tape into a spiral. The tape may or may not be housed in a plastic container depending on the application. This type of core does not require laminated steel punchings. Since the core is one continuous length of metal, flux leakage is kept to a minimum. The tape-wound core is one of the most efficient core designs available.

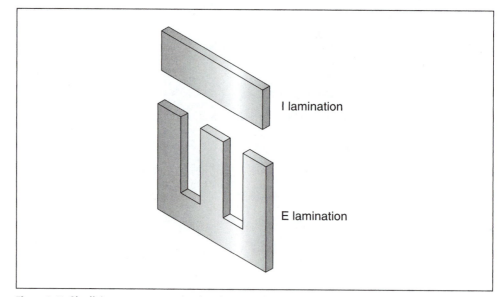

Figure 4-11 Shell-type cores are made of E and I laminations.

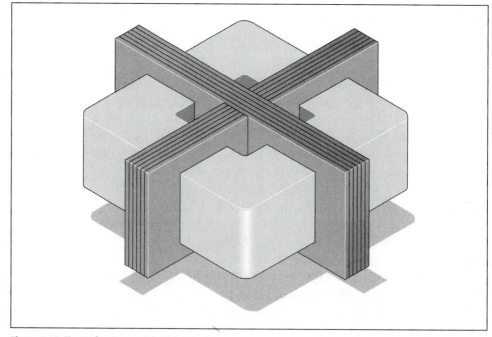

Figure 4-12 Transformer with H-type core.

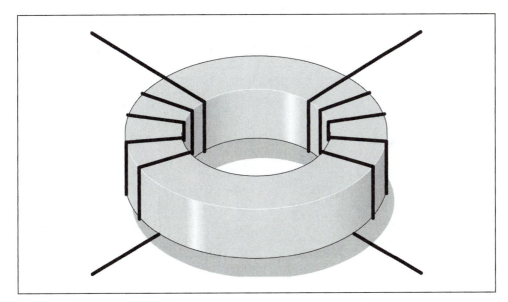

Figure 4-13 Toroid transformer.

Basic Operating Principles

In *Figure 4-14,* one winding of the transformer has been connected to an alternating-current supply, and the other winding has been connected to a load. As current increases from zero to its peak positive point, a magnetic field expands outward around the coil. When the current decreases from its peak positive point toward zero, the magnetic field collapses. When the current increases toward its negative peak, the magnetic field again expands, but with an opposite polarity of that previously. The field again collapses when the current decreases from its negative peak toward zero. This continually expanding and collapsing magnetic field cuts the windings of the primary and induces a voltage into it. This induced voltage opposes the applied voltage and limits the current flow of the primary. When a coil induces a voltage into itself, it is known as *self-induction.* It is this induced voltage, inductive reactance, that limits the flow of current in the primary winding. If the resistance of the primary winding is measured with an ohmmeter, it will indicate only the resistance of the wire used to construct the winding and will not give an indication of the actual current-limiting effect of the winding. Most transformers with a large kVA rating will appear to be almost a short circuit when measured with an ohmmeter. When connected to power, however, the actual no-load current is generally relatively small.

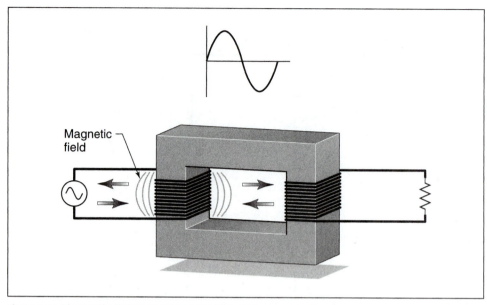

Figure 4-14 Magnetic field produced by alternating current.

Excitation Current

There will always be some amount of current flow in the primary of any voltage transformer even if there is no load connected to the secondary. This is called the **excitation current** of the transformer. The excitation current is the amount of current required to magnetize the core of the transformer. The excitation current remains constant from no load to full load. As a general rule, the excitation current is such a small part of the full-load current it is often omitted when making calculations.

Mutual Induction

Since the secondary windings of an isolation transformer are wound on the same core as the primary, the magnetic field produced by the primary winding cuts the windings of the secondary also *(Figure 4-15)*. This continually changing magnetic field induces a voltage into the secondary winding. The ability of one coil to induce a voltage into another coil is called **mutual induction**. The amount of voltage induced in the secondary is determined by the ratio of the number of turns of wire in the secondary to those in the primary. For example, assume the primary has 240 turns of wire and is connected to 120 VAC. This gives the transformer a **volts-per-turn ratio** of 0.5 (120 V/240 turns = 0.5 volt-per-turn). Now assume that the secondary winding

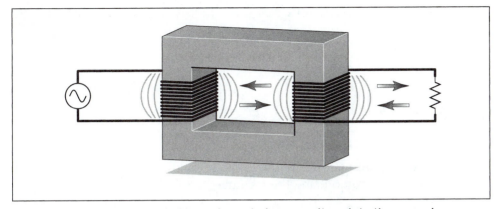

Figure 4-15 The magnetic field of the primary induces a voltage into the secondary.

contains 100 turns of wire. Since the transformer has a volts-per-turn ratio of 0.5, the secondary voltage will be 50 V (100 × 0.5 = 50).

Transformer Calculations

In the following examples, values of voltage, current, and turns for different transformers will be computed.

Example 1

Assume that the isolation transformer shown in *Figure 4-2* has 240 turns of wire on the primary and 60 turns of wire on the secondary. This is a ratio of 4:1 (240/60 = 4). Now assume that 120 V is connected to the primary winding. What is the voltage of the secondary winding?

$$\frac{E_P}{E_S} = \frac{N_P}{N_S}$$

$$\frac{120}{E_S} = \frac{240}{60}$$

$$E_S = 30 \text{ V}$$

The transformer in this example is known as a **step-down transformer** because its secondary voltage is lower than its primary voltage.

Now assume that the load connected to the secondary winding has an impedance of 5 Ω. The next problem is to calculate the current flow in the secondary and primary windings. The current flow of the secondary

step-down transformer

can be computed using Ohm's law since the voltage and impedance are known.

$$I = \frac{E}{Z}$$

$$I = \frac{30}{5}$$

$$I = 6 \text{ amps}$$

Now that the amount of current flow in the secondary is known, the primary current can be computed using the formula

$$\frac{E_P}{E_S} = \frac{I_S}{I_P}$$

$$\frac{120}{30} = \frac{6}{I_P}$$

$$120 \, I_P = 180$$

$$I_P = 1.5 \text{ A}$$

Notice that the primary voltage is higher than the secondary voltage, but the primary current is much less than the secondary current. *A good rule for any type of transformer is that power in must equal power out.* If the primary voltage and current are multiplied together, it should equal the product of the voltage and current of the secondary.

Primary	Secondary
120 x 1.5 = 180 volt amps	30 x 6 = 180 volt amps

Example 2

In the next example, assume that the primary winding contains 240 turns of wire and the secondary contains 1200 turns of wire. This is a turns ratio of 1:5 (1200/240 = 5). Now assume that 120 V is connected to the primary winding. Compute the voltage output of the secondary winding.

$$\frac{E_P}{E_S} = \frac{N_P}{N_S}$$

$$\frac{120}{E_S} = \frac{240}{1200}$$

$$240 \, E_S = 144,000$$

$$E_S = 600 \text{ V}$$

Notice that the secondary voltage of this transformer is higher than the primary voltage. This type of transformer is known as a **step-up transformer**.

step-up transformer

Now assume that the load connected to the secondary has an impedance of 2400 Ω. Find the amount of current flow in the primary and secondary windings. The current flow in the secondary winding can be computed using Ohm's law.

$$I = \frac{E}{Z}$$

$$I = \frac{600}{2400}$$

$$I = 0.25 \text{ A}$$

Now that the amount of current flow in the secondary is known, the primary current can be computed using the formula

$$\frac{E_P}{E_S} = \frac{I_S}{I_P}$$

$$\frac{120}{600} = \frac{0.25}{I_P}$$

$$120 \, I_P = 150$$

$$I_P = 1.25 \text{ A}$$

Notice that the amount of power input equals the amount of power output.

Primary	Secondary
120 x 1.25 = 150 VA	600 x 0.25 = 150 VA

Calculating Transformer Values Using the Turns Ratio

As illustrated in the previous examples, transformer values of voltage, current, and turns can be computed using formulas. It is also possible to compute these same values using the turns ratio. There are several ways in which turns ratios can be expressed. One method is to use a whole number value such as 13:5 or 6:21. The first ratio indicates that one winding has 13 turns of wire for every 5 turns of wire in the other winding. The second ratio indicates that there are 6 turns of wire in one winding for every 21 turns in the other.

A second method is to use the number 1 as a base. When using this method, the number 1 is always assigned to the winding with the lowest voltage rating. The ratio is found by dividing the higher voltage by the lower voltage. The number on the left side of the ratio represents the primary winding and the number on the right of the ratio represents the secondary

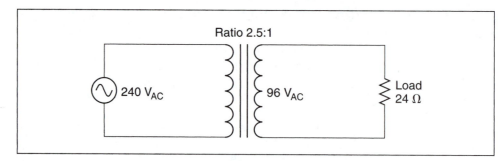

Figure 4-16 Computing transformer values using the turns ratio.

winding. For example, assume a transformer has a primary rated at 240 volts and a secondary rated at 96 volts *(Figure 4-16)*. The turns ratio can be computed by dividing the higher voltage by the lower voltage.

$$\text{Ratio} = \frac{240}{96}$$

$$\text{Ratio} = 2.5{:}1$$

Notice in this example that the primary winding has the higher voltage rating and the secondary has the lower. Therefore, the 2.5 is placed on the left and the base unit, 1, is placed on the right. This ratio indicates that there are 2.5 turns of wire in the primary winding for every 1 turn of wire in the secondary.

Now assume that a resistance of 24 Ω is connected to the secondary winding. The amount of secondary current can be found using Ohm's law.

$$I_S = \frac{96}{24}$$

$$I_S = 4 \text{ amps}$$

The primary current can be found using the turns ratio. Recall that the volt amps of the primary must equal the volt amps of the secondary. Since the primary voltage is greater, the primary current will have to be less than the secondary current. Therefore, the secondary current will be divided by the turns ratio.

$$I_P = \frac{I_S}{\text{turns ratio}}$$

$$I_P = \frac{4}{2.5}$$

$$I_P = 1.6 \text{ A}$$

To check the answer, find the volt amps of the primary and secondary.

Primary	Secondary
240 x 1.6 = 384	96 x 4 = 384

Now assume that the secondary winding contains 150 turns of wire. The primary turns can be found by using the turns ratio also. Since the primary voltage is higher than the secondary voltage, the primary must have more turns of wire. The secondary turns will be multiplied by the turns ratio.

$$N_P = N_S \times \text{turns ratio}$$

$$N_P = 150 \times 2.5$$

$$N_P = 375 \text{ turns}$$

In the next example, assume that a transformer has a primary voltage of 120 volts and a secondary voltage of 500 volts. The secondary has a load impedance of 1200 Ω. The secondary contains 800 turns of wire *(Figure 4-17)*.

The turns ratio can be found by dividing the higher voltage by the lower voltage.

$$\text{ratio} = \frac{500}{120}$$

$$\text{ratio} = 1{:}4.17$$

The secondary current can be found using Ohm's law.

$$I_S = \frac{500}{1200}$$

$$I_S = 0.417 \text{ amps}$$

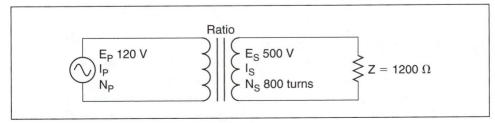

Figure 4-17 Calculating transformer values.

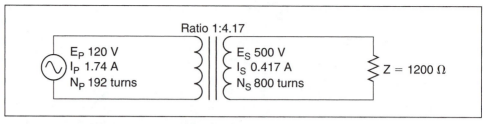

Figure 4-18 Transformer with completed values.

In this example, the primary voltage is lower than the secondary voltage. Therefore, the primary current must be higher. To find the primary current, multiply the secondary current by the turns ratio.

$$I_P = I_S \times \text{turns ratio}$$

$$I_P = 0.417 \times 4.17$$

$$I_P = 1.74 \text{ amps}$$

To check this answer, compute the volt amps of both windings.

Primary	Secondary
120 x 1.74 = 208.8	500 x 0.417 = 208.5

The slight difference in answers is caused by rounding off values.

Since the primary voltage is less than the secondary voltage, the turns of wire in the primary will be less also. The primary turns will be found by dividing the turns of wire in the secondary by the turns ratio.

$$N_P = \frac{N_S}{\text{turns ratio}}$$

$$N_P = \frac{800}{4.17}$$

$$N_P = 192 \text{ turns}$$

Figure 4-18 shows the transformer with all completed values.

Multiple-Tapped Windings

It is not uncommon for isolation transformers to be designed with windings that have more than one set of lead wires connected to the primary or secondary. The transformer shown in *Figure 4-19* contains a secondary winding rated at 24 V. The primary winding contains several taps, however. One of the primary lead wires is labeled C, and is the common for the other leads. The

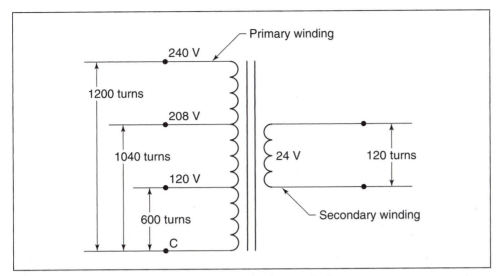

Figure 4-19 Transformer with multiple-tapped primary winding.

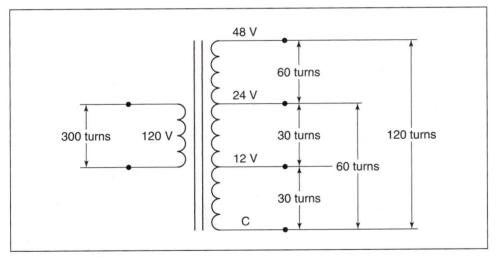

Figure 4-20 Transformer secondary with multiple taps.

other leads are labeled 120, 208, and 240. This transformer is designed in such a manner that it can be connected to different primary voltages without changing the value of the secondary voltage. In this example, it is assumed that the secondary winding has a total of 120 turns of wire. To maintain the proper turns ratio, the primary would have 600 turns of wire between C and 120, 1040 turns between C and 208, and 1200 turns between C and 240.

The transformer shown in *Figure 4-20* contains a single primary winding. The secondary winding, however, has been tapped at several points. One of

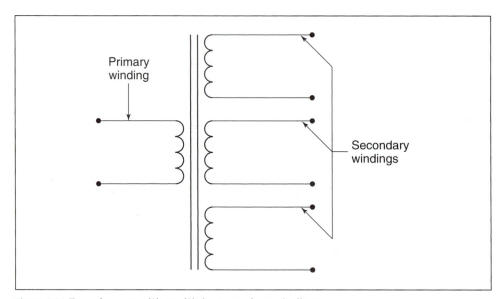

Figure 4-21 Transformer with multiple secondary windings.

the secondary lead wires is labeled C and is common to the other lead wires. When rated voltage is applied to the primary, voltages of 12, 24, and 48 V can be obtained at the secondary. It should also be noted that this arrangement of taps permits the transformer to be used as a center-tapped transformer for two of the voltages. If a load is placed across the lead wires labeled C and 24, the lead wire labeled 12 becomes a center tap. If a load is placed across the C and 48 lead wires, the 24 lead wire becomes a center tap.

In this example, it is assumed that the primary winding has 300 turns of wire. In order to produce the proper turns ratio, it would require 30 turns of wire between C and 12, 60 turns of wire between C and 24, and 120 turns of wire between C and 48.

The transformer shown in *Figure 4-21* is similar to the transformer in *Figure 4-20.* The transformer in *Figure 4-21,* however, has multiple secondary windings instead of a single secondary winding with multiple taps. The advantage of the transformer in *Figure 4-21* is that the secondary windings are electrically isolated from each other. These secondary windings can be either step-up or step-down depending on the application of the transformer.

Computing Values for Isolation Transformers with Multiple Secondaries

When computing the values of a transformer with multiple secondary windings, each secondary must be treated as a different transformer. For

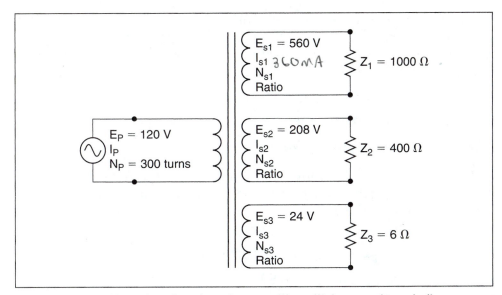

Figure 4-22 Computing values for a transformer with multiple secondary windings.

example, the transformer in *Figure 4-22* contains one primary winding and three secondary windings. The primary is connected to 120 VAC and contains 300 turns of wire. One secondary has an output voltage of 560 V and a load impedance of 1000 Ω. The second secondary has an output voltage of 208 V and a load impedance of 400 Ω, and the third secondary has an output voltage of 24 V and a load impedance of 6 Ω. The current, turns of wire, and ratio for each secondary and the current of the primary will be found.

The first step will be to compute the turns ratio of the first secondary. The turns ratio can be found by dividing the smaller voltage into the larger.

$$ratio = \frac{E_{S1}}{E_P}$$

$$ratio = \frac{560}{120}$$

ratio = 1:4.67

The current flow in the first secondary can be computed using Ohm's law.

$$I_{S1} = \frac{560}{1000}$$

I_{S1} = 0.56 amps

The number of turns of wire in the first secondary winding will be found using the turns ratio. Since this secondary has a higher voltage than the primary, it must have more turns of wire. The number of primary turns will be multiplied by the turns ratio.

$$N_{S1} = N_P \times \text{turns ratio}$$

$$N_{S1} = 300 \times 4.67$$

$$N_{S1} = 1401 \text{ turns}$$

The amount of primary current needed to supply this secondary winding can be found using the turns ratio also. Since the primary has less voltage, it will require more current. The primary current can be determined by multiplying the secondary current by the turns ratio.

$$I_{P(\text{First Secondary})} = I_{S1} \times \text{turns ratio}$$

$$I_{P(\text{First Secondary})} = 0.56 \times 4.67$$

$$I_{P(\text{First Secondary})} = 2.61 \text{ amps}$$

The turns ratio of the second secondary winding will be found by dividing the higher voltage by the lower.

$$\text{ratio} = \frac{208}{120}$$

$$\text{ratio} = 1:1.73$$

The amount of current flow in this secondary can be determined using Ohm's law.

$$I_{S2} = \frac{208}{400}$$

$$I_{S2} = 0.52 \text{ amps}$$

Since the voltage of this secondary is greater than that of the primary, it will have more turns of wire than the primary. The turns of this secondary will be found using the turns ratio.

$$N_{S2} = N_P \times \text{turns ratio}$$

$$N_{S2} = 300 \times 1.73$$

$$N_{S2} = 519 \text{ turns}$$

The voltage of the primary is less than that of this secondary. The primary will, therefore, require a greater amount of current. The amount of current required to operate this secondary will be computed by multiplying the secondary current by the turns ratio.

$$I_{P(Second\ Secondary)} = I_{S2} \times turns\ ratio$$

$$I_{P(Second\ Secondary)} = 0.52 \times 1.732$$

$$I_{P(Second\ Secondary)} = 0.9\ amps$$

The turns ratio of the third secondary winding will be computed in the same way as the other two. The larger voltage will be divided by the smaller.

$$ratio = \frac{120}{24}$$

$$ratio = 5:1$$

The primary current will be found using Ohm's law.

$$I_{S3} = \frac{24}{6}$$

$$I_{S3} = 4\ amps$$

The output voltage of the third secondary is less than that of the primary. The number of turns of wire will, therefore, be smaller than the primary turns. To find the number of secondary turns, divide the number of primary turns by the turns ratio.

$$N_{S3} = \frac{N_P}{turns\ ratio}$$

$$N_{S3} = \frac{300}{5}$$

$$N_{S3} = 60\ turns$$

The primary has a higher voltage than that of this secondary. The primary current will therefore be less by the amount of the turns ratio.

$$I_{P(Third\ Secondary)} = \frac{I_{S3}}{turns\ ratio}$$

$$I_{P(Third\ Secondary)} = \frac{4}{5}$$

$$I_{P(Third\ Secondary)} = 0.8\ amps$$

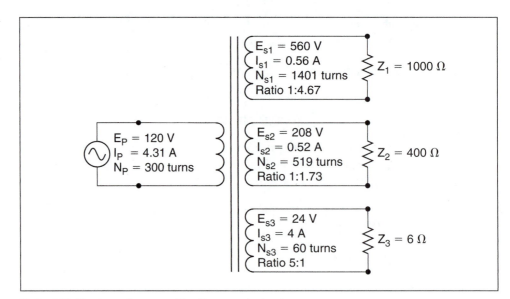

Figure 4-23 The transformer with all computed values.

The primary must supply current to each of the three secondary windings. Therefore, the total amount of primary current will be the sum of the currents required to supply each secondary.

$$I_{P(Total)} = I_{P1} + I_{P2} + I_{P3}$$

$$I_{P(Total)} = 2.61 + 0.9 + 0.8$$

$$I_{P(Total)} = 4.31 \text{ amps}$$

The transformer with all computed values is shown in *Figure 4-23*.

Distribution Transformers

distribution transformer

A very common type of isolation transformer is the **distribution transformer** *(Figure 4-24)*. This transformer is used to supply power to most homes and many businesses. In this example, it is assumed that the primary is connected to a 7200-volt line. The secondary is 240 volts with a center tap. The center tap is grounded and becomes the **neutral conductor**. If voltage is measured across the entire secondary, a voltage of 240 volts will be seen. If voltage is measured from either line to the center tap, half of the secondary voltage, or 120 volts, will be seen *(Figure 4-25)*. The reason is that the voltages between the two secondary lines are in phase with each other. If a

neutral conductor

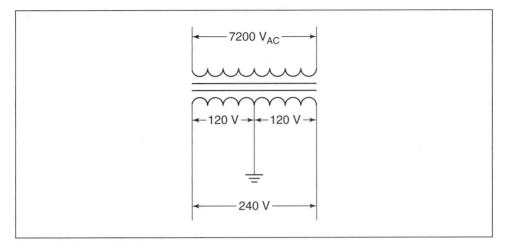

Figure 4-24 Distribution transformer.

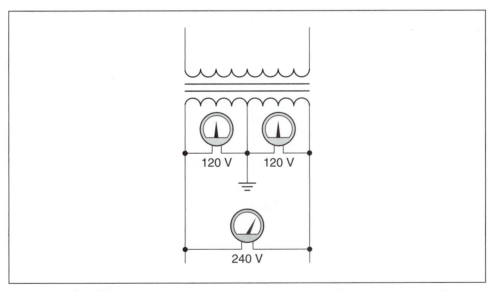

Figure 4-25 The voltage from either line to neutral is 120 volts. The voltage across the entire secondary winding is 240 volts.

vector diagram is drawn to illustrate this condition, it will be seen that the grounded neutral conductor is connected to the junction point of the two voltage vectors *(Figure 4-26)*. Loads that are intended to operate on 240 volts, such as water heaters, electric-resistance heating units, and central air conditioners, are connected directly across the lines of the secondary *(Figure 4-27)*.

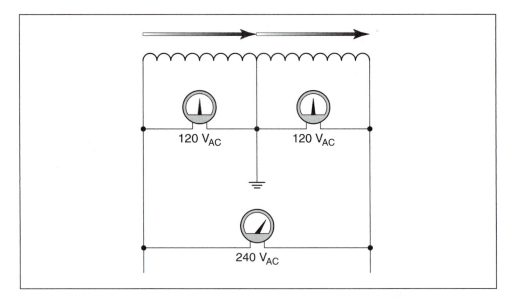

Figure 4-26 The voltages across the secondary are in phase with each other.

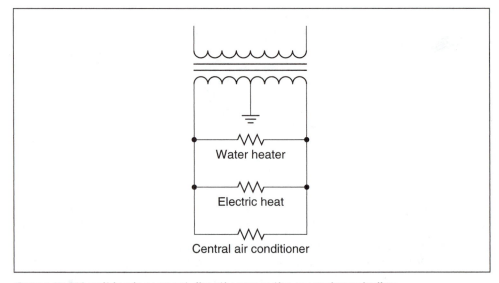

Figure 4-27 240-volt loads connect directly across the secondary winding.

Loads intended to operate on 120 volts connect from the center tap, or neutral, to one of the secondary lines. The function of the neutral is to carry the difference in current between the two secondary lines and maintain a balanced voltage. In the example shown in *Figure 4-28,* it is assumed that one of the secondary lines has a current flow of 30 amperes and

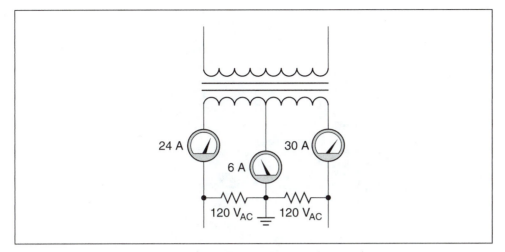

Figure 4-28 The neutral carries the sum of the unbalanced current.

3	5	10	15	25
37.5	50	75	100	167
250	333	500	833	1250
1667	2500	3333	5000	6667
8333	10,000	12,500	16,667	20,000
25,000	33,333			

Table 4-1 Common kVA Values for Single-Phase Transformers.

the other has a current flow of 24 amperes. The neutral will conduct the sum of the unbalanced load. In this example, the neutral current will be 6 A (30 − 24 = 6). Common kVA ratings for single-phase distribution transformers are given in *Table 4-1*.

Control Transformers

Another common type of isolation transformer found throughout industry is the **control transformer** *(Figure 4-29)*. The control transformer is used to reduce the line voltage to the value needed to operate control circuits. The most common type of control transformer contains two primary windings and one secondary. The primary windings are generally rated at

**control
transformer**

Figure 4-29 Control transformer with fuse protection added to the secondary winding (courtesy of Hevi-Duty Electric).

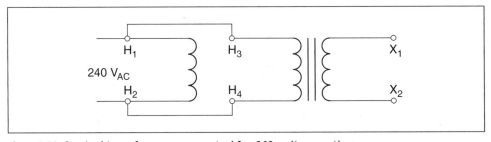

Figure 4-30 Control transformer connected for 240-volt operation.

240 volts each, and the secondary is rated at 120 volts. This provides a 2:1 turns ratio between each of the primary windings and the secondary. For example, assume that each of the primary windings contains 200 turns of wire. The secondary will contain 100 turns of wire.

One of the primary windings is labeled H_1 and H_2. The other is labeled H_3 and H_4. The secondary winding is labeled X_1 and X_2. If the primary of the transformer is to be connected to 240 volts, the two primary windings will be connected in parallel by connecting H_1 and H_3 together, and H_2 and H_4 together *(Figure 4-30)*. When the primary windings are connected in parallel, the same voltage is applied across both windings. This has the same effect as using one primary winding with a total of 200 turns of wire. A turns ratio of 2:1 is maintained and the secondary voltage will be 120 volts.

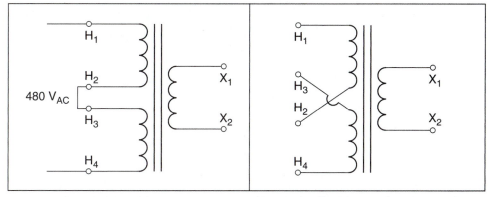

Figure 4-31 Control transformer connected   **Figure 4-32** The primary windings of a con-
for 480-volt operation.   trol transformer are crossed.

If the transformer is to be connected to 480 volts, the two primary windings
will be connected in series by connecting H_2 and H_3 together *(Figure 4-31)*.
The incoming power is connected to H_1 and H_4. Series-connecting the primary windings has the effect of increasing the number of turns in the primary
to 400. This produces a turns ratio of 4:1. When 480 volts is connected to the
primary, the secondary voltage will remain at 120.

The primary leads of a control transformer are generally cross-connected
as shown in *Figure 4-32*. This is done so that metal links can be used to connect the primary for 240 or 480 volt operation. If the primary is to be connected for 240 volt operation, the metal links will be connected under
screws as shown in *Figure 4-33*. Notice that leads H_1 and H_3 are connected
together and leads H_2 and H_4 are connected together. Compare this connection with the connection shown in *Figure 4-30*.

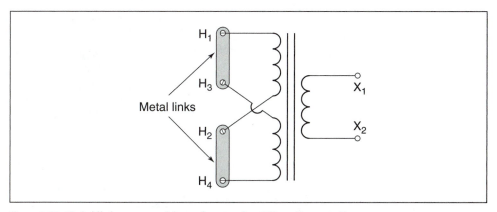

Figure 4-33 Metal links connect transformer for 240-volt operation.

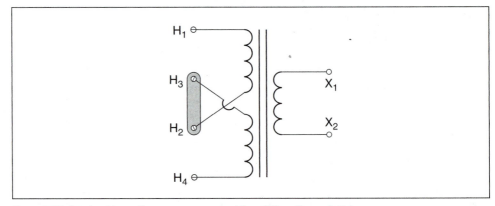

Figure 4-34 Control transformer connected for 480-volt operation.

If the transformer is to be connected for 480 volt operation, terminals H_2 and H_3 are connected as shown in *Figure 4-34*. Compare this connection with the connection shown in *Figure 4-31*.

Potential Transformer

The potential transformer is basically an isolation transformer with a high-voltage rating on the primary winding, and that is intended for use in metering circuits. They generally have low power ratings, such as 100 to 500 VA and have a standard secondary voltage rating of 120 volts. Potential transformers are commonly used to operate meters that require a voltage input such as voltmeters, wattmeters, varmeters, and so forth *(Figure 4-35)*. Since the

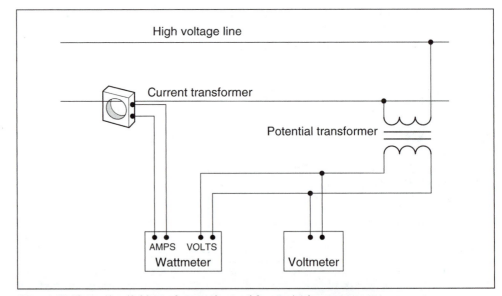

Figure 4-35 The potential transformer is used for metering purposes.

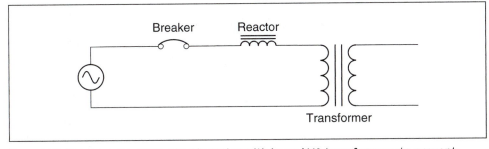

Figure 4-36 Reactors are connected in series with large kVA transformers to prevent excessive inrush current.

secondary voltage will be proportional to the voltage applied to the primary, they can be used to accurately measure high voltages without the necessity of trying to insulate a meter for operation at a high voltage. Potential transformers intended for connection to voltages of 25,000 volts or less can be designed for indoor and outdoor usage. For voltage above 25,000 volts they are generally designed for outdoor connection only.

Transformer Inrush Current

Although transformers and reactors are both inductive devices, there is a great difference in their operating characteristics. Reactors (chokes) are often connected in series with the primary winding of large kVA transformers to prevent inrush current from becoming excessive when a circuit is first turned on *(Figure 4-36)*. Transformers can produce extremely high inrush currents when power is first applied to the primary winding. The type of core used when constructing inductors and transformers is primarily responsible for this difference in characteristics.

Magnetic Domains

Magnetic materials contain tiny magnetic structures in their molecular material known as **magnetic domains**. These domains can be affected by outside sources of magnetism. *Figure 4-37* illustrates a magnetic domain that has not been polarized by an outside magnetic source.

magnetic domains

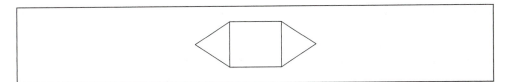

Figure 4-37 Magnetic domain in neutral position.

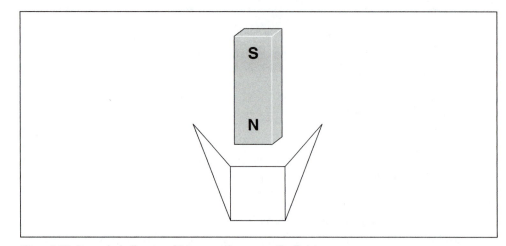

Figure 4-38 Domain influenced by a north magnetic field.

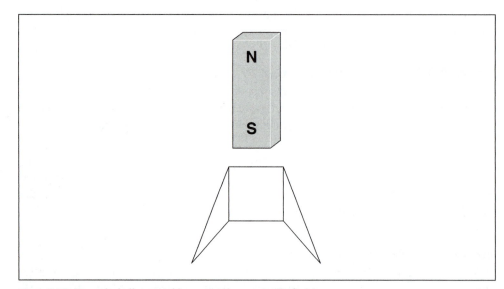

Figure 4-39 Domain influenced by a south magnetic field.

Now assume that the north pole of a magnet is placed toward the top of the material that contains the magnetic domains *(Figure 4-38)*. Notice that the structure of the domain has changed to realign the molecules in the direction of the outside magnetic field. If the polarity of the magnetic pole is changed *(Figure 4-39)*, the molecular structure of the domain will change to realign itself with the new magnetic lines of flux. This external influence can be produced by an electromagnet as well as a permanent magnet.

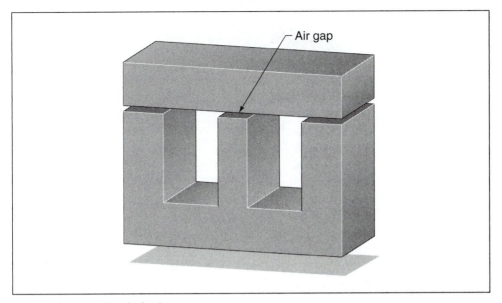

Figure 4-40 Core of an inductor.

In certain types of cores, the molecular structure of the domain will snap back to its neutral position when the magnetizing force is removed. This type of core is used in the construction of reactors or chokes *(Figure 4-40).* A core of this type is constructed by separating sections of the steel laminations with an air gap. This air gap breaks the magnetic path through the core material and is responsible for the domains returning to their neutral position once the magnetizing force is removed.

The core construction of a transformer, however, does not contain an air gap. The steel laminations are connected together in such a manner as to produce a very low-reluctance path for the magnetic lines of flux. In this type of core, the domains remain in their set position once the magnetizing force has been removed. This type of core "remembers" where it was last set. This was the principle of operation of the core memory of early computers. It is also the reason that transformers can have extremely high inrush currents when they are first connected to the power line.

The amount of inrush current in the primary of a transformer is limited by three factors:

1. Amount of applied voltage
2. Resistance of the wire in the primary winding
3. Flux change of the magnetic field in the core

The amount of flux change determines the amount of inductive reactance produced in the primary winding when power is applied. *Figure 4-41*

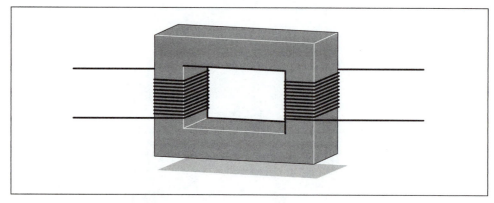

Figure 4-41 Isolation transformer.

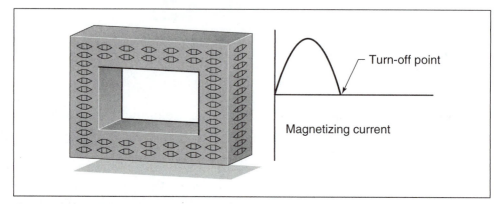

Figure 4-42 Magnetic domains are left in the neutral position.

illustrates a simple isolation-type transformer. The alternating current applied to the primary winding produces a magnetic field around the winding. As the current changes in magnitude and direction, the magnetic lines of flux change also. Since the lines of flux in the core are continually changing polarity, the magnetic domains in the core material are changing also. As stated previously, the magnetic domains in the core of a transformer remember their last set position. For this reason, the point on the waveform at which current is disconnected from the primary winding can have a great bearing on the amount of inrush current when the transformer is re-connected to power. For example, assume the power supplying the primary winding is disconnected at the zero crossing point *(Figure 4-42)*. In this instance, the magnetic domains would be set at the neutral point. When power is restored to the primary winding, the core material can be

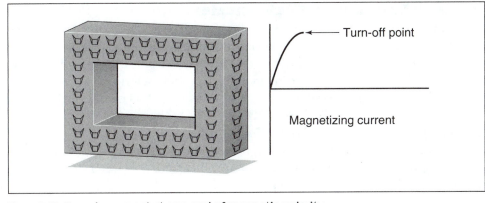

Figure 4-43 Domains are set at one end of magnetic polarity.

magnetized by either magnetic polarity. This permits a change of flux, which is the dominant current-limiting factor. In this instance, the amount of inrush current would be relatively low.

If the power supplying current to the primary winding is interrupted at the peak point of the positive or negative half-cycle, however, the domains in the core material will be set at that position. *Figure 4-43* illustrates this condition. It is assumed that the current was stopped as it reached its peak positive point. If the power is reconnected to the primary winding during the positive half-cycle, only a very small amount of flux change can take place. Since the core material is saturated in the positive direction, the primary winding of the transformer is essentially an air-core inductor, which greatly decreases the inductive characteristics of the winding. The inrush current in this situation would be limited by the resistance of the winding and a very small amount of inductive reactance.

This characteristic of transformers can be demonstrated with a clamp-on ammeter that has a "peak hold" capability. If the ammeter is connected to one of the primary leads and power is switched on and off several times, the amount of inrush current will vary over a wide range.

Transformer Polarities

To understand what is meant by transformer polarity, one must consider the voltage produced across a winding during some point in time. In a 60-Hz AC circuit, the voltage changes polarity 120 times per second. When discussing transformer polarity, it is necessary to consider the relationship between the different windings at the same point in time. It will, therefore, be assumed that this point in time is when the peak positive voltage is being produced across the winding.

Polarity Markings on Schematics

When a transformer is shown on a schematic diagram it is common practice to indicate the polarity of the transformer windings by placing a dot beside one end of each winding, as shown in *Figure 4-44*. These dots signify that the polarity is the same at that point in time for each winding. For example, assume the voltage applied to the primary winding is at its peak positive value at the terminal indicated by the dot. The voltage at the dotted lead of the secondary will be at its peak positive value at the same time.

This same type of polarity notation is used for transformers that have more than one primary or secondary winding. An example of a transformer with a multiple secondary is shown in *Figure 4-45*.

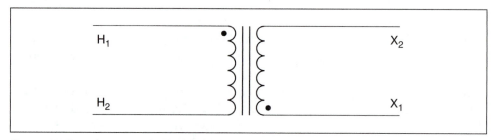

Figure 4-44 Transformer polarity dots.

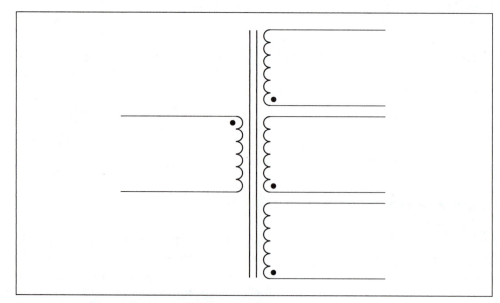

Figure 4-45 Polarity marks for multiple secondaries.

Additive and Subtractive Polarities

The polarity of transformer windings can be determined by connecting one lead of the primary to one lead of the secondary and testing for an increase or decrease in voltage. This is often referred to as a **buck** or **boost** connection *(Figure 4-46)*. The transformer shown in the example has a primary voltage rating of 120 volts and a secondary voltage rating of 24 volts. This same circuit has been redrawn in *Figure 4-47* to show the connection more clearly. Notice that the secondary winding has been connected in series with the primary winding. When 120 volts is applied to the primary winding, the voltmeter connected across the secondary will indicate either the *sum* of the two voltages or the *difference* between the two voltages. If this voltmeter indicates 144 volts (120 + 24 = 144) the windings are connected additive (boost), and polarity dots can be placed as shown in *Figure 4-48*. Notice in this connection that the secondary voltage is added to the primary voltage.

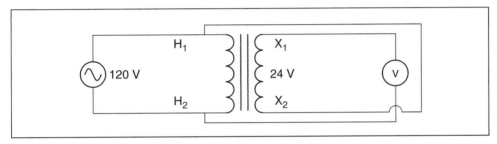

Figure 4-46 Connecting the secondary and primary windings.

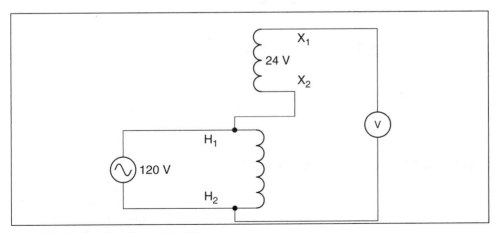

Figure 4-47 Redrawing the connection.

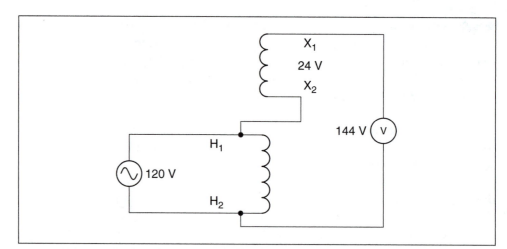

Figure 4-48 Placing polarity dots to indicate additive polarity.

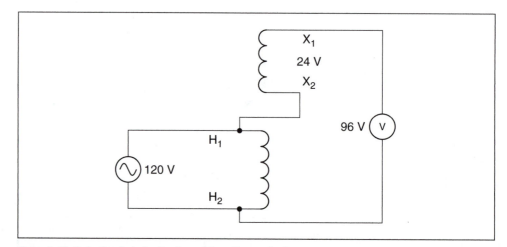

Figure 4-49 Polarity dots indicate subtractive polarity.

If the voltmeter connected to the secondary winding indicates a voltage of 96 volts (120 − 24 = 96), the windings are connected subtractive (buck), and polarity dots are placed as shown in *Figure 4-49*.

Using Arrows to Place Dots

To help in the understanding of additive and subtractive polarity, arrows can be used to indicate a direction of greater than or less than values. In *Figure 4-50*, arrows have been added to indicate the direction in which the dot is to be placed. In this example, the transformer is connected additive, or boost, and both of the arrows point in the same direction. Notice that the

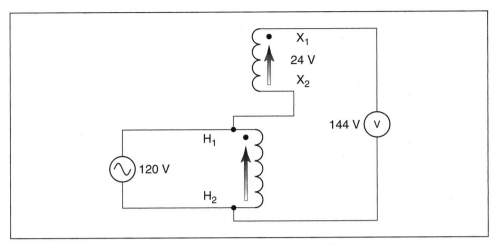

Figure 4-50 Arrows help indicate the placement of the polarity dots.

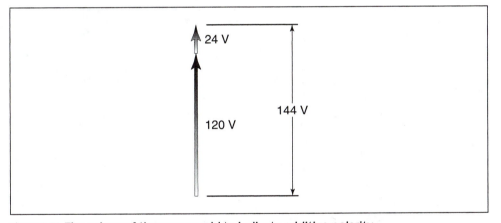

Figure 4-51 The values of the arrows add to indicate additive polarity.

arrow points to the dot. In *Figure 4-51*, it is seen that values of the two arrows add to produce 144 volts.

In *Figure 4-52*, arrows have been added to a subtractive, or buck, connection. In this instance, the arrows point in opposite directions, and the voltage of one tries to cancel the voltage of the other. The result is that the smaller value is eliminated, and the larger value is reduced as shown in *Figure 4-53*.

Voltage and Current Relationships in a Transformer

When the primary of a transformer is connected to power but there is no load connected to the secondary, current is limited by the inductive reactance of the primary. At this time, the transformer is essentially an inductor, and the excitation current is lagging the applied voltage by 90°

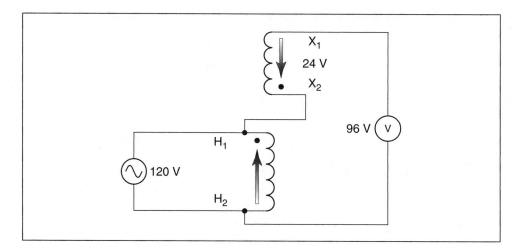

Figure 4-52 The arrows help indicate subtractive polarity.

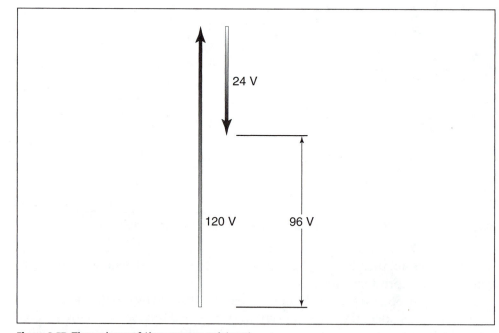

Figure 4-53 The values of the arrows subtract.

(Figure 4-54). The primary current induces a voltage in the secondary. This induced voltage is proportional to the rate of change of current. The secondary voltage will be at maximum during the periods that the primary current is changing the most (0°, 180°, and 360°), and it will be zero when the primary current is not changing (90° and 270°). If the primary current and

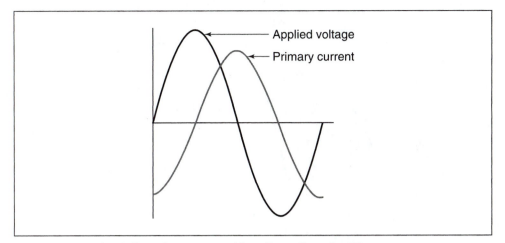

Figure 4-54 At no load, the primary current lags the voltage by 90°.

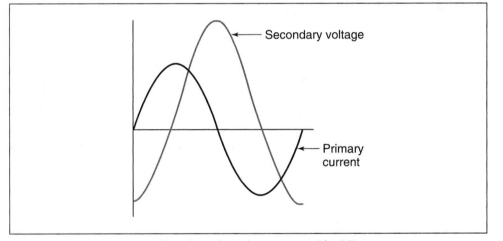

Figure 4-55 The secondary voltage lags the primary current by 90°.

secondary voltage are plotted, it will be seen that the secondary voltage lags the primary current by 90° *(Figure 4-55)*. Since the secondary voltage lags the primary current by 90° and the applied voltage leads the primary current by 90°, the secondary voltage is 180° out of phase with the applied voltage and in phase with the counter induced in the primary.

Adding Load to the Secondary

When a load is connected to the secondary, current begins to flow. Because the transformer is an inductive device, the secondary current lags the secondary voltage by 90°. Since the secondary voltage lags the primary

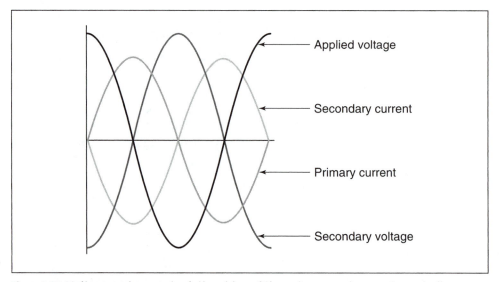

Figure 4-56 Voltage and current relationships of the primary and secondary windings.

current by 90°, the secondary current is 180° out of phase with the primary current *(Figure 4-56)*.

The current of the secondary induces a counter voltage in the secondary windings that is in opposition to the counter voltage induced in the primary. The counter voltage of the secondary weakens the counter voltage of the primary and permits more primary current to flow. As secondary current increases, primary current increases proportionally.

Since the secondary current causes a decrease in the counter voltage produced in the primary, the current of the primary is limited less by inductive reactance and more by the resistance of the windings as load is added to the secondary. If a wattmeter were connected to the primary, you would see that the true power would increase as load was added to the secondary.

Testing the Transformer

There are several tests that can be made to determine the condition of the transformer. A simple test for grounds, shorts, or opens can be made with an ohmmeter *(Figure 4-57)*. Ohmmeter A is connected to one lead of the primary and one lead of the secondary. This test checks for shorted windings between the primary and secondary. The ohmmeter should indicate infinity. If there is more than one primary or secondary winding, all isolated windings should be tested for shorts. Ohmmeter B illustrates testing the windings for grounds.

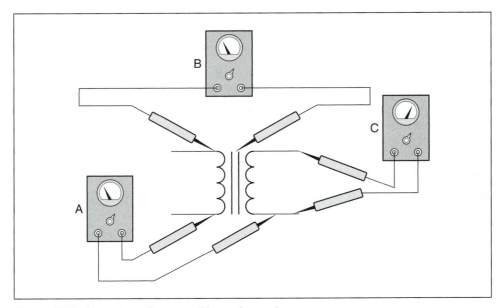

Figure 4-57 Testing a transformer with an ohmmeter.

One lead of the ohmmeter is connected to the case of the transformer, and the other is connected to the winding. All windings should be tested for grounds, and the ohmmeter should indicate infinity for each winding. Ohmmeter C illustrates testing the windings for continuity. The wire resistance of the winding should be indicated by the ohmmeter. Each winding should be tested for continuity.

If the transformer appears to be in good condition after the ohmmeter test, it should then be tested for shorts and grounds with a megohmmeter. A MEGGER® will reveal problems of insulation breakdown that an ohmmeter will not. Large oil-filled transformers should have the condition of the dielectric oil tested at periodic intervals. This involves taking a sample of the oil and performing tests for dielectric strength and contamination.

Transformer Ratings

Most transformers contain a **nameplate** that lists information concerning the transformer. The information listed is generally determined by the size, type, and manufacturer. Almost all nameplates will list the primary voltage, secondary voltage, and kVA rating. Transformers are rated in kilo-volt-amps and not kilowatts because the true power is determined by the power factor of the load. Other information that may be listed includes frequency, temperature rise in C°, percent impedance (%Z), type of insulating oil, gallons of

nameplate

insulating oil, serial number, type number, model number, and whether the transformer is single-phase or three-phase.

Determining Maximum Current

Notice that the nameplate does not list the current rating of the windings. Since power input must equal power output, the current rating for a winding can be determined by dividing the kVA rating by the winding voltage. For example, assume a transformer has a kVA rating of 0.5 kVA, a primary voltage of 480 volts, and a secondary voltage of 120 volts. To determine the maximum current that can be supplied by the secondary, divide the kVA rating by the secondary voltage.

$$I_S = \frac{kVA}{E_S}$$

$$I_S = \frac{500}{120}$$

$$I_S = 4.16 \text{ amps}$$

The primary current can be computed in the same way.

$$I_P = \frac{kVA}{E_P}$$

$$I_P = \frac{500}{480}$$

$$I_P = 1.04 \text{ amps}$$

Transformers with multiple secondary windings will generally have the current rating listed with the voltage rating.

Transformer Impedance

Transformer impedance is determined by the physical construction of the transformer. Factors such as the amount and type of core material, wire size used to construct the windings, number of turns, and the degree of magnetic coupling between the windings greatly affect the transformer's impedance. Impedance is expressed as a percentage and is measured by connecting a short circuit across the low-voltage winding of the transformer and then connecting a variable-voltage source to the high-voltage winding *(Figure 4-58)*. The variable voltage is then increased until rated current flows in the low-voltage winding. The transformer impedance is determined by calculating

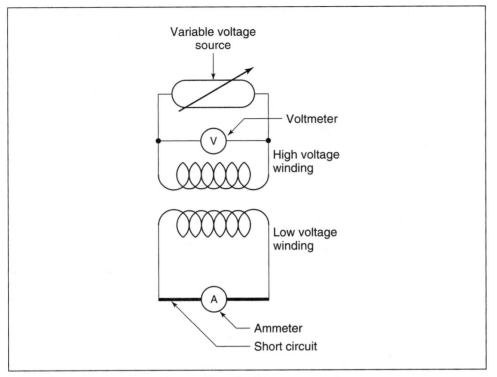

Figure 4-58 Determining transformer impedance.

the percentage of variable voltage compared to the rated voltage of the high-voltage winding.

Example

Assume that the transformer shown in *Figure 4-58* is a 2400/480 volt 15-kVA transformer. To determine the impedance of the transformer, first compute the full-load current rating of the secondary winding.

$$I = \frac{VA}{E}$$

$$I = \frac{15,000}{480}$$

$$I = 31.25A$$

Next, increase the source voltage connected to the high-voltage winding until a current of 31.25 amperes flows in the low-voltage winding. For the purpose of this example, assume that voltage value is 138 volts. Finally, determine the percentage of applied voltage as compared to the rated voltage.

$$\%Z = \frac{\text{Source Voltage}}{\text{Rated Voltage}} \times 100$$

$$\%Z = \frac{138}{2400} \times 100$$

$$\%Z = 0.0575 \times 100$$

$$\%Z = 5.75$$

The impedance of this transformer is 5.75%.

Transformer impedance is a major factor in determining the amount of voltage drop a transformer will exhibit between no load and full load and in determining the amount of current flow in a short-circuit condition. Short-circuit current can be computed using the formulas

$$\text{(Single-Phase) } I_{SC} = \frac{VA}{E \times \%Z}$$

$$\text{(Three-Phase) } I_{SC} = \frac{VA}{E \times \sqrt{3} \times \%Z}$$

One of the formulas for determining current in a single-phase circuit is:

$$I = \frac{VA}{E}$$

One of the formulas for determining current in a three-phase circuit is:

$$I = \frac{VA}{E \times \sqrt{3}}$$

The preceding formulas for determining short-circuit current can be modified to show that the short-circuit current can be computed by dividing the rated secondary current by the %Z.

$$I_{SC} = \frac{I_{Rated}}{\%Z}$$

Transformer Markings

The *American National Standards Institute* (ANSI) sets standardization rules concerning the way in which the terminals of transformers are to be marked. According to ANSI rules, the high-voltage leads of a transformer are to be marked H_1 – H_2, and so forth, and the low-voltage leads are marked X_1 – X_2, and so forth. The order is to be such that if the H_1 and X_1 terminals are connected together and voltage is applied to the primary of the transformer, the voltage measured between the highest numbered H lead and the highest numbered X lead should be less than the voltage of the high-voltage winding *(Figure 4-59)*. When the leads are marked in this manner, the transformer is considered to have subtractive polarity when the H_1 and X_1 leads are adjacent to each other. If the H_1 and X_1 leads are located diagonally, the transformer is considered to be additive polarity *(Figure 4-60)*.

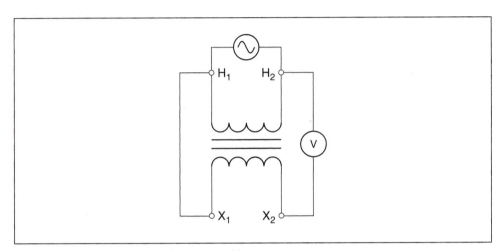

Figure 4-59 Determining transformer markings.

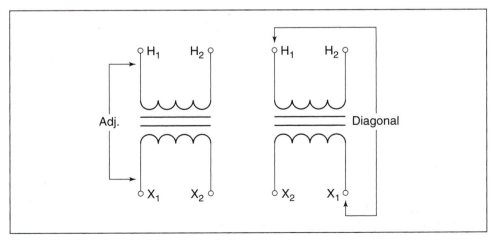

Figure 4-60 Additive and subtractive polarity.

Standard markings can become very important when it is necessary to connect transformers together. It is sometimes necessary to use single-phase transformers to form a three-phase bank. (This will be discussed fully later in this text.) Another instance is when transformers are to be connected in series or parallel. The secondary windings of transformers are sometimes connected in series to increase the output voltage or to form a center-tap connection. When this is done, the primary windings are connected in parallel to the power source *(Figure 4-61)*. Series connection of the secondary windings does not present a problem with balance because the same current must flow through both windings.

Connecting the secondary windings of transformers in parallel is generally avoided, but is sometimes necessary. When this connection is made, the primary windings are connected in parallel to the power source *(Figure 4-62)*. When transformers are connected in parallel, their characteristics must be the same or problems of imbalance can occur, causing one transformer to begin supplying current to the other. This causes one transformer to produce

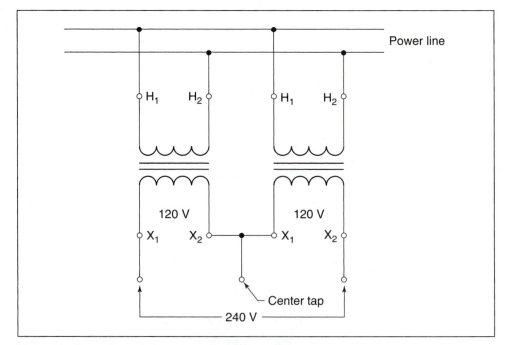

Figure 4-61 Secondary windings connected in series.

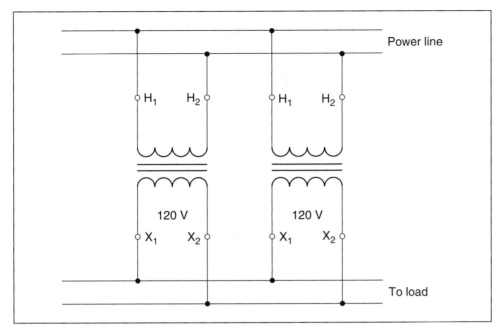

Figure 4-62 Secondary windings connected in parallel.

excessive current and the other to produce almost no current. Transformers intended for parallel connection should have the same voltage ratings, kVA ratings, and impedance.

Transformer Losses

Although transformers are probably the most efficient machines known, they are not perfect. A transformer operating at 90% efficiency has a power loss of 10%. Some of these losses are I^2R losses, eddy current losses, hysteresis losses, and magnetic flux leakage. Most of these losses result in heat production. Recall that I^2R is one of the formulas for finding power or wattage. In the case of a transformer, it describes the power loss associated with heat due to the resistance of the wire in both primary and secondary windings.

Eddy currents are currents that are induced into the metal-core material by the changing magnetic field as alternating current produces a changing flux. Eddy currents are so named because they circulate inside the metal in a manner similar to the swirling eddies in a river *(Figure 4-63)*. These swirling currents produce heat, which is a power loss. Transformers are constructed with laminated cores to help reduce eddy currents. The surface of each

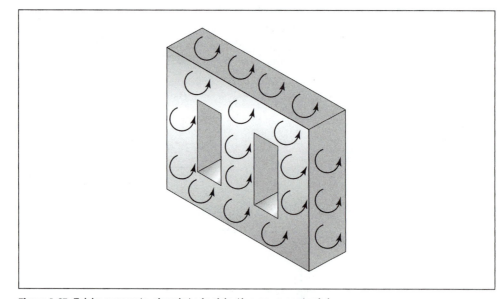

Figure 4-63 Eddy currents circulate inside the core material.

lamination forms a layer of iron oxide, which acts as an insulator to help prevent the formation of eddy currents.

Hysteresis losses are due to molecular friction. As discussed previously, the reversal of the direction of current flow causes the molecules of iron in the core to realign each time the current changes direction. The molecules of iron are continually rubbing against each other as they realign magnetically. The friction of the molecules rubbing together causes heat, which is a power loss. Hysteresis loss is proportional to frequency. The higher the frequency, the greater the loss. A special steel called *silicon steel* is often used in transformer cores to help reduce hysteresis loss. Power loss due to hysteresis and eddy currents is often called core losses.

Magnetic flux leakage does not produce heat, but does constitute a power loss. Flux leakage is caused by magnetic lines of flux radiating away from the transformer and not cutting the secondary windings. Flux leakage can be reduced by better core designs.

Transformer Voltage Regulation

The voltage regulation of a transformer is expressed as a percentage based on the difference between the secondary voltage at no load compared to full load. To determine regulation, load is added to the secondary until rated secondary voltage is applied across the load. The load is then removed, which will cause the secondary voltage to increase. The voltage regulation is the percentage of voltage increase compared to the rated voltage. The voltage regulation is proportional to the impedance of the transformer. Transformers with a lower-percent impedance will have better voltage regulation characteristics.

Constant-Current Transformers

A very special type of isolation transformer is the *constant-current transformer*, or *current regulator*. Constant-current transformers are designed to deliver a constant current, generally 6.6 amperes, under varying load conditions. They are most often used to provide the power for series-connected street lights. Street lights are often connected in series instead of parallel because of the savings in wire. Series-connected lights require one conductor run from lamp to lamp instead of two conductors *(Figure 4-64)*. When series-connected lights are installed, a device must be used to provide a connection if one of the lamps should fail. Some lights use a reactor coil connected in parallel with the lamp *(Figure 4-65)*. If the lamp should fail,

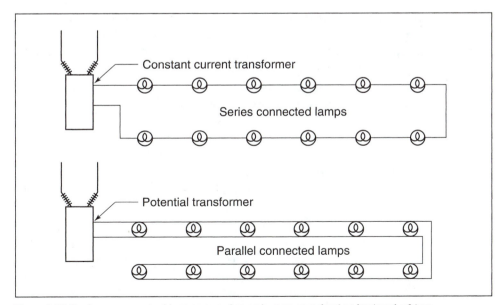

Figure 4-64 Series-connected lamps require only one conductor instead of two.

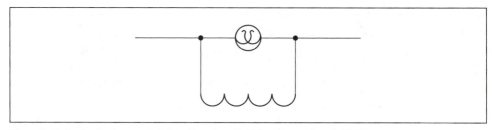

Figure 4-65 An inductor maintains the circuit if the lamp should fail.

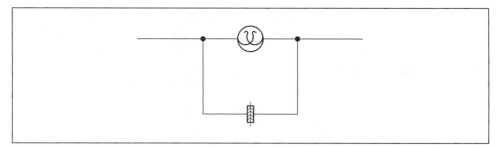

Figure 4-66 A film cut-out device shorts and maintains the circuit if the lamp should fail.

the circuit is continued through the reactor coil. Others use a film cut-out device *(Figure 4-66)*. The film cut-out device consists of two pieces of metal separated by an insulating film designed to puncture at a predetermined voltage. As long as the lamp is in operation, the voltage drop across the lamp is insufficient to rupture the film. If the lamp should burn out and become an open circuit, the entire circuit voltage is dropped across the film cut-out.

Constant-current transformers contain primary and secondary windings that are movable with respect to each other. Either winding can be made movable. Both windings are mounted on the same core material *(Figure 4-67)*. Some transformers attach a counterweighted lever to the moving coil to help balance the weight of the movable coil.

The constant-current regulator operates by producing a magnetic field in the secondary windings that is in opposition to the magnetic field produced in the primary winding. If the load current of the secondary increases, the magnetic repulsion increases and the two coils move farther apart *(Figure 4-68)*. This produces a greater amount of flux leakage between the two windings and causes a reduction in the secondary voltage. If the secondary

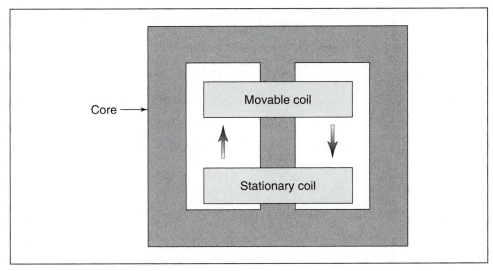

Figure 4-67 Constant-current transformers contain both a stationary and a movable coil.

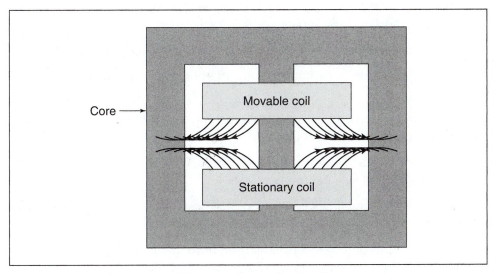

Figure 4-68 The magnetic flux of the two coils repel each other.

current should decrease, the magnetic field of the secondary decreases also. This permits the two windings to move closer together, which reduces the flux leakage and secondary voltage increases. Most moving coil-type current regulators contain a dashpot mechanism to reduce sudden changes in the spacing between the two coils. This helps reduce any "hunting" action between the two coils.

Summary

1. All values of voltage, current, and impedance in a transformer are proportional to the turns ratio.

2. Transformers can change values of voltage, current, and impedance, but cannot change the frequency.

3. The primary winding of a transformer is connected to the power line.

4. The secondary winding is connected to the load.

5. A transformer that has a lower secondary voltage than primary voltage is a step-down transformer.

6. A transformer that has a higher secondary voltage than primary voltage is a step-up transformer.

7. An isolation transformer has its primary and secondary windings electrically and mechanically separated from each other.

8. When a coil induces a voltage into itself, this is known as self-induction.

9. When a coil induces a voltage into another coil, it is known as mutual induction.

10. Transformers can have very high inrush current when first connected to the power line because of the magnetic domains in the core material.

11. Inductors provide an air gap in their core material that causes the magnetic domains to reset to a neutral position.

12. Inductors are sometimes connected in series with the primary winding of large transformers to reduce initial inrush current.

13. Either winding of a transformer can be used as the primary or secondary as long as its voltage or current ratings are not exceeded.

14. Isolation transformers help filter voltage and current spikes between the primary and secondary side.

15. Polarity dots are often added to schematic diagrams to indicate transformer polarity.

16. Transformers can be connected as additive or subtractive polarity.

17. Transformer impedance is expressed as a percentage.

18. Voltage regulation is a percentage based on the change in voltage between the voltage at full load and the voltage at no load.

19. The high-voltage leads of a transformer are marked with an H, and the low-voltage leads are marked with an X.

Review Questions

1. What is a transformer?

2. What are common efficiencies for transformers?

3. What is an isolation transformer?

4. All values of a transformer are proportional to its _____ _____ .

5. A transformer has a primary voltage of 480 volts and a secondary voltage of 20 volts. What is the turns ratio of the transformer?

6. If the secondary of the transformer in question 5 supplies a current of 9.6 amperes to a load, what is the primary current? (Disregard excitation current.)

7. Explain the difference between a step-up and a step-down transformer.

8. A transformer has a primary voltage of 240 volts and a secondary voltage of 48 volts. What is the turns ratio of this transformer?

9. A transformer has an output of 750 VA. The primary voltage is 120 V. What is the primary current?

10. A transformer has a turns ratio of 1:6. The primary current is 18 amperes. What is the secondary current?

11. What do the dots shown beside the terminal leads of a transformer represent on a schematic?

12. A transformer has a primary voltage rating of 240 volts and a secondary voltage rating of 80 volts. If the windings were connected subtractive, what voltage would appear across the entire connection?

13. If the windings of the transformer in question 12 were to be connected additive, what voltage would appear across the entire winding?

14. The primary leads of a transformer are labeled 1 and 2. The secondary leads are labeled 3 and 4. If polarity dots are placed beside leads 1 and 4, which secondary lead would be connected to terminal 2 to make the connection additive?

Problems

Refer to *Figure 4-69* to answer the following questions. Find all the missing values.

1.

E_P 120	E_S 24
I_P _____	I_S _____
N_P 300	N_S _____
Ratio _____	$Z = 3\ \Omega$

2.

E_P 240	E_S 320
I_P _____	I_S _____
N_P _____	N_S 280
Ratio _____	$Z = 500\ \Omega$

3.

E_P _____	E_S 160
I_P _____	I_S _____
N_P _____	N_S 80
Ratio 1:2.5	$Z = 12\ \Omega$

4.

E_P 48	E_S 240
I_P _____	I_S _____
N_P 220	N_S _____
Ratio _____	$Z = 360\ \Omega$

5.

E_P _____	E_S _____
I_P 16.5	I_S 3.25
N_P _____	N_S 450
Ratio _____	$Z = 56\ \Omega$

6.

E_P 480	E_S _____
I_P _____	I_S _____
N_P 275	N_S 525
Ratio _____	$Z = 1.2\ k\Omega$

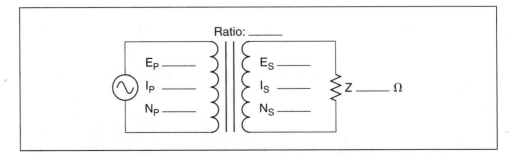

Figure 4-69 Practice problems 1 through 6.

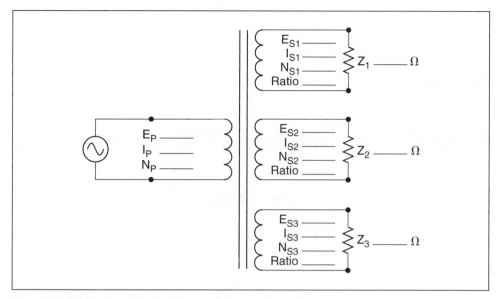

Figure 4-70 Practice problems 7 through 8.

Refer to *Figure 4-70* to answer the following questions. Find all the missing values.

7.

E_P 208	E_{S1} 320	E_{S2} 120	E_{S3} 24
I_P _____	I_{S1} _____	I_{S2} _____	I_{S3} _____
N_P 800	N_{S1} _____	N_{S2} _____	N_{S3} _____
	Ratio 1:	Ratio 2:	Ratio 3:
	R_1 12 kΩ	R_2 6 Ω	R_3 8 Ω

8.

E_P 277	E_{S1} 480	E_{S2} 208	E_{S3} 120
I_P _____	I_{S1} _____	I_{S2} _____	I_{S3} _____
N_P 350	N_{S1} _____	N_{S2} _____	N_{S3} _____
	Ratio 1:	Ratio 2:	Ratio 3:
	R_1 200 Ω	R_2 60 Ω	R_3 24 Ω

UNIT 5

Autotransformers

Objectives

After studying this unit, you should be able to:

- Discuss the operation of an autotransformer.
- List differences between isolation transformers and autotransformers.
- Compute values of voltage, current, and turns ratios for autotransformers.
- Connect an autotransformer for operation.

The word *auto* means self. An autotransformer is literally a *self-transformer*. It uses the same winding as both primary and secondary transformers. Recall that the definition of a primary winding is a winding that is connected to the source of power, and the definition of a secondary winding is a winding that is connected to a load. Autotransformers have very high efficiencies, most in the range of 95% to 98%.

In *Figure 5-1*, the entire winding is connected to the power source, and part of the winding is connected to the load. All the turns of wire form the primary while part of the turns form the secondary. Since the secondary part of the winding contains fewer turns than the primary section, the secondary will produce less voltage. This autotransformer is a step-down transformer.

In *Figure 5-2*, the primary section is connected across part of a winding and the secondary is connected across the entire winding. The secondary section contains more windings than the primary. This type of transformer is

114

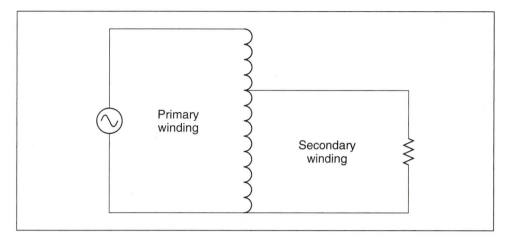

Figure 5-1 Autotransformer used as a step-down transformer.

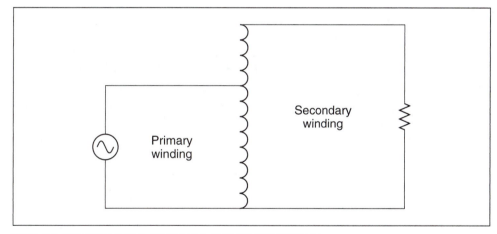

Figure 5-2 Autotransformer used as a step-up transformer.

a step-up transformer. (Notice that autotransformers, like isolation transformers, can be used as step-up or step-down transformers.)

DETERMINING VOLTAGE VALUES

Autotransformers are not limited to a single secondary winding. Many autotransformers have multiple taps to provide different voltages as shown in *Figure 5-3*. In this example, there are 40 turns of wire between taps A and B, 80 turns of wire between taps B and C, 100 turns of wire between taps C and D, and 60 turns of wire between taps D and E. The primary section of the windings is connected between taps B and E. The primary is connected to a 120-volt source.

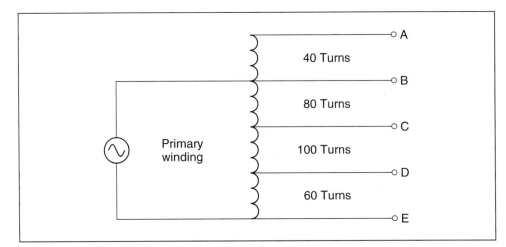

Figure 5-3 Autotransformer with multiple taps.

There is generally more than one method that can be employed to determine values of a transformer. Since the number of turns between each tap is known, the volts-per-turn method will be used in this example. *The volts-per-turn for any transformer is determined by the primary winding.* In *Figure 5-3*, the primary winding is connected across taps B and E. The number of primary turns are, therefore, the sum of the turns between taps B and E (80 + 100 + 60 = 240 turns). Since 120 volts is connected across 240 turns, this transformer will have a volts-per-turn ratio of 0.5 (240 turns/120 volts = 0.5 volts-per-turn). To determine the amount of voltage between each set of taps, multiply the number of turns by the volts-per-turn, as shown here:

A–B: (40 turns x 0.5 = 20 volts)
A–C: (120 turns x 0.5 = 60 volts)
A–D: (220 turns x 0.5 = 110 volts)
A–E: (280 turns x 0.5 = 140 volts)
B–C: (80 turns x 0.5 = 40 volts)
B–D: (180 turns x 0.5 = 90 volts)
B–E: (240 turns x 0.5 = 120 volts)
C–D: (100 turns x 0.5 = 100 volts)
C–E: (160 turns x 0.5 = 80 volts)
D–E: (60 turns x 0.5 = 30 volts)

USING TRANSFORMER FORMULAS

The values of voltage and current for autotransformers can also be determined by using standard transformer formulas. The primary winding of the transformer shown in *Figure 5-4* is between points B and N, and has

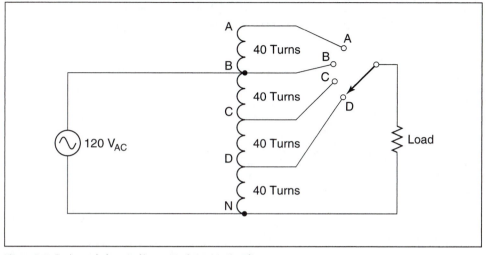

Figure 5-4 Determining voltage and current values.

120 volts applied to it. There are 120 turns of wire between points B and N. Now assume that the selector switch is set to point D. The load is now connected between points D and N. The secondary of this transformer contains 40 turns of wire. If the amount of voltage applied to the load is to be computed, the following formula can be used:

$$\frac{E_P}{E_S} = \frac{N_P}{N_S}$$

$$\frac{120}{E_S} = \frac{120}{40}$$

$$120\,E_S = 4800$$

$$E_S = 40 \text{ volts}$$

Assume that the load connected to the secondary has an impedance of 10 Ω. The amount of current flow in the secondary circuit can be computed using the formula

$$I = \frac{E}{Z}$$

$$I = \frac{40}{10}$$

$$I = 4 \text{ amps}$$

The primary current can be computed by using the same formula that was used to compute primary current for an isolation-type transformer, as follows:

$$\frac{E_P}{E_S} = \frac{I_S}{I_P}$$

$$\frac{120}{40} = \frac{4}{I_P}$$

$$I_P = 1.333 \text{ amps}$$

The amount of power input and output for the autotransformer must also be the same.

Primary	Secondary
120 x 1333 = 160 volt amps	40 x 4 = 160 volt amps

Now assume that the rotary switch is connected to point A. The load is now connected to 160 turns of wire. The voltage applied to the load can be computed by

$$\frac{E_P}{E_S} = \frac{N_P}{N_S}$$

$$\frac{120}{E_S} = \frac{120}{60}$$

$$120\ E_S = 19{,}200$$

$$E_S = 160 \text{ volts}$$

The amount of secondary current can be computed using the formula

$$I = \frac{E}{Z}$$

$$I = \frac{160}{10}$$

$$I = 16 \text{ amps}$$

The primary current can be computed using the formula

$$\frac{E_P}{E_S} = \frac{I_S}{I_P}$$

$$\frac{120}{160} = \frac{16}{I_P}$$

$$120\ I_P = 2560$$

$$I_P = 21.333 \text{ amps}$$

The answers can be checked by determining if the power in and power out are the same.

Primary	Secondary
120 x 21.333 = 2560 volt amps	160 x 16 = 2560 volt amps

CURRENT RELATIONSHIPS

An autotransformer with a 2:1 turns ratio is shown in *Figure 5-5*. A voltage of 480 volts is connected across the entire winding. Since the transformer has a turns ratio of 2:1, a voltage of 240 volts will be supplied to the load. Ammeters connected in series with each winding indicate the current flow in the circuit. It is assumed that the load produces a current flow of 4 amperes on the secondary. Note that a current flow of 2 amperes is supplied to the primary.

$$I_{Primary} = \frac{I_{Secondary}}{Ratio}$$

$$I_P = \frac{4}{2}$$

$$I_P = 2 \text{ amperes}$$

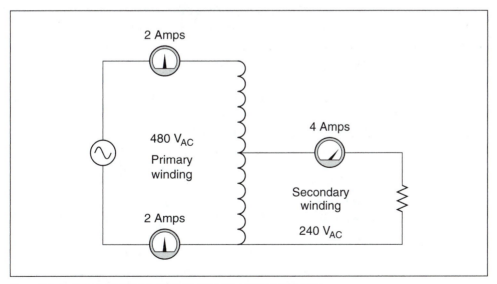

Figure 5-5 Current divides between primary and secondary.

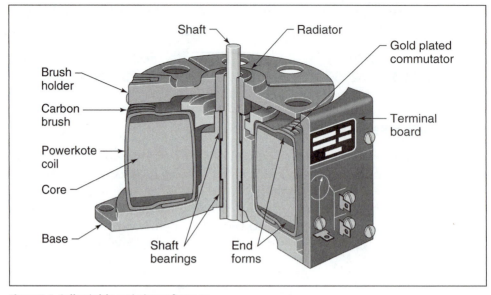

Figure 5-6 Adjustable autotransformer.

If the rotary switch shown in *Figure 5-4* were to be removed and replaced with a sliding tap that made contact directly with the transformer winding, the turns ratio could be adjusted continuously. This type of transformer is commonly referred to as a Variac or Powerstat depending on the manufacturer. A variable autotransformer is shown in *Figure 5-6*. The windings are wrapped around a tape-wound toroid core inside a plastic case. The top of the windings has been milled flat, similar to a commutator. A carbon brush makes contact with the windings. When the brush is moved across the windings the turns ratio changes, which changes the output voltage. This type of autotransformer provides a very efficient means of controlling AC voltage.

Autotransformers are often used by power companies to provide a small increase or decrease to line voltage. They help provide voltage regulation to large power lines. A 600-MVA (mega-volt-amp) three-phase autotransformer is shown in *Figure 5-7*. Notice the cooling fins on the side of the transformer. These fins are actually hollow tubes through which dielectric oil is circulated to provide cooling.

The autotransformer does have one disadvantage: since the load is connected to one side of the power line, there is no line isolation between the incoming power and the load. This can cause problems with certain types of equipment and must be a consideration when designing a power system.

Figure 5-7 600 MVA autotransformer (courtesy of Reliant Energy).

Summary

1. The autotransformer has only one winding, which is used as both the primary and secondary.

2. Autotransformers have efficiencies that range from about 95% to 98%.

3. Values of voltage, current, and turns can be computed in the same manner as an isolation transformer.

4. Autotransformers can be step-up or step-down transformers.

5. Autotransformers can be made to provide a variable-output voltage by connecting a sliding tap to the windings.

6. Autotransformers have the disadvantage of no line isolation between primary and secondary.

7. One of the simplest ways of computing values of voltage for an autotransformer when the turns are known is to use the volts-per-turn method.

Review Questions

1. An AC power source is connected across 325 turns of an autotransformer, and the load is connected across 260 turns. What is the turns ratio of this transformer?

2. Is the transformer in question 1 a step-up or step-down transformer?

3. An autotransformer has a turns ratio of 3.2:1. A voltage of 208 volts is connected across the primary. What is the voltage of the secondary?

4. A load impedance of 52 Ω is connected to the secondary winding of the transformer in question 3. How much current will flow in the secondary?

5. How much current will flow in the primary of the transformer in question 4?

6. The autotransformer shown in *Figure 5-3* has the following number of turns between windings: A–B (120 turns), B–C (180 turns), C–D (250 turns), and D–E (300 turns). A voltage of 240 volts is connected across B and E. Find the voltages between each of the following points:

A–B _____   A–C _____   A–D _____   A–E _____   B–C _____   B–D _____

B–E _____   C–D _____   C–E _____   D–E _____

Problems

Refer to the transformer shown in *Figure 5-8* to answer the following questions.

1. Assume that a voltage of 208 volts is applied across terminals B and E. How much voltage is across each of the following terminals?

A–B _____   A–C _____   A–D _____   A–E _____   B–C _____   B–D _____

B–E _____   C–D _____   C–E _____   D–E _____

2. Assume a voltage of 120 volts is connected across terminals B and E, and that a load impedance of 20 Ω is connected across terminals A and C. What is the secondary and primary current?

I_P _____ I_S _____

3. What is the turns ratio of the winding between points B and E as compared to the winding between points D and E?

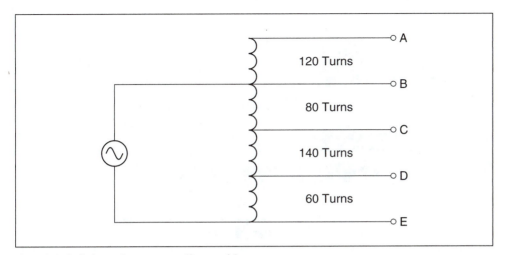

Figure 5-8 Autotransformer practice problems.

4. Assume that a voltage of 480 volts is connected across terminals A and E. What is the voltage across terminals C and D?

5. The transformer shown in *Figure 5-8* is supplying 325 volt amperes to a load. The primary voltage is 240 volts. What is the primary current?

UNIT **6**

Current Transformers

After studying this unit, you should be able to:

- Discuss the operation of a current transformer.
- Describe how current transformers differ from voltage transformers.
- Discuss safety precautions that should be observed when using current transformers.
- Connect a current transformer in a circuit.

Current transformers differ from voltage transformers in that the primary winding is generally part of the power line. The primary winding of a current transformer must be connected in series with the load *(Figure 6-1)*. Current transformers are used to change the full-scale range of AC ammeters. Most in-line ammeters (ammeters that must be connected directly into the line) that have multiple-range values use a current transformer to provide the different ranges *(Figure 6-2)*. The full-scale value of the ammeter is changed by adjusting the turns ratio. Assume that the ammeter in *Figure 6-2* is to provide range values of 5 A, 2.5 A, 1 A, and 0.5 A. Also, assume that the meter movement requires a current flow of 100 mA (0.100) to deflect the meter full scale and that the primary of the current transformer contains 5 turns of wire. Transformer formulas can be used to determine the number of secondary turns needed to produce the desired ranges.

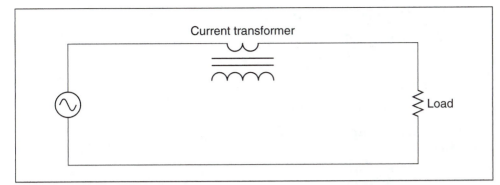

Figure 6-1 The primary winding of a current transformer is connected in series with a load.

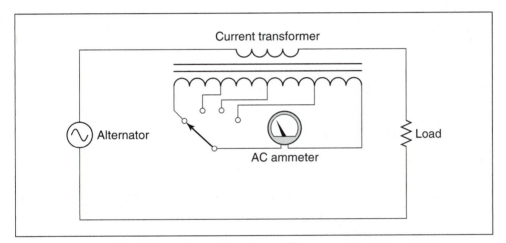

Figure 6-2 A current transformer is used to change the range of an AC ammeter.

The number of turns needed for a full-scale range of 5 amperes can be calculated as follows:

$$\frac{N_P}{N_S} = \frac{I_S}{I_P}$$

$$\frac{5}{N_S} = \frac{0.1}{5}$$

$$0.1\, N_S = 25$$

$$N_S = 250 \text{ turns}$$

To calculate the number of turns needed for a full-scale range of 2.5 amperes:

$$\frac{5}{N_S} = \frac{0.1}{2.5}$$

$$0.1\ N_S = 12.5$$

$$N_S = 125 \text{ turns}$$

The number of turns needed for a full-scale range of 1 ampere can be determined by

$$\frac{5}{N_S} = \frac{0.1}{1}$$

$$0.1\ N_S = 5$$

$$N_S = 50 \text{ turns}$$

Use the following formula to find the number of turns needed for a full-scale range of 0.5 ampere:

$$\frac{5}{N_S} = \frac{0.1}{0.5}$$

$$0.1\ N_S = 2.5$$

$$N_S = 25 \text{ turns}$$

An in-line ammeter that can be set for different scale values is shown in *Figure 6-3*.

When a large amount of AC current must be measured, a different type of current transformer is connected in the power line. The ratios of these transformers range from 200:5 to several thousand to 5. This type of current transformer, generally referred to in industry as "CT," has a standard secondary-current rating of 5 amps AC. They are designed to be operated with a 5-amp AC ammeter connected directly to their secondary winding, which produces a short circuit. CTs are designed to operate with the secondary winding shorted. *The secondary winding of a CT should never be opened when there is power applied to the primary. This will cause the transformer to produce a step-up in voltage that could be high enough to kill anyone who comes in contact with it.*

A current transformer of this type is basically a toroid transformer, a transformer constructed with a hollow core, similar to a donut in that it has a hole in the middle *(Figure 6-4)*. When current transformers are used, the main power line is inserted through the opening in the transformer *(Figure 6-5)*. The power line acts as the primary of the transformer and is considered to be one turn.

Figure 6-3 In-line ammeter.

Figure 6-4 Current transformer (CT) with a ratio of 1500:5 (courtesy of Square D Company).

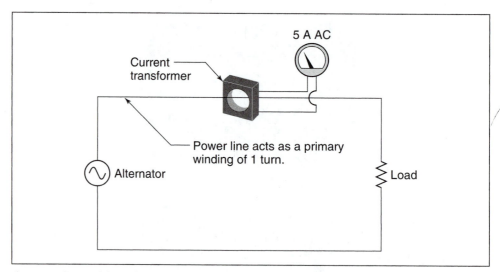

Figure 6-5 Current transformer used to change the scale factor of an AC ammeter.

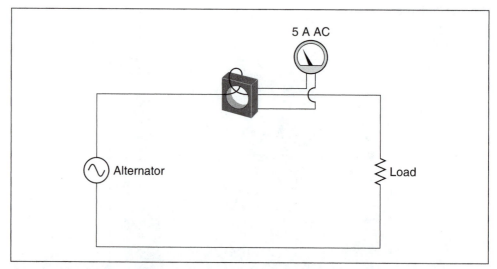

Figure 6-6 The primary conductor loops through the CT to produce a second turn, changing the turns ratio.

The turns ratio of the transformer can be changed by looping the power wire through the opening in the transformer to produce a primary winding of more than one turn. For example, assume that a current transformer has a ratio of 600:5. If the primary power wire is inserted through the opening, it will require a current of 600 amps to deflect the meter full scale. If the primary power conductor is looped around and inserted through the window a second time, the primary now contains two turns of wire instead of one *(Figure 6-6)*.

It now requires 300 amps of current flow in the primary to deflect the meter full scale. If the primary conductor is looped through the opening a third time, it would require only 200 amps of current flow to deflect the meter full scale.

CLAMP-ON AMMETERS

Many electricians use the clamp-on type of AC ammeter *(Figure 6-7)*. To use this meter, clamp the jaw of the meter around one of the conductors supplying power to the load *(Figure 6-8)*. Clamp the meter around only one of the lines. If the meter is clamped around more than one line, the magnetic fields of the wires cancel each other and the meter indicates zero.

This type of meter uses a current transformer to operate. The jaw of the meter is part of the core material of the transformer. When the meter is connected around the current-carrying wire, the changing magnetic field produced by the AC current induces a voltage into the current transformer. The strength of the magnetic field and its frequency determine the amount of voltage induced in the current transformer. Since 60 Hz is a standard frequency throughout the United States and Canada, the amount of induced voltage is proportional to the strength of the magnetic field.

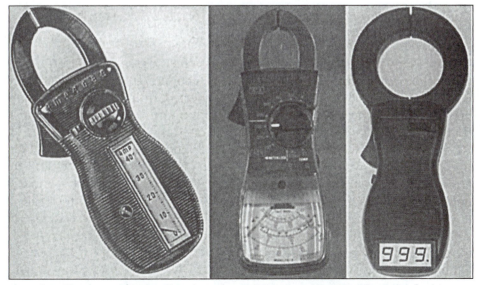

Figure 6-7 Clamp-on AC ammeters (courtesy of Advanced Test Products, Inc.).

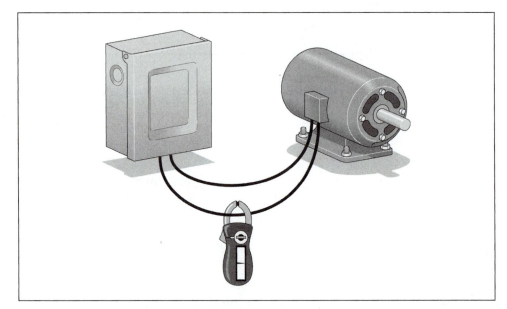

Figure 6-8 The clamp-on ammeter connects around only one conductor.

The clamp-on ammeter can have different range settings by changing the turns ratio of the secondary of the transformer just as the in-line ammeter does. The primary of the transformer is the conductor that the movable jaw is connected around. If the ammeter is connected around one wire, the primary has one turn of wire as compared to the turns of the secondary. The turns ratio can be changed in the same manner as changing the ratio of the CT. If two turns of wire are wrapped around the jaw of the ammeter, *Figure 6-9*, the primary winding now contains two turns instead of one, and the turns ratio of the transformer is changed. The ammeter will now indicate double the amount of the current in the circuit. The reading on the scale of the meter would have to be divided by two to get the correct reading. The ability to change the turns ratio of a clamp-on ammeter can be very useful for measuring low current.

Changing the turns ratio is not limited to wrapping two turns of wire around the jaw of the ammeter. Any number of turns can be wrapped around the jaw. Divide the reading by that number. A problem with many clamp-on ammeters is that their lowest scale value is too high to accurately measure low-current values. An ammeter with a full-scale range of 0–6 amperes would not be able to measure accurately a current flow of 0.2 amperes. The answer to this problem is to wrap multiple turns of wire around the jaw of the ammeter to change the secondary-scale value. If 10 turns of wire are wrapped around the jaw of a 0–6 ampere meter, the meter will indicate a full-scale

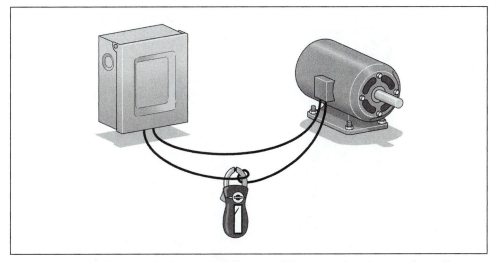

Figure 6-9 Looping the conductor around the jaw of the ammeter changes the ratio.

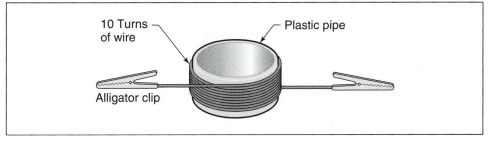

Figure 6-10 A simple scale divider for clamp-on ammeters.

value of 0.6 ampere. The meter could now accurately measure a current of 0.2 ampere. In the field, however, it is not often that there is enough slack wire to make 10 wraps around the jaw of an ammeter. A simple device can be constructed to overcome this problem. Assume that 10 turns of wire is wound around a piece of nonconductive material, such as a piece of plastic pipe *(Figure 6-10)*. Plastic tape is then used to prevent the wire from slipping off the plastic core. If alligator clips are attached to the ends of the wire, the device can be inserted in series with the load in a similar manner to using an inline ammeter. The jaw of the ammeter can then be inserted through the opening in the plastic pipe *(Figure 6-11)*. The current value is read by moving the decimal one place to the left. A full-scale value of 0–6 amperes becomes a full-scale value of 0–0.6 ampere, or a 0–30 ampere scale becomes a 0–3 ampere scale.

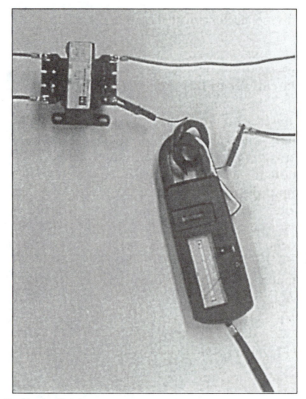

Figure 6-11 Using the scale divider.

Summary

1. Current transformers have their primary winding connected in series with a load.

2. Current transformers are often used to provide multiple-scale values for in-line AC ammeters.

3. Current transformers are often referred to as CTs.

4. CTs are used to measure large amounts of AC current.

5. CTs have a standard secondary-current value of 5 amperes.

6. CTs are designed to be operated with their secondary winding shorted.

7. The short circuit connected across the secondary of the CT should never be removed when power is connected to the circuit because the secondary voltage can become very high.

8. Many clamp-on AC ammeters operate on the principle of a current transformer.

9. The movable jaw of the clamp-on ammeter is the core of the transformer.

10. The secondary-current value of a current transformer can be changed by changing the turns of wire of the primary.

Review Questions

1. Explain the difference in connection between the primary winding of a voltage transformer and the primary winding of a current transformer.

2. What is the standard current rating for the secondary winding of a CT?

3. Why should the secondary winding of a CT never be disconnected from its load when there is current flow in the primary?

4. A current transformer has a ratio of 600:5. If three loops of wire are wound through the transformer core, how much primary current is required to produce 5 amperes of current in the secondary winding?

5. Assume that a primary current of 75 amperes flows through the windings of the transformer in question 4. How much current will flow in the secondary winding?

6. What type of core is generally used in the construction of a CT?

7. A current transformer has 4 turns of wire in its primary winding. How many turns of wire are needed in the secondary winding to produce a current of 2 amperes when a current of 60 amperes flows in the primary winding?

8. A 1500:5 CT develops 3 volts across the primary winding. If the secondary is disconnected from its load, how much voltage would be developed across the secondary terminals?

9. A CT has a current flow of 80 amperes in its primary winding and a current of 2 amperes in its secondary winding. What is the ratio of the CT?

10. What is the most common use for a CT?

UNIT 7

Three-Phase Circuits

Objectives

After studying this unit, you should be able to:

- Discuss the differences between three-phase and single-phase voltages.

- Discuss the characteristics of delta and wye connections.

- Compute voltage and current values for delta and wye circuits.

- Connect delta and wye circuits and make measurements with measuring instruments.

- Compute the amount of capacitance needed to correct the power factor of a three-phase motor.

Most of the electrical power generated in the world today is three-phase. Three-phase power was first conceived by Nikola Tesla. In the early days of electrical power generation, Tesla not only led the battle concerning whether the nation should be powered with low-voltage direct current or high-voltage alternating current, but he also proved that three-phase power was the most efficient way that electricity could be produced, transmitted, and consumed.

THREE-PHASE CIRCUITS

There are several reasons why three-phase power is superior to single-phase power.

1. The horsepower rating of three-phase motors and the kilovolt-amp rating of three-phase transformers are about 150% greater than for single-phase motors or transformers with a similar frame size.
2. The power delivered by a single-phase system pulsates *(Figure 7-1)*. The power falls to zero three times during each cycle. The power delivered by a three-phase circuit pulsates, but it never falls to zero *(Figure 7-2)*. In a three-phase system, the power delivered to the load is the same at any instant. This produces superior operating characteristics for three-phase motors.
3. In a balanced three-phase system, the conductors need be only about 75% of the size of conductors for a single-phase two-wire system of the

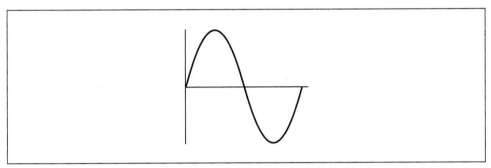

Figure 7-1 Single-phase power falls to zero three times each cycle.

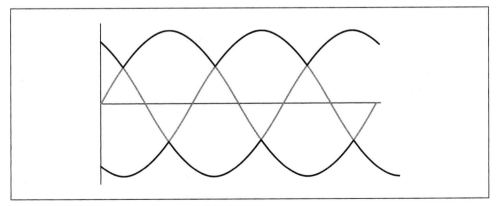

Figure 7-2 Three-phase power never falls to zero.

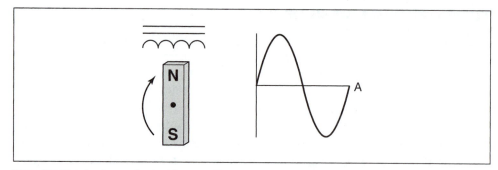

Figure 7-3 Producing a single-phase voltage.

same kVA (kilovolt-amp) rating. This savings helps offset the cost of supplying the third conductor required by three-phase systems.

A single-phase alternating voltage can be produced by rotating a magnetic field through the conductors of a stationary coil, as shown in *Figure 7-3*.

Since alternate polarities of the magnetic field cut through the conductors of the stationary coil, the induced voltage will change polarity at the same speed as the rotation of the magnetic field. The alternator shown in *Figure 7-3* is single-phase because it produces only one AC voltage.

If three separate coils are spaced 120° apart, as shown in *Figure 7-4*, three voltages 120° out of phase with each other will be produced when the magnetic field cuts through the coils. This is the manner in which a three-phase voltage is produced. There are two basic three-phase connections: the wye, or star, and the delta.

WYE CONNECTIONS

wye connection
star connection
phase voltage

line voltage

The **wye**, or **star, connection** is made by connecting one end of each of the three-phase windings together, as shown in *Figure 7-5*. The voltage measured across a single winding, or phase, is known as the **phase voltage** as shown in *Figure 7-6*. The voltage measured between the lines is known as the line-to-line voltage, or simply as the **line voltage**.

In *Figure 7-7*, ammeters have been placed in the phase winding of a wye-connected load and in the line that supplies power to the load. Voltmeters have been connected across the input to the load and across the phase. A line voltage of 208 V has been applied to the load. Notice that the voltmeter connected across the lines indicates a value of 208 V, but the voltmeter connected across the phase indicates a value of 120 V.

In a wye-connected system, the line voltage is higher than the phase voltage by a factor of the square root of 3 (1.732). Two formulas used to

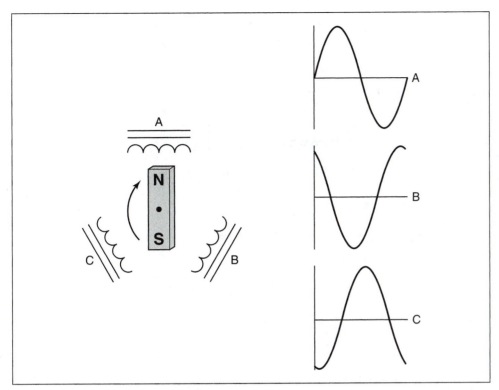

Figure 7-4 The voltages of a three-phase system are 120° out of phase with each other.

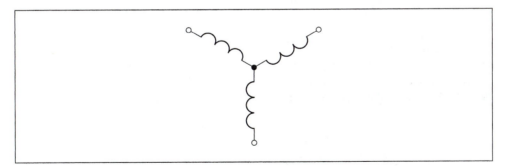

Figure 7-5 A wye connection is formed by joining one end of each winding together.

compute the voltage in a wye-connected system are

$$E_{Line} = E_{Phase} \times 1.732$$

and

$$E_{Phase} = \frac{E_{Line}}{1.732}$$

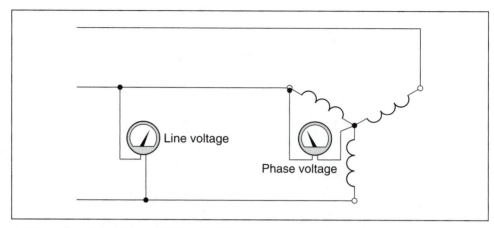

Figure 7-6 Line and phase voltages are different in a wye connection.

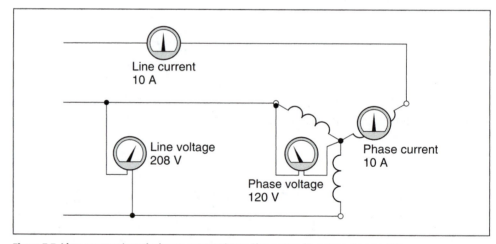

Figure 7-7 Line current and phase current are the same in a wye connection.

Notice in *Figure 7-7* that 10 A of current flow in both the phase and the line. *In a wye-connected system, phase current and line current are the same.*

$$I_{Line} = I_{Phase}$$

Voltage Relationships in a Wye Connection

Many students of electricity have difficulty at first understanding why the line voltage of the wye connection used in this illustration is 208 V instead of

240 V. Since line voltage is measured across two phases that have a voltage of 120 V each, it would appear that the sum of the two voltages should be 240 V. One cause of this misconception is that many students are familiar with the 240/120-V connection supplied to most homes. If voltage is measured across the two incoming lines, a voltage of 240 V will result. If voltage is measured from either of the two lines to the neutral, a voltage of 120 V will be seen. The reason for this is that this is a single-phase connection derived from the center tap of a transformer *(Figure 7-8)*. If the center tap is used as a common point, the two line voltages on either side of it will be in phase with each other *(Figure 7-9)*. The vector sum of these two voltages would be 240 V.

Three-phase voltages are 120° apart, not in phase. If the three voltages are drawn 120° apart, the vector sum of these voltages is 208 V *(Figure 7-10)*.

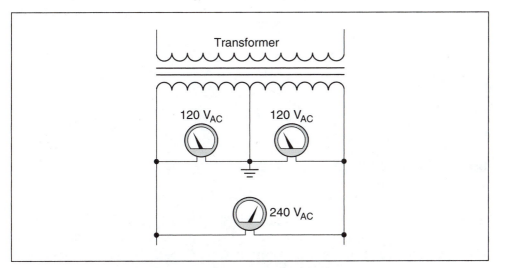

Figure 7-8 Single-phase transformer with grounded center tap.

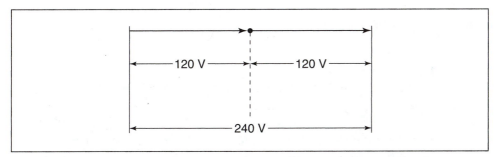

Figure 7-9 The voltages of a single-phase system are in phase with each other.

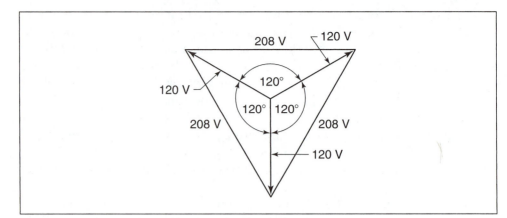

Figure 7-10 Vector sum of the voltages in a three-phase wye connection.

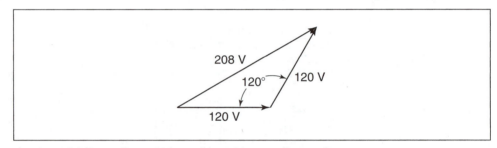

Figure 7-11 Adding voltage vectors of two-phase voltage values.

Another illustration of vector addition is shown in *Figure 7-11*. In this illustration, two phase-voltage vectors are added, and the resultant is drawn from the starting point of one vector to the end point of the other. The parallelogram method of vector addition for the voltages in a wye-connected three-phase system is shown in *Figure 7-12*.

DELTA CONNECTIONS

delta connection

    In *Figure 7-13*, three separate inductive loads have been connected to form a **delta connection**. This connection receives its name from the fact that a schematic diagram of this connection resembles the Greek letter delta (Δ). In *Figure 7-14*, voltmeters have been connected across the lines and across the phase. Ammeters have been connected in the line and in the phase. *In a delta connection, line voltage and phase voltage are the same.* Notice that both voltmeters indicate a value of 480 V.

$$E_{Line} = E_{Phase}$$

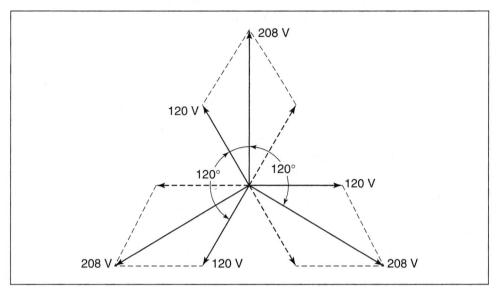

Figure 7-12 The parallelogram method of adding three-phase vectors.

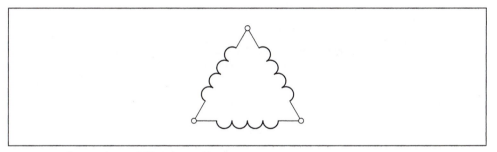

Figure 7-13 Three-phase delta connection.

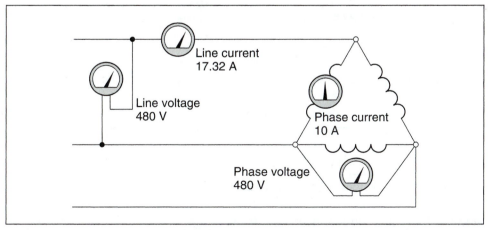

Figure 7-14 Voltage and current relationships in a delta connection.

The line current and phase current, however, are different. *The line current of a delta connection is higher than the phase current by a factor of the square root of 3 (1.732).* In the example shown, it is assumed that each of the phase windings has a current flow of 10 A. The current in each of the lines, however, is 17.32 A. The reason for this difference in current is that current flows through different windings at different times in a three-phase circuit. During some periods of time, current will flow between two lines only. At other times, current will flow from two lines to the third *(Figure 7-15)*. The delta connection is similar to a parallel connection because there is always more than one path for current flow. Since these currents are 120° out of phase with each other, vector addition must be used when finding the sum of the currents *(Figure 7-16)*. Formulas for determining the current in a delta connection are

$$I_{Line} = I_{Phase} \times 1.732$$

and

$$I_{Phase} = \frac{I_{Line}}{1.732}$$

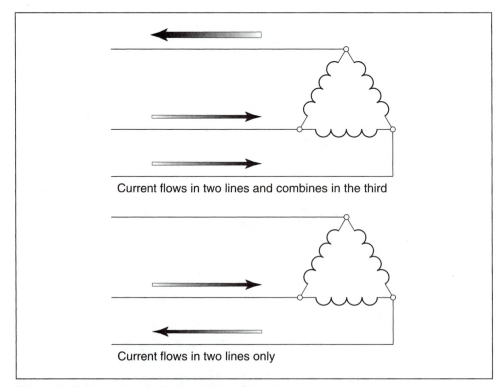

Current flows in two lines and combines in the third

Current flows in two lines only

Figure 7-15 Division of currents in a delta connection.

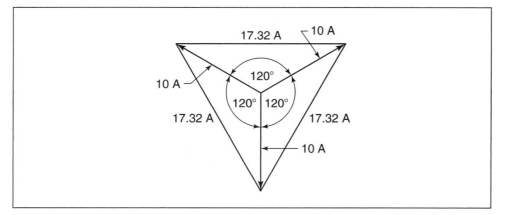

Figure 7-16 Vector addition is used to compute the sum of the currents in a delta connection.

THREE-PHASE POWER

Students sometimes become confused when computing values of power in three-phase circuits. One reason for this confusion is there are actually two formulas that can be used. If *line* values of voltage and current are known, the apparent power of the circuit can be computed using the formula

$$VA = \sqrt{3} \times E_{LINE} \times I_{LINE}$$

If the *phase* values of voltage and current are known, the apparent power can be computed using the formula

$$VA = 3 \times E_{Phase} \times I_{Phase}$$

Notice that in the first formula, the line values of voltage and current are multiplied by the square root of 3. In the second formula, the phase values of voltage and current are multiplied by 3. The first formula is used more because it is generally more convenient to obtain line values of voltage and current since they can be measured with a voltmeter and clamp-on ammeter.

WATTS AND VARS

Watts and VARs can be computed in a similar manner. **Three-phase watts** can be computed by multiplying the apparent power by the power factor

three-phase watts

$$P = \sqrt{3} \times E_{Line} \times I_{Line} \times PF$$

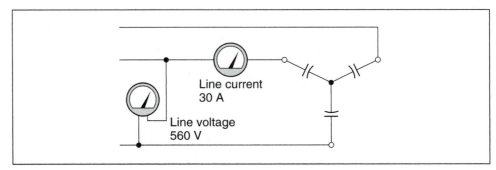

Figure 7-17 Pure capacitive three-phase load.

or

$$P = 3 \times E_{Phase} \times I_{Phase} \times PF$$

(**Note:** When computing the power of a pure resistive load, the voltage and current are in phase with each other and the power factor is 1.)

three-phase VARs

Three-phase VARs can be computed in a similar manner, except that voltage and current values of a pure reactive load are used. For example, a pure capacitive load is shown in *Figure 7-17*. In this example, it is assumed that the line voltage is 480 V and the line current is 30 A. Capacitive VARs can be computed using the formula

$$VARS_C = \sqrt{3} \times E_{Line\ (Capacitive)} \times I_{Line\ (Capacitive)}$$

$$VARS_C = 1.732 \times 560 \times 30$$

$$VARS_C = 29,097.6$$

THREE-PHASE CIRCUIT CALCULATIONS

In the following examples, values of line and phase voltage, line and phase current, and power will be computed for different types of three-phase connections.

Example 1

A wye-connected three-phase alternator supplies power to a delta-connected resistive load *(Figure 7-18)*. The alternator has a line voltage of

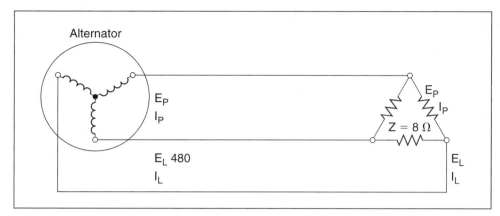

Figure 7-18 Computing three-phase values. Example problem 1.

480 V. Each resistor of the delta load has 8 Ω of resistance. Find the following values:

$E_{L(Load)}$ — line voltage of the load

$E_{P(Load)}$ — phase voltage of the load

$I_{P(Load)}$ — phase current of the load

$I_{L(Load)}$ — line current to the load

$I_{L(Alt)}$ — line current delivered by the alternator

$I_{P(Alt)}$ — phase current of the alternator

$E_{P(Alt)}$ — phase voltage of the alternator

P — true power

Solution

The load is connected directly to the alternator. Therefore, the line voltage supplied by the alternator is the line voltage of the load.

$$E_{L(Load)} = 480 \text{ V}$$

The three resistors of the load are connected in a delta connection. In a delta connection, the phase voltage is the same as the line voltage.

$$E_{P(Load)} = E_{L(Load)}$$

$$E_{P(Load)} = 480 \text{ V}$$

Each of the three resistors in the load is one phase of the load. Now that the phase voltage is known (480 V), the amount of phase current can be computed using Ohm's law.

$$I_{P(Load)} = \frac{E_{P(Load)}}{Z}$$

$$I_{P(LOAD)} = \frac{480}{8}$$

$$I_{P(Load)} = 60 \text{ A}$$

The three load resistors are connected as a delta with 60 A of current flow in each phase. The line current supplying a delta connection must be 1.732 times greater than the phase current.

$$I_{L(Load)} = I_{P(Load)} \times 1.732$$

$$I_{L(Load)} = 60 \times 1.732$$

$$I_{L(Load)} = 103.92 \text{ A}$$

The alternator must supply the line current to the load or loads to which it is connected. In this example, only one load is connected to the alternator. Therefore, the line current of the load will be the same as the line current of the alternator.

$$I_{L(Alt)} = 103.92 \text{ A}$$

The phase windings of the alternator are connected in a wye connection. In a wye connection, the phase current and line current are equal. The phase current of the alternator will, therefore, be the same as the alternator line current.

$$I_{P(Alt)} = 103.92 \text{ A}$$

The phase voltage of a wye connection is less than the line voltage by a factor of the square root of 3. The phase voltage of the alternator will be

$$E_{P(Alt)} = \frac{E_{L(Alt)}}{1.732}$$

$$E_{P(ALT)} = \frac{480}{1.732}$$

$$E_{P(Alt)} = 277.13 \text{ V}$$

In this circuit, the load is pure resistive. The voltage and current are in phase with each other, which produces a unity power factor of 1. The true power in this circuit will be computed using the formula

$$P = 1.732 \times E_{L(Alt)} \times I_{L(Alt)} \times PF$$

$$P = 1.732 \times 480 \times 103.92 \times 1$$

$$P = 86{,}394.93 \text{ W}$$

Example 2

A delta-connected alternator is connected to a wye-connected resistive load *(Figure 7-19)*. The alternator produces a line voltage of 240 V and the resistors have a value of 6 Ω each. Find the following values:

$E_{L(Load)}$ — line voltage of the load

$E_{P(Load)}$ — phase voltage of the load

$I_{P(Load)}$ — phase current of the load

$I_{L(Load)}$ — line current to the load

$I_{L(Alt)}$ — line current delivered by the alternator

$I_{P(Alt)}$ — phase current of the alternator

$E_{P(Alt)}$ — phase voltage of the alternator

P — true power

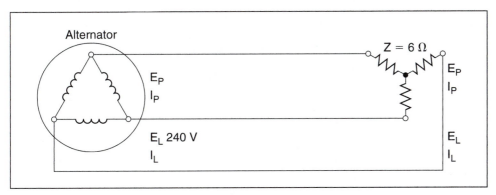

Figure 7-19 Computing three-phase values. Example problem 2.

Solution

As was the case in the previous example, the load is connected directly to the output of the alternator. The line voltage of the load must, therefore, be the same as the line voltage of the alternator.

$$E_{L(Load)} = 240 \text{ V}$$

The phase voltage of a wye connection is less than the line voltage by a factor of 1.732.

$$E_{P(Load)} = \frac{240}{1.732}$$

$$E_{P(Load)} = 138.57 \text{ V}$$

Each of the three 6-Ω resistors is one phase of the wye-connected load. Since the phase voltage is 138.57 V, this voltage is applied to each of the three resistors. The amount of phase current can now be determined using Ohm's law.

$$I_{P(Load)} = \frac{E_{P(Load)}}{Z}$$

$$I_{P(Load)} = \frac{138.57}{6}$$

$$I_{P(Load)} = 23.1 \text{ A}$$

The amount of line current needed to supply a wye-connected load is the same as the phase current of the load.

$$I_{L(Load)} = 23.1 \text{ A}$$

Only one load is connected to the alternator. The line current supplied to the load is the same as the line current of the alternator.

$$I_{L(Alt)} = 23.1 \text{ A}$$

The phase windings of the alternator are connected in delta. In a delta connection, the phase current is less than the line current by a factor of 1.732.

$$I_{P(Alt)} = \frac{I_{L(Alt)}}{1.732}$$

$$I_{P(Alt)} = \frac{23.1}{1.732}$$

$$I_{P(Alt)} = 13.34 \text{ A}$$

The phase voltage of a delta is the same as the line voltage.

$$E_{P(Alt)} = 240 \text{ V}$$

Since the load in this example is pure resistive, the power factor has a value of unity, or 1. Power will be computed by using the line values of voltage and current.

$$P = 1.732 \times E_L \times I_L \times PF$$

$$P = 1.732 \times 240 \times 23.1 \times 1$$

$$P = 9,602.21 \text{ W}$$

Example 3

The phase windings of an alternator are connected in wye. The alternator produces a line voltage of 440 V and supplies power to two resistive loads. One load contains resistors with a value of 4 Ω each, connected in wye. The second load contains resistors with a value of 6 Ω each, connected in delta *(Figure 7-20)*. Find the following circuit values:

$E_{L(Load\ 2)}$ — line voltage of load 2

$E_{P(Load\ 2)}$ — phase voltage of load 2

$I_{P(Load\ 2)}$ — phase current of load 2

$I_{L(Load\ 2)}$ — line current to load 2

$E_{P(Load\ 1)}$ — phase voltage of load 1

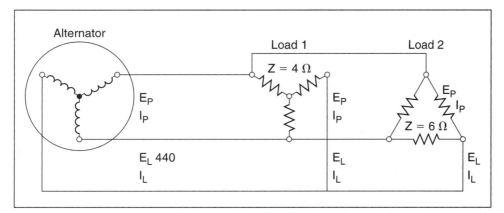

Figure 7-20 Computing three-phase values. Example problem 3.

$I_{P(Load\ 1)}$ — phase current of load 1

$I_{L(Load\ 1)}$ — line current to load 1

$I_{L(Alt)}$ — line current delivered by the alternator

$I_{P(Alt)}$ — phase current of the alternator

$E_{P(Alt)}$ — phase voltage of the alternator

P   — true power

Solution

Both loads are connected directly to the output of the alternator. The line voltage for both loads 1 and 2 will be the same as the line voltage of the alternator.

$$E_{L(Load\ 2)} = 440\ V$$

$$E_{L(Load\ 1)} = 440\ V$$

Load 2 is connected as a delta. The phase voltage will be the same as the line voltage.

$$E_{P(Load\ 2)} = 440\ V$$

Each of the resistors that constitutes a phase of load 2 has a value of 6 Ω. The amount of phase current can be found using Ohm's law.

$$I_{P(Load\ 2)} = \frac{E_{P(Load\ 2)}}{Z}$$

$$I_{P(Load\ 2)} = \frac{440}{6}$$

$$I_{P(Load\ 2)} = 73.33\ A$$

The line current supplying a delta-connected load is 1.732 times greater than the phase current. The amount of line current needed for load 2 can be computed by increasing the phase current value by 1.732.

$$I_{L(Load\ 2)} = I_{P(Load\ 2)} \times 1.732$$

$$I_{L(Load\ 2)} = 73.33 \times 1.732$$

$$I_{L(Load\ 2)} = 127.01\ A$$

The resistors of load 1 are connected to form a wye. The phase voltage of a wye connection is less than the line voltage by a factor of 1.732.

$$E_{P(Load\ 1)} = \frac{E_{L(Load\ 1)}}{1.732}$$

$$E_{P(Load\ 1)} = \frac{440}{1.732}$$

$$E_{P(Load\ 1)} = 254.04\ V$$

Now that the voltage applied to each of the 4-Ω resistors is known, the phase current can be computed using Ohm's law.

$$I_{P(Load\ 1)} = \frac{E_{P(Load\ 1)}}{Z}$$

$$I_{P(Load\ 1)} = \frac{254.04}{4}$$

$$I_{P(Load\ 1)} = 63.51\ A$$

The line current supplying a wye-connected load is the same as the phase current. Therefore, the amount of line current needed to supply load 1 is

$$I_{L(Load\ 1)} = 63.51\ A$$

The alternator must supply the line current needed to operate both loads. In this example, both loads are resistive. The total line current supplied by the alternator will be the sum of the line currents of the two loads.

$$I_{L(Alt)} = I_{L(Load\ 1)} + I_{L(Load\ 2)}$$

$$I_{L(Alt)} = 63.51 + 127.01$$

$$I_{L(Alt)} = 190.52\ A$$

Since the phase windings of the alternator in this example are connected in a wye, the phase current will be the same as the line current.

$$I_{P(Alt)} = 190.52\ A$$

The phase voltage of the alternator will be less than the line voltage by a factor of 1.732.

$$E_{P(Alt)} = \frac{440}{1.732}$$

$$E_{P(Alt)} = 254.04\ V$$

Both of the loads in this example are resistive and have a unity power factor of 1. The total power in this circuit can be found by using the line voltage and total line current supplied by the alternator.

$$P = 1.732 \times E_L \times I_L \times PF$$

$$P = 1.732 \times 440 \times 190.52 \times 1$$

$$P = 145{,}191.48 \text{ W}$$

Example 4

A wye-connected three-phase alternator with a line voltage of 560 V supplies power to three different loads *(Figure 7-21)*. The first load is formed by three resistors with a value of 6 Ω each, connected in a wye. The second load comprises three inductors with an inductive reactance of 10 Ω each, connected in delta, and the third load comprises three capacitors with a capacitive reactance of 8 Ω each, connected in wye. Find the following circuit values.

$E_{L(Load\ 3)}$ — line voltage of load 3 (capacitive)

$E_{P(Load\ 3)}$ — phase voltage of load 3 (capacitive)

$I_{P(Load\ 3)}$ — phase current of load 3 (capacitive)

$I_{L(Load\ 3)}$ — line current to load 3 (capacitive)

$E_{L(Load\ 2)}$ — line voltage of load 2 (inductive)

$E_{P(Load\ 2)}$ — phase voltage of load 2 (inductive)

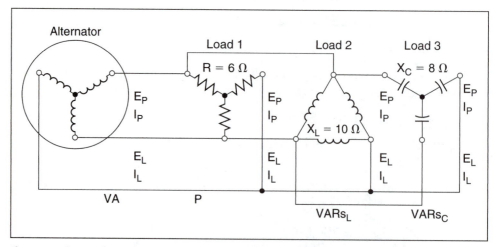

Figure 7-21 Computing three-phase values. Example problem 4.

$I_{P(Load\ 2)}$ — phase current of load 2 (inductive)

$I_{L(Load\ 2)}$ — line current to load 2 (inductive)

$E_{L(Load\ 1)}$ — line voltage of load 1 (resistive)

$E_{P(Load\ 1)}$ — phase voltage of load 1 (resistive)

$I_{P(Load\ 1)}$ — phase current of load 1 (resistive)

$I_{L(Load\ 1)}$ — line current to load 1 (resistive)

$I_{L(Alt)}$ — line current delivered by the alternator

$E_{P(Alt)}$ — phase voltage of the alternator

P — true power

$VARS_L$ — reactive power of the inductive load

$VARS_C$ — reactive power of the capacitive load

VA — apparent power

PF — power factor

Solution

All three loads are connected to the output of the alternator. The line voltage connected to each load is the same as the line voltage of the alternator.

$$E_{L(Load\ 3)} = 560\ V$$

$$E_{L(Load\ 2)} = 560\ V$$

$$E_{L(Load\ 1)} = 560\ V$$

LOAD 3 CALCULATIONS

Load 3 is formed from three capacitors with a capacitive reactance of 8 Ω each, connected in a wye. Since this load is wye-connected, the phase voltage will be less than the line voltage by a factor of 1.732.

$$E_{P(Load\ 3)} = \frac{E_{L(Load\ 3)}}{1.732}$$

$$E_{P(Load\ 3)} = \frac{560}{1.732}$$

$$E_{P(Load\ 3)} = 323.33\ V$$

Now that the voltage applied to each capacitor is known, the phase current can be computed using Ohm's law.

$$I_{P(Load\ 3)} = \frac{E_{P(Load\ 3)}}{X_C}$$

$$I_{P(Load\ 3)} = \frac{323.33}{8}$$

$$I_{P(Load\ 3)} = 40.42\ A$$

The line current required to supply a wye-connected load is the same as the phase current.

$$I_{L(Load\ 3)} = 40.42\ A$$

The reactive power of load 3 can be found using a formula similar to the formula for computing apparent power. Since load 3 is pure capacitive, the current and voltage are 90° out of phase with each other, and the power factor is zero.

$$VARS_C = 1.732 \times E_{L(Load\ 3)} \times I_{L(Load\ 3)}$$

$$VARS_C = 1.732 \times 560 \times 40.42$$

$$VARS_C = 39{,}204.17$$

LOAD 2 CALCULATIONS

Load 2 comprises three inductors connected in a delta with an inductive reactance of 10 Ω each. Since the load is connected in delta, the phase voltage will be same as the line voltage.

$$E_{L(Load\ 2)} = 560\ V$$

The phase current can be computed by using Ohm's law.

$$I_{P(Load\ 2)} = \frac{E_{P(Load\ 2)}}{X_L}$$

$$I_{P(Load\ 2)} = \frac{560}{10}$$

$$I_{P(Load\ 2)} = 56\ A$$

The amount of line current needed to supply a delta-connected load is 1.732 times greater than the phase current of the load.

$$I_{L(Load\ 2)} = I_{P(Load\ 2)} \times 1.732$$

$$I_{L(Load\ 2)} = 56 \times 1.732$$

$$I_{L(Load\ 2)} = 96.99\ A$$

Since load 2 is made up of inductors, the reactive power can be computed using the line values of voltage and current supplied to the load.

$$VARS_L = 1.732 \times E_{L(Load\ 2)} \times I_{L(Load\ 2)}$$

$$VARS_L = 1.732 \times 560 \times 96.99$$

$$VARS_L = 94,072.54$$

LOAD 1 CALCULATIONS

Load 1 consists of three resistors with a resistance of 6 Ω each, connected in wye. In a wye connection, the phase voltage is less than the line voltage by a factor of 1.732. The phase voltage for load 1 will be the same as the phase voltage for load 3.

$$E_{P(Load\ 1)} = 323.33\ V$$

The amount of phase current can now be computed using the phase voltage and the resistance of each phase.

$$I_{P(Load\ 1)} = \frac{E_{P(Load\ 1)}}{R}$$

$$I_{P(Load\ 1)} = \frac{323.33}{6}$$

$$I_{P(Load\ 1)} = 53.89\ A$$

Since the resistors of load 1 are connected in a wye, the line current will be the same as the phase current.

$$I_{L(Load\ 1)} = 53.89\ A$$

Since load 1 is pure resistive, true power can be computed using the line- and phase-current values.

$$P = 1.732 \times E_{L(Load\ 1)} \times I_{L(Load\ 1)}$$

$$P = 1.732 \times 560 \times 53.89$$

$$P = 52,267\ W$$

ALTERNATOR CALCULATIONS

The alternator must supply the line current for each of the loads. In this problem, however, the line currents are out of phase with each other. To find the total line current delivered by the alternator, vector addition must be used. The current flow in load 1 is resistive and in phase with the line voltage. The current flow in load 2 is inductive and lags the line voltage by 90°. The current flow in load 3 is capacitive and leads the line voltage by 90°. A formula similar to the formula used to find total current flow in an RLC (Resistive, Inductive, Capacitive) parallel circuit can be employed to find the total current delivered by the alternator.

$$I_{L(Alt)} = \sqrt{I_{L(Load\ 1)}{}^2 + (I_{L(Load\ 2)} - I_{L(Load\ 3)})^2}$$

$$I_{L(Alt)} = \sqrt{53.89^2 + (96.99 - 40.42)^2}$$

$$I_{L(Alt)} = \quad 78.13\ A$$

The apparent power can now be found using the line voltage and current values of the alternator.

$$VA = 1.732 \times E_{L(Alt)} \times I_{L(Alt)}$$

$$VA = 1.732 \times 560 \times 78.13$$

$$VA = 75,779.85$$

The circuit power factor is the ratio of apparent power and true power.

$$PF = \frac{W}{VA}$$

$$PF = \frac{52,267}{75,779.85}$$

$$PF = 69\%$$

POWER FACTOR CORRECTION

Correcting the power factor of a three-phase circuit is similar to the procedure used to correct the power factor of a single-phase circuit.

Example 5

A three-phase motor is connected to a 480-V, 60-Hz line. A clamp-on ammeter indicates a running current of 68 A at full load, and a three-phase wattmeter indicates a true power of 40,277 W. Compute the motor power factor first. Then find the amount of capacitance needed to correct the power factor to 95%. Assume that the capacitors used for power factor correction are to be connected in wye, and the capacitor bank is then to be connected in parallel with the motor.

Solution

First, find the amount of apparent power in the circuit.

$$VA = 1.732 \times E_L \times I_L$$

$$VA = 1.732 \times 480 \times 68$$

$$VA = 56,532.48$$

The motor power factor can be computed by dividing the true power by the apparent power.

$$PF = \frac{P}{VA}$$

$$PF = \frac{40,277}{56,532.48}$$

$$PF = 71.2\%$$

The inductive VARs in the circuit can be computed using the formula

$$VARS_L = \sqrt{VA^2 - P^2}$$

$$VARS_L = \sqrt{56,532.48^2 - 40,277^2}$$

$$VARS_L = 39,669.69$$

If the power factor is to be corrected to 95%, the apparent power at 95% power factor must be found. This can be done using the formula

$$VA = \frac{P}{PF}$$

$$VA = \frac{40,277}{0.95}$$

$$VA = 42,396.84$$

The amount of inductive VARs needed to produce an apparent power of 42,396.84 VA can be found using the formula

$$VARS_L = \sqrt{VA^2 - P^2}$$

$$VARS_L = \sqrt{42{,}396.84^2 - 40{,}277^2}$$

$$VARS_L = 13{,}238.4$$

To correct the power factor to 95%, the inductive VARs must be reduced from 39,669.69 to 13,238.4. This can be done by connecting a bank of capacitors in the circuit that will produce a total of 26,431.29 capacitive VARs (39,669.69 − 13,238.4 = 26,431.29). This amount of capacitive VARs will reduce the inductive VARs to the desired amount.

Now that the amount of capacitive VARs needed to correct the power factor is known, the amount of line current supplying the capacitor bank can be computed using the formula

$$I_L = \frac{VARS_C}{E_L \times 1.732}$$

$$I_L = \frac{26{,}431.29}{480 \times 1.732}$$

$$I_L = 31.79 \text{ A}$$

The capacitive load bank is to be connected in a wye. Therefore, the phase current will be the same as the line current. The phase voltage, however, will be less than the line voltage by a factor of 1.732, or 277.14 V. Ohm's law can be used to find the amount of capacitive reactance needed to produce a phase current of 31.79 A with an applied voltage of 277.14 V.

$$X_C = \frac{E_P}{I_P}$$

$$X_C = \frac{277.14}{31.79}$$

$$X_C = 8.72 \ \Omega$$

The amount of capacitance needed to produce a capacitive reactance of 8.72 Ω can now be computed.

$$C = \frac{1}{2\pi F X_C}$$

$$C = \frac{40}{377 \times 8.72}$$

$$C = 304.2 \ \mu F$$

When a bank of wye-connected capacitors with a value of 304.2 µF each is connected in parallel with the motor, the power factor will be corrected to 95%.

Summary

1. The voltages of a three-phase system are 120° out of phase with each other.

2. The two types of three-phase connections are wye and delta.

3. Wye connections are characterized by the fact that one terminal of each of the devices is connected together.

4. In a wye connection, the phase voltage is less than the line voltage by a factor of 1.732. The phase current and line current are the same.

5. In a delta connection, the phase voltage is the same as the line voltage. The phase current is less than the line current by a factor of 1.732.

Review Questions

1. How many degrees out of phase with each other are the voltages of a three-phase system?

2. What are the two main types of three-phase connections?

3. A wye-connected load has a voltage of 480 V applied to it. What is the voltage dropped across each phase?

4. A wye-connected load has a phase current of 25 A. How much current is flowing through the lines supplying the load?

5. A delta connection has a voltage of 560 V connected to it. How much voltage is dropped across each phase?

6. A delta connection has 30 A of current flowing through each phase winding. How much current is flowing through each of the lines supplying power to the load?

7. A three-phase load has a phase voltage of 240 V and a phase current of 18 A. What is the apparent power of this load?

8. If the load in question 7 is connected in a wye, what would be the line voltage and line current supplying the load?

9. An alternator with a line voltage of 2400 V supplies a delta-connected load. The line current supplied to the load is 40 A. Assuming the load is a balanced three-phase load, what is the impedance of each phase?

10. What is the apparent power of the circuit in question 9?

Problems

1. Refer to the circuit shown in *Figure 7-18* to answer the following questions. Assume that the alternator has a line voltage of 240 V and the load has an impedance of 12 Ω per phase. Find all the missing values.

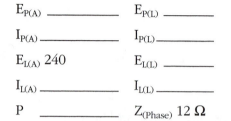

$E_{P(A)}$ _____ $E_{P(L)}$ _____

$I_{P(A)}$ _____ $I_{P(L)}$ _____

$E_{L(A)}$ 240 $E_{L(L)}$ _____

$I_{L(A)}$ _____ $I_{L(L)}$ _____

P _____ $Z_{(Phase)}$ 12 Ω

2. Refer to the circuit shown in *Figure 7-19* to answer the following questions. Assume that the alternator has a line voltage of 4160 V, and the load has an impedance of 60 Ω per phase. Find all the missing values.

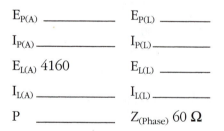

$E_{P(A)}$ _____ $E_{P(L)}$ _____

$I_{P(A)}$ _____ $I_{P(L)}$ _____

$E_{L(A)}$ 4160 $E_{L(L)}$ _____

$I_{L(A)}$ _____ $I_{L(L)}$ _____

P _____ $Z_{(Phase)}$ 60 Ω

3. Refer to the circuit shown in *Figure 7-20* to answer the following questions. Assume that the alternator has a line voltage of 560 V. Load 1 has an impedance of 5 Ω per phase, and load 2 has an impedance of 8 Ω per phase. Find all the missing values.

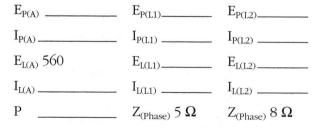

$E_{P(A)}$ _____ $E_{P(L1)}$ _____ $E_{P(L2)}$ _____

$I_{P(A)}$ _____ $I_{P(L1)}$ _____ $I_{P(L2)}$ _____

$E_{L(A)}$ 560 $E_{L(L1)}$ _____ $E_{L(L2)}$ _____

$I_{L(A)}$ _____ $I_{L(L1)}$ _____ $I_{L(L2)}$ _____

P _____ $Z_{(Phase)}$ 5 Ω $Z_{(Phase)}$ 8 Ω

4. Refer to the circuit shown in *Figure 7-21* to answer the following questions. Assume that the alternator has a line voltage of 480 V. Load 1 has a resistance of 12 Ω per phase. Load 2 has an inductive reactance of 16 Ω per phase, and load 3 has a capacitive reactance of 10 Ω per phase. Find all the missing values.

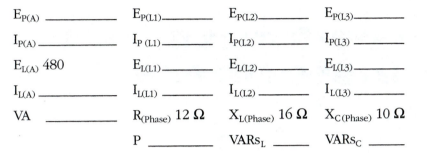

$E_{P(A)}$ _____ $E_{P(L1)}$_____ $E_{P(L2)}$_____ $E_{P(L3)}$_____

$I_{P(A)}$ _____ $I_{P\,(L1)}$_____ $I_{P(L2)}$ _____ $I_{P(L3)}$ _____

$E_{L(A)}$ 480 $E_{L(L1)}$_____ $E_{L(L2)}$_____ $E_{L(L3)}$_____

$I_{L(A)}$ _____ $I_{L(L1)}$ _____ $I_{L(L2)}$ _____ $I_{L(L3)}$ _____

VA _____ $R_{(Phase)}$ 12 Ω $X_{L(Phase)}$ 16 Ω $X_{C(Phase)}$ 10 Ω

                      P _____ $VARs_L$ _____ $VARs_C$ _____

UNIT 8

Three-Phase Transformers

Objectives

After studying this unit, you should be able to:

- Discuss the operation of three-phase transformers.

- Connect three single-phase transformers to form a three-phase bank.

- Calculate voltage and current values for a three-phase transformer connection.

- Connect two single-phase transformers to form a three-phase open-delta connection.

- Discuss the characteristics of an open-delta connection.

Three-phase transformers are used throughout industry to change values of three-phase voltage and current. Since three-phase power is the major way in which power is produced, transmitted, and used, an understanding of how three-phase transformer connections are made is essential. This unit will discuss different types of three-phase transformer connections and present examples of how values of voltage and current for these connections are computed.

THREE-PHASE TRANSFORMERS

A three-phase transformer is constructed by winding three single-phase transformers on a single core *(Figure 8-1)*. A photograph of a three-phase transformer is shown in *Figure 8-2*. The transformer is shown before it is mounted in an enclosure, which will be filled with a **dielectric oil**. The dielectric oil performs several functions. Since it is a dielectric, it provides electrical insulation between the windings and the case. It is also used to help provide cooling and to prevent the formation of moisture, which can deteriorate the winding insulation.

dielectric oil

Three-Phase Transformer Connections

Three-phase transformers are connected in delta or wye configurations. A **wye-delta** transformer has its primary winding connected in a wye and its secondary winding connected in a delta *(Figure 8-3)*. A **delta-wye** transformer has its primary winding connected in delta and its secondary connected in wye *(Figure 8-4)*.

wye-delta
delta-wye

Connecting Single-Phase Transformers into a Three-Phase Bank

If three-phase transformation is needed and a three-phase transformer of the proper size and turns ratio is not available, three single-phase transformers can be connected to form a **three-phase bank**. When three single-phase transformers are used to make a three-phase transformer bank, their primary and secondary windings are connected in a wye or delta connection. The three transformer windings in *Figure 8-5* have been labeled A, B,

three-phase bank

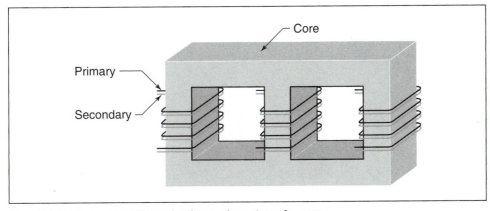

Figure 8-1 Basic construction of a three-phase transformer.

Figure 8-2 Three-phase pyranol-filled transformer.

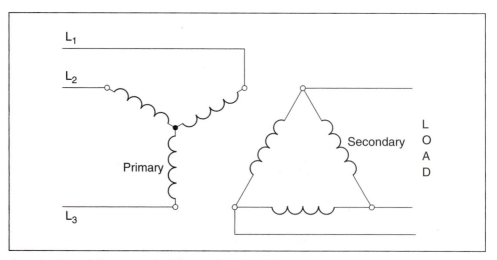

Figure 8-3 Wye-delta connected three-phase transformer.

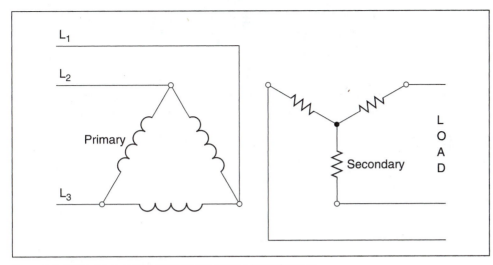

Figure 8-4 Delta-wye connected three-phase transformer.

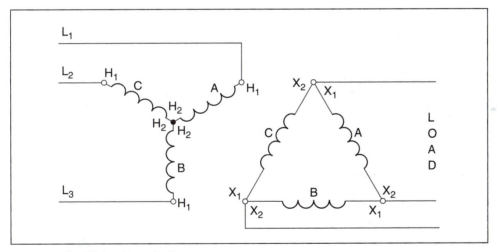

Figure 8-5 Identifying the windings.

and C. One end of each primary lead is labeled H_1, and the other end is labeled H_2. One end of each secondary lead is labeled X_1, and the other end is labeled X_2.

Figure 8-6 shows three single-phase transformers labeled A, B, and C. The primary leads of each transformer have been labeled H_1 and H_2, and the secondary leads are labeled X_1 and X_2. The schematic diagram of *Figure 8-5* will be used to connect the three single-phase transformers into a three-phase wye-delta connection as shown in *Figure 8-7*.

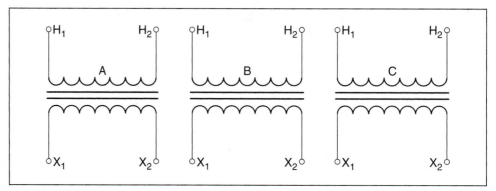

Figure 8-6 Three single-phase transformers.

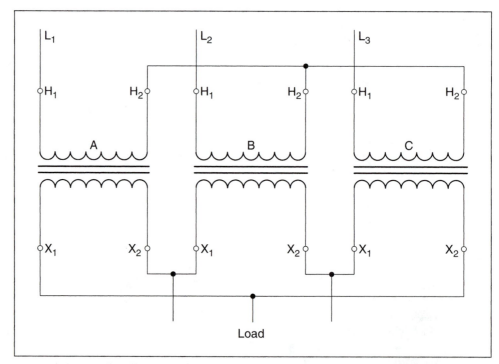

Figure 8-7 Connecting three single-phase transformers to form a wye-delta three-phase bank.

The primary winding will be tied into a wye connection first. The schematic in *Figure 8-5* shows that the H_2 leads of primary windings are connected together, and the H_1 lead of each winding is open for connection to the incoming power line. Notice in *Figure 8-7* that the H_2 leads of each primary winding have been connected together, and the H_1 lead of each winding has been connected to the incoming power line.

Figure 8-5 shows that the X_1 lead of transformer A is connected to the X_2 lead of transformer C. Notice that this same connection has been made in *Figure 8-7.* The X_1 lead of transformer B is connected to the X_2 lead of transformer A, and the X_1 lead of transformer C is connected to the X_2 lead of transformer B. The load is connected to the points of the delta connection.

Although *Figure 8-5* illustrates the proper schematic symbology for a three-phase transformer connection, some electrical schematics and wiring diagrams do not illustrate three-phase transformer connections in this manner. One type of diagram, called the **one-line diagram**, would illustrate a delta-wye connection as shown in *Figure 8-8.* These diagrams are generally used to show the main power distribution system of a large industrial plant. The one-line diagram in *Figure 8-9* shows the main power to the plant and the transformation of voltages to different subfeeders. Notice that each transformer shows the secondary voltage of the subfeeder, and whether the primary and secondary are connected as a wye or delta.

one-line diagram

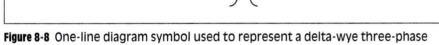

Figure 8-8 One-line diagram symbol used to represent a delta-wye three-phase transformer connection.

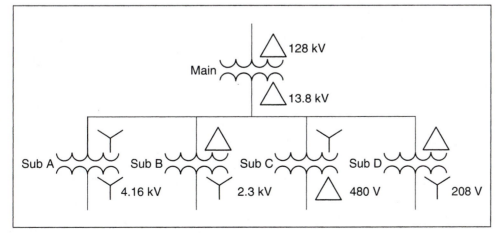

Figure 8-9 One-line diagrams are generally used to show the main power distribution of a plant.

CLOSING A DELTA

Delta connections should be checked for proper polarity before making the final connection and applying power. If the phase winding of one transformer is reversed, an extremely high current will flow when power is applied. Proper phasing can be checked with a voltmeter, as shown in *Figure 8-10*. If power is applied to the transformer bank before the delta connection is closed, the voltmeter should indicate 0 volts. If one phase winding has been reversed, however, the voltmeter will indicate double the amount of voltage. For example, assume that the output voltage of a delta secondary is 240 volts. If the voltage is checked before the delta is closed, the voltmeter should indicate a voltage of 0 volts if all windings have been phased properly. If one winding has been reversed, however, the voltmeter will indicate a voltage of 480 volts (240 + 240). This test will confirm whether a phase winding has been reversed, but it will not indicate whether the reversed winding is located in the primary or secondary. If either primary or secondary windings have been reversed, the voltmeter will indicate double the output voltage.

It should be noted, however, that a voltmeter is a high-impedance device. It is not unusual for a voltmeter to indicate some amount of voltage before the delta is closed, especially if the primary has been connected as a wye and the secondary as a delta. When this is the case, however, the voltmeter will generally indicate close to the normal output voltage if

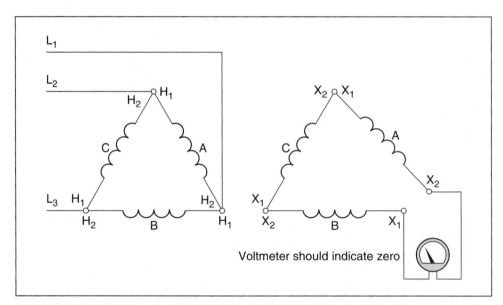

Figure 8-10 Testing for proper transformer polarity before closing the delta.

the connection is correct and double the output voltage if the connection is incorrect.

THREE-PHASE TRANSFORMER CALCULATIONS

When the values of voltage and current for three-phase transformers are computed, the formulas used for making transformer calculations and three-phase calculations must be followed. Another very important rule that must be understood is that *only phase values of voltage and current can be used when computing transformer values.* When three-phase transformers are connected as a wye or delta, the primary and secondary windings themselves become the phases of a three-phase connection. This is true whether a three-phase transformer is used or whether three single-phase transformers are employed, to form a three-phase bank. Refer to transformer A in *Figure 8-6*. All transformation of voltage and current takes place between the primary and secondary windings. Since these windings form the phase values of the three-phase connection, only phase values— not line values—can be used when calculating transformed voltages and currents.

Example 1

A three-phase transformer connection is shown in *Figure 8-11*. Three single-phase transformers have been connected to form a wye-delta bank. The primary is connected to a three-phase line of 13,800 volts, and the secondary voltage is 480. A three-phase resistive load with an impedance of 2.77 Ω per

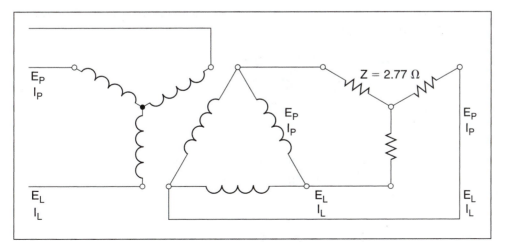

Figure 8-11 Example 1: Three-phase transformer calculation.

phase is connected to the secondary of the transformer. The following values will be computed for this circuit.

$E_{P(Primary)}$ — phase voltage of the primary

$E_{P(Secondary)}$ — phase voltage of the secondary

Ratio — turns ratio of the transformer

$E_{P(Load)}$ — phase voltage of the load bank

$I_{P(Load)}$ — phase current of the load bank

$I_{L(Secondary)}$ — line current of the secondary

$I_{P(Secondary)}$ — phase current of the secondary

$I_{P(Primary)}$ — phase current of the primary

$I_{L(Primary)}$ — line current of the primary

The primary windings of the three single-phase transformers have been connected to form a wye connection. In a wye connection, the phase voltage is less than the line voltage by a factor of 1.732. Therefore, the phase value of the primary voltage can be computed using the formula

$$E_{P(Primary)} = \frac{E_L}{1.732}$$

$$E_{P(PRIMARY)} = \frac{13{,}800}{1.732}$$

$$E_{P(Primary)} = 7967.67 \text{ volts}$$

The secondary windings are connected as a delta. In a delta connection, the phase voltage and line voltage are the same.

$$E_{P(Secondary)} = E_{L(Secondary)}$$

$$E_{P(Secondary)} = 480 \text{ volts}$$

The turns ratio can be computed by comparing the phase voltage of the primary with the phase voltage of the secondary.

$$Ratio = \frac{\text{primary-phase voltage}}{\text{secondary-phase voltage}}$$

$$Ratio = \frac{7967.67}{480}$$

$$Ratio = 16.6{:}1$$

The load bank is connected in a wye connection. The voltage across the phase of the load bank will be less than the line voltage by a factor of 1.732.

$$E_{P(Load)} = \frac{E_{L(Load)}}{1.732}$$

$$E_{P(Load)} = \frac{480}{1.732}$$

$$E_{P(Load)} = 277 \text{ volts}$$

Now that the voltage across each of the load resistors is known, the current flow through the phase of the load can be computed using Ohm's law.

$$I_{P(Load)} = \frac{E_{P(Load)}}{R}$$

$$I_{P(Load)} = \frac{277}{2.77}$$

$$I_{P(Load)} = 100 \text{ amperes}$$

Since the load is connected as a wye connection, the line current will be the same as the phase current. Therefore, the line current supplied by the secondary of the transformer is equal to the phase current of the load.

$$I_{L(Secondary)} = 100 \text{ A}$$

The secondary of the transformer bank is connected as a delta. The phase current of the delta is less than the line current by a factor of 1.732.

$$I_{P(Secondary)} = \frac{I_{L(Secondary)}}{1.732}$$

$$I_{P(Secondary)} = \frac{100}{1.732}$$

$$I_{P(Secondary)} = 57.74 \text{ amps}$$

The amount of current flow through the primary can be computed using the turns ratio. Since the primary has a higher voltage than the secondary, it will have a lower current. (Volts × amps input must equal volts × amps output.)

$$I_{P(Primary)} = \frac{I_{P(Secondary)}}{Ratio}$$

$$I_{P(Primary)} = \frac{57.74}{16.6}$$

$$I_{P(Primary)} = 3.48 \text{ amperes}$$

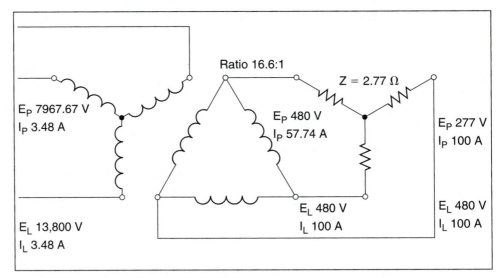

Figure 8-12 Example 1 with all missing values.

Recall that all transformed values of voltage and current take place across the phases. The primary has a phase current of 3.48 amps. In a wye connection, the phase current is the same as the line current.

$$I_{L(Primary)} = 3.48 \text{ amps}$$

The transformer connection with all computed values is shown in *Figure 8-12*.

Example 2

In the next example, a three-phase transformer is connected in a delta-delta configuration *(Figure 8-13)*. The load is connected as a wye, and each phase has an impedance of 7 Ω. The primary is connected to a line voltage

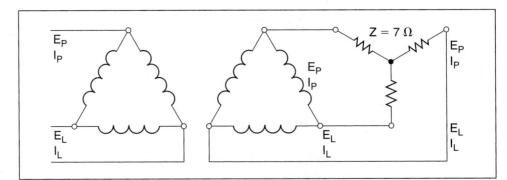

Figure 8-13 Example 2: Three-phase transformer calculation.

of 4160 volts, and the secondary line voltage is 440 volts. The following values will be found:

$E_{P(Primary)}$ — phase voltage of the primary

$E_{P(Secondary)}$ — phase voltage of the secondary

Ratio — turns ratio of the transformer

$E_{P(Load)}$ — phase voltage of the load bank

$I_{P(Load)}$ — phase current of the load bank

$I_{L(Secondary)}$ — line current of the secondary

$I_{P(Secondary)}$ — phase current of the secondary

$I_{P(Primary)}$ — phase current of the primary

$I_{L(Primary)}$ — line current of the primary

The primary is connected as a delta. The phase voltage will be the same as the applied line voltage.

$$E_{P(Primary)} = E_{L(Primary)}$$
$$E_{P(Primary)} = 4160 \text{ volts}$$

The secondary of the transformer is connected as a delta also. Therefore, the phase voltage of the secondary will be the same as the line voltage of the secondary.

$$E_{P(Secondary)} = 440 \text{ volts}$$

All transformer values must be computed using phase values of voltage and current. The turns ratio can be found by dividing the phase voltage of the primary by the phase voltage of the secondary.

$$Ratio = \frac{\text{Primary-phase voltage}}{\text{Secondary-phase voltage}}$$

$$Ratio = \frac{4160}{440}$$

$$Ratio = 9.45:1$$

The load is connected directly to the output of the secondary. The line voltage applied to the load must, therefore, be the same as the line voltage of the secondary.

$$E_{L(Load)} = 440 \text{ volts}$$

The load is connected in a wye. The voltage applied across each phase will be less than the line voltage by a factor of 1.732.

$$E_{P(Load)} = \frac{E_{L(Load)}}{1.732}$$

$$E_{P(Load)} = \frac{440}{1.732}$$

$$E_{P(Load)} = 254 \text{ volts}$$

The phase current of the load can be computed using Ohm's law.

$$I_{P(Load)} = \frac{E_{P(Load)}}{Z}$$

$$I_{P(Load)} = \frac{254}{7}$$

$$I_{P(Load)} = 36.29 \text{ amps}$$

The amount of line current supplying a wye-connected load will be the same as the phase current of the load.

$$I_{L(Load)} = 36.29 \text{ amps}$$

Since the secondary of the transformer is supplying current to only one load, the line current of the secondary will be the same as the line current of the load.

$$I_{L(Secondary)} = 36.29 \text{ amps}$$

The phase current in a delta connection is less than the line current by a factor of 1.732.

$$I_{P(Secondary)} = \frac{I_{L(Secondary)}}{1.732}$$

$$I_{P(Secondary)} = \frac{36.29}{1.732}$$

$$I_{P(Secondary)} = 20.95 \text{ amps}$$

The phase current of the transformer primary can now be computed using the phase current of the secondary and the turns ratio.

$$I_{P(Primary)} = \frac{I_{P(Secondary)}}{ratio}$$

$$I_{P(Primary)} = \frac{20.95}{9.45}$$

$$I_{P(Primary)} = 2.27 \text{ amps}$$

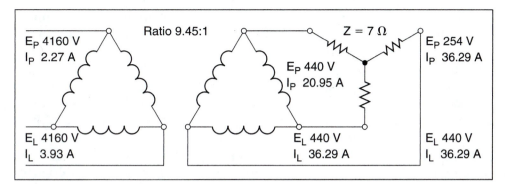

Figure 8-14 Example 2 with all missing values.

In this example, the primary of the transformer is connected as a delta. The line current supplying the transformer will be higher than the phase current by a factor of 1.732.

$$I_{L(Primary)} = I_{P(Primary)} \times 1.732$$

$$I_{L(Primary)} = 2.27 \times 1.732$$

$$I_{L(Primary)} = 3.93 \text{ amps}$$

The circuit with all computed values is shown in *Figure 8-14*.

OPEN-DELTA CONNECTION

The **open-delta** transformer connection can be made with only two transformers instead of three *(Figure 8-15)*. This connection is often used when the amount of three-phase power needed is not excessive, such as in a small

open-delta

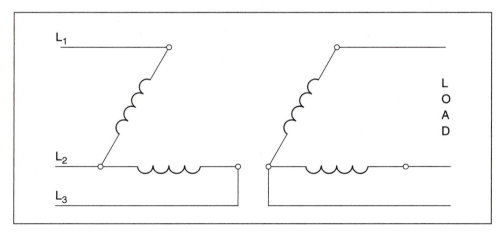

Figure 8-15 Open-delta connection.

business. It should be noted that the output power of an open-delta connection is only 87% of the rated power of the two transformers. For example, assume two transformers, each having a capacity of 25 kVA (kilovolt amperes), are connected in an open-delta connection. The total output power of this connection is 43.5 kVA (50 kVA × 0.87 = 43.5 kVA).

Another figure given for this calculation is 58%. This percentage assumes a closed-delta bank containing three transformers. If three 25 kVA transformers were connected to form a closed-delta connection, the total output power would be 75 kVA (3 × 25 kVA = 75 kVA). If one of these transformers were removed and the transformer bank operated as an open-delta connection, the output power would be reduced to 58% of its original capacity of 75 kVA. The output capacity of the open-delta bank is 43.5 kVA (75 kVA × 0.58 = 43.5 kVA).

The voltage and current values of an open-delta connection are computed in the same manner as a standard delta-delta connection when three transformers are employed. The voltage and current rules for a delta connection must be used when determining line and phase values of voltage and current.

T-Connected Transformers

**T-Connection
main trans-
former**

Another connection involving the use of two transformers to supply three-phase power is the **T-connection** *(Figure 8-16)*. In this connection, one transformer is generally referred to as the **main transformer** and the other is called the teaser transformer. The main transformer must contain a center or 50% tap for both the primary and secondary windings, and it is preferred that the teaser transformer contain an 86.6% voltage tap for both the primary and secondary windings. Although the 86.6% tap is preferred, the connection can be made with a teaser transformer that has the same voltage rating as the main transformer. In this instance the teaser transformer is operated at reduced flux *(Figure 8-17)*. This connection permits two transformers to be connected T instead of open delta in the event that one transformer of a delta-delta bank should fail.

Transformers intended for use as T-connected transformers are often specially wound for the purpose, and both transformers are often contained in the same case. When making the T-connection, the main transformer is connected directly across the power line. One primary lead of the teaser transformer is connected to the center tap of the main transformer, and the 86.6% tap is connected to the power line. The same basic connection is made for the secondary. A vector diagram illustrating the voltage relationships of the T-connection is shown in *Figure 8-18*. The greatest advantage of the T-connection over the open-delta connection is that it maintains a better phase balance. The greatest disadvantage of the T-connection is that one transformer must contain a center tap of both its primary and secondary windings.

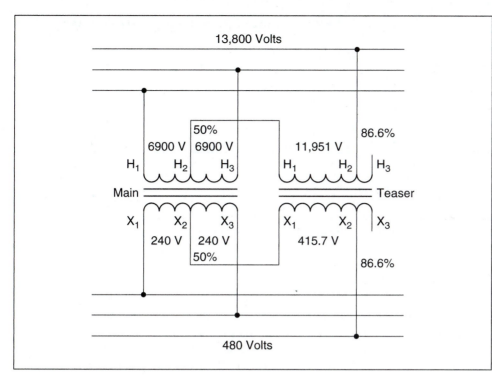

Figure 8-16 T-connected transformers.

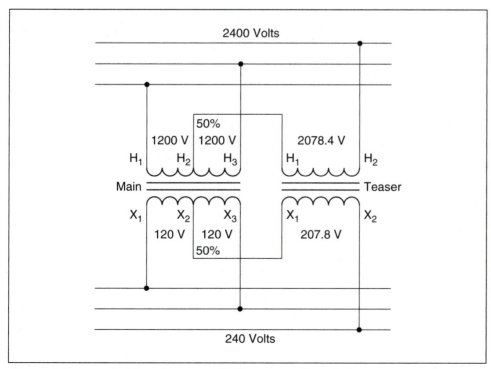

Figure 8-17 T-connected transformers with the same voltage ratings.

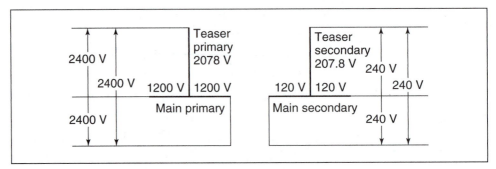

Figure 8-18 Vector voltage relationships of a T-connection.

Scott Connection

   The **Scott connection** is used to convert three-phase power into two-phase power using two single-phase transformers. The Scott connection is very similar to the T-connection in that one transformer, called the main transformer, must have a center or 50% tap, and the second, or teaser transformer, must have an 86.6% tap on the primary side. The difference between the Scott and T connections lies in the connection of the secondary windings (*Figure 8-19*). In the Scott connection, the secondary windings of each transformer provide the phases of a two-phase system. The voltages of the secondary windings are 90° out of phase with each other. The Scott connection is generally used to provide two-phase power for the operation of two-phase motors.

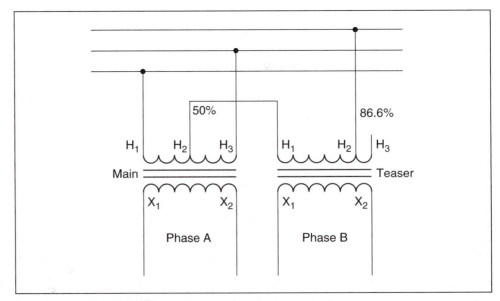

Figure 8-19 Scott connection.

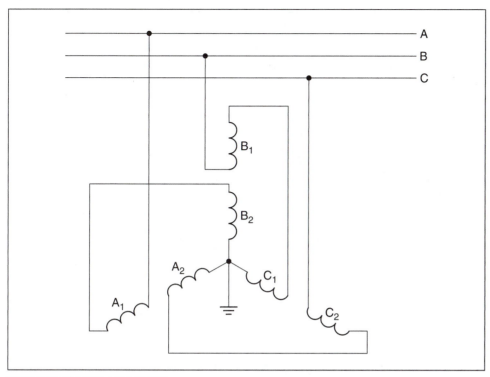

Figure 8-20 Zig-zag connection.

Zig-Zag Connection

The **zig-zag** or **interconnected-wye** transformer is used primarily for grounding purposes to establish a neutral point for the grounding of fault currents. The zig-zag connection is basically a three-phase autotransformer with windings divided into six equal parts *(Figure 8-20)*. In the event of a fault current, the zig-zag connection forces the current to flow equally in the three legs of the autotransformer, offering minimum impedance to the flow of fault current. A schematic diagram of the zig-zag connection is shown in *Figure 8-21*.

zig-zag
interconnected-
wye

Three-Phase to Six-Phase Connections

There are some instances when it is desirable to have a power system with more than three phases. A good example of this is when it is necessary to convert or rectify alternating current into direct current with a minimum amount of ripple (pulsations of voltage). Power supplies that produce a low amount of ripple require less filtering. One of the most common three-phase to six-phase connections is the **diametrical connection** *(Figure 8-22)*. The diametrical connection is preferred because it requires only one low-voltage winding on each transformer. If these windings are center tapped, a neutral conductor can

diametrical
connection

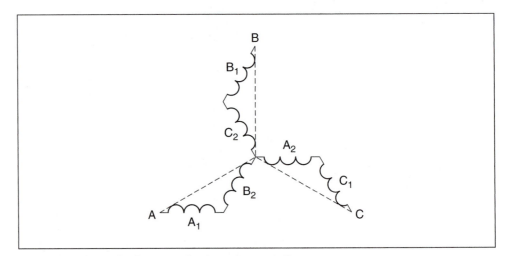

Figure 8-21 Schematic diagram of a zig-zag connection.

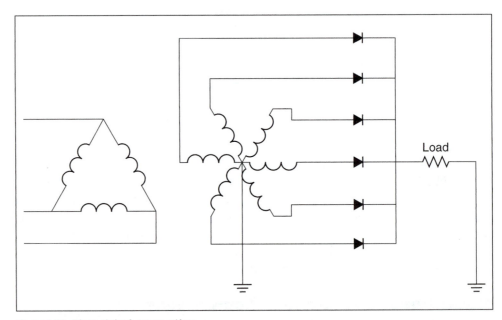

Figure 8-22 Diametrical connection.

be provided for the six-phase output, permitting half-wave rectification to be used. The high-voltage windings can be connected in wye or delta, but delta is preferred because it helps to reduce harmonics in the secondary winding. A schematic diagram of a diametrical connection with a delta-connected primary and three-phase half-wave rectifier is shown in *Figure 8-23*.

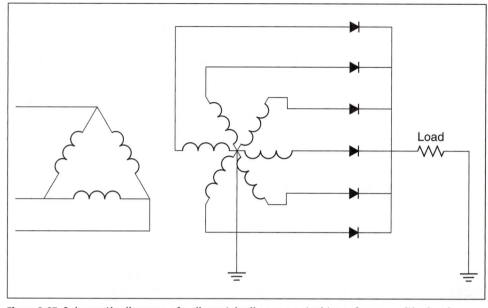

Figure 8-23 Schematic diagram of a diametrically connected transformer with six-phase half-wave rectifier.

A connection diagram for converting three-phase into six-phase is shown in *Figure 8-24*. The conversion is made by connecting the center taps of three single-phase center-tapped transformers together.

Double-Delta Connection

Another three-phase to six-phase connection is the **double-delta**. The transformers used in this connection must have two secondary windings that provide equal voltage. The secondary windings are connected in such a manner that two delta connections are formed. The two delta windings are reverse connected so they have an angular difference of 180° with respect to each other *(Figure 8-25)*. A schematic drawing showing the double-delta connection is shown in *Figure 8-26*. The primary windings may be connected in a wye or delta.

double-delta

Double-Wye Connection

The **double-wye** connection is very similar to the double-delta connection in that the transformers must have two secondary windings that produce equal voltage *(Figure 8-27)*. This connection is made by connecting the secondary windings in two different wyes. The windings are reverse connected so a phase displacement of 180° is formed between the two wye connections. A schematic of this connection is shown in *Figure 8-28*. Notice that

double-wye

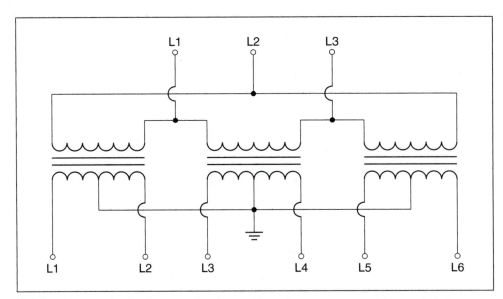

Figure 8-24 Three single-phase center-tapped transformers are used to convert three-phase into six phase.

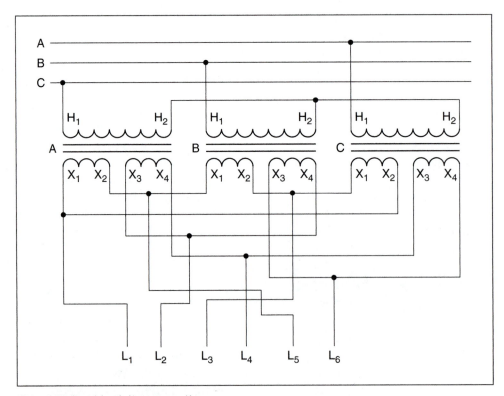

Figure 8-25 Double-delta connection.

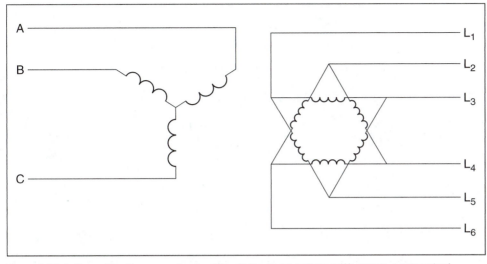

Figure 8-26 Schematic diagram of a double-delta connection with a wye-connected primary.

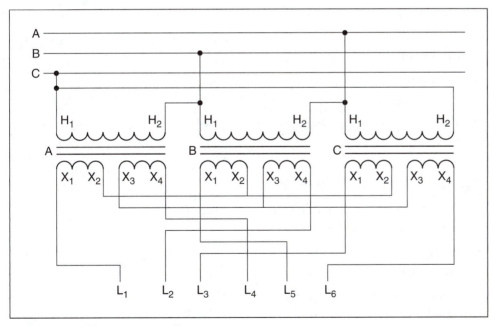

Figure 8-27 Double-wye connection.

this connection is very similar to the diametrical connection except that no center tap connection is possible. The primary windings may be connected wye or delta. The delta connection is preferred because it provides better phase and voltage stability.

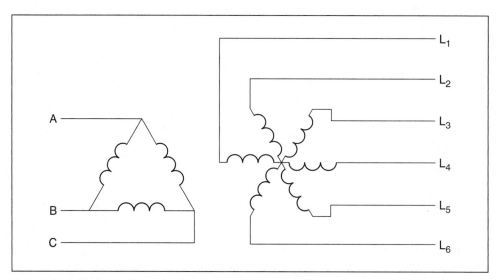

Figure 8-28 Schematic diagram of a double-wye connection.

Summary

1. Three-phase transformers are constructed by winding three separate transformers on the same core material.

2. Single-phase transformers can be used as a three-phase transformer bank by connecting their primary and secondary windings as either wyes or deltas.

3. When computing three-phase transformer values, follow the rules for three-phase circuits as well as the rules for transformers.

4. Phase values of voltage and current must be used when computing the values associated with the transformer.

5. The total power output of a three-phase transformer bank is the sum of the ratings of the three transformers.

6. An open-delta connection can be made with the use of only two transformers.

7. When an open-delta connection is used, the total output power is 87% of the sum of the power rating of both transformers.

8. Two transformers connected in T can be used to supply three-phase power.

9. When making a T-connection, the main transformer must contain a center tap, and it is preferred that the teaser transformer contain a 86.6% voltage tap.

10. The Scott connection is used to change three-phase power into two-phase power.

11. Three-phase transformers may be connected to produce six-phase power.

12. The most common three-phase to six-phase connection is the diametrical connection.

13. The diametrical connection is favored because only one secondary winding is needed on each transformer, and if the secondary windings are center tapped, the six-phase output can have a center tap for use as a neutral conductor.

14. The double-delta and double-wye connections require transformers with two secondary windings on each transformer.

15. A center-tap winding is not possible with the double-delta or double-wye connection.

Review Questions

1. How many transformers are needed to make an open-delta connection?

2. Two transformers rated at 100 kVA each are connected in an open-delta connection. What is the total output power that can be supplied by this bank?

3. When computing values of voltage and current for a three-phase transformer, should the line values of voltage and current be used or the phase values?

Refer to *Figure 8-29* to answer the following questions.

4. Assume that a line voltage of 2400 volts is connected to the primary of the three-phase transformer and the line voltage of the secondary is 240 volts. What is the turns ratio of the transformer?

5. Assume the load has an impedance of 3.5 Ω per phase. What is the line current provided by the transformer secondary?

6. How much current is flowing through the secondary winding?

7. How much current is flowing through the primary winding?

Refer to *Figure 8-30* to answer the following questions.

8. Assume that a line voltage of 12,470 volts is connected to the primary of the transformer and the line voltage of the secondary is 480 volts. What is the turns ratio of the transformer?

9. Assume the load has an impedance of 6 Ω per phase. What is the secondary line current?

10. How much current is flowing in the secondary winding?

11. How much current is flowing in the primary winding?

12. What is the line current of the primary?

Problems

Refer to the transformer shown in *Figure 8-29* to fill in the following values.

1.

Primary	Secondary	Load
E_P _____	E_P _____	E_P _____
I_P _____	I_P _____	I_P _____
E_L 4160	E_L 440	E_L _____
I_L _____	I_L _____	I_L _____
Ratio	$Z = 3.5 \, \Omega$	

2.

Primary	Secondary	Load
E_P _____	E_P _____	E_P _____
I_P _____	I_P _____	I_P _____
E_L 7200	E_L 240	E_L _____
I_L _____	I_L _____	I_L _____
Ratio	$Z = 4 \, \Omega$	

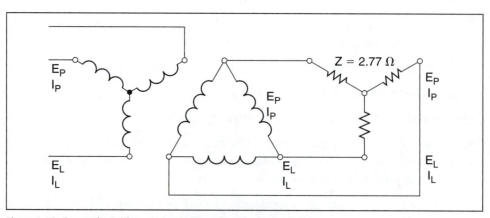

Figure 8-29 Example 1: Three-phase transformer calculation.

Refer to the transformer connection shown in *Figure 8-30* to fill in the following values.

3.

	Primary		Secondary		Load
E_P	_____	E_P	_____	E_P	_____
I_P	_____	I_P	_____	I_P	_____
E_L	13,800	E_L	480	E_L	_____
I_L	_____	I_L	_____	I_L	_____
Ratio		$Z = 2.5\ \Omega$			

4.

	Primary		Secondary		Load
E_P	_____	E_P	_____	E_P	_____
I_P	_____	I_P	_____	I_P	_____
E_L	23,000	E_L	208	E_L	_____
I_L	_____	I_L	_____	I_L	_____
Ratio		$Z = 3\ \Omega$			

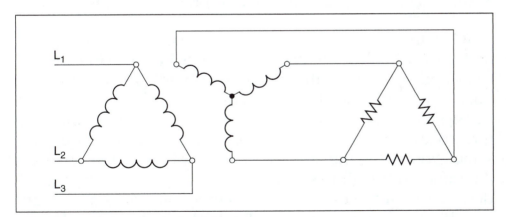

Figure 8-30 Practice problems circuit 2.

UNIT 9

Single-Phase Loads for Three-Phase Transformers

Objectives

After studying this unit, you should be able to:

- Compute values of voltage and current for single-phase loads connected to a three-phase transformer bank.

- Discuss different types of three-phase transformer connections generally used to supply single-phase loads.

- Connect single-phase loads to a three-phase transformer.

When true three-phase loads are connected to a three-phase transformer bank, there are no problems in balancing the currents and voltages of the individual phases. *Figure 9-1* illustrates this condition. In this circuit, a delta-wye three-phase transformer bank is supplying power to a wye-connected three-phase load in which the impedance of each phase is the same. Notice that the amount of current flow in the phases is the same. Although this is the ideal condition and is certainly desired for all three-phase transformer loads, it is not always possible to obtain a balanced load. Three-phase transformer connections are often used to supply **single-phase loads**, which tend to unbalance the system.

single-phase loads

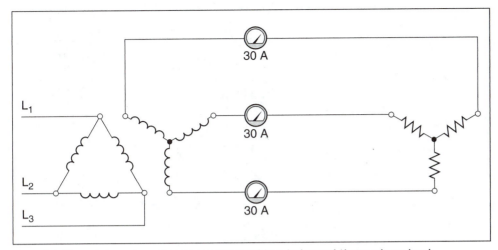

Figure 9-1 Three-phase transformer connected to a balanced three-phase load.

Open-Delta Connection Supplying a Single-Phase Load

The type of three-phase transformer connection to use is generally determined by the amount of power needed. When a transformer bank must supply both three-phase and single-phase loads, the utility company will often provide an open-delta connection with one transformer center-tapped as shown in *Figure 9-2*. In this connection, it is assumed that the amount of three-phase power needed is 20 kVA, and the amount of single-phase power needed is 30 kVA. Notice that the transformer that has been center-tapped must supply power to both the three-phase and single-phase loads. Since this is an open-delta connection, the transformer bank can be loaded to only 87% of its full capacity when supplying a three-phase load. The rating of the three-phase transformer bank must therefore be 23 kVA (20 kVA/0.87 = 23 kVA). Since the rating of the two transformers can be added to obtain a total output power rating, one transformer is rated at only half the total amount of power needed, or 12 kVA (23 kVA/2 = 11.5 kVA). The transformer that is used to supply power to the three-phase load only is rated at 12 kVA. The transformer that has been center-tapped must supply power to both the single-phase and three-phase load. Its capacity will, therefore, be 42 kVA (12 kVA + 30 kVA). Therefore, a 45-kVA transformer will be used.

Voltage Values

The connection shown in *Figure 9-2* has a line-to-line voltage of 240 volts. The three voltmeters V_1, V_2, and V_3 have all been connected across the three-phase lines and should indicate 240 volts each. Voltmeters V_4 and V_5 have

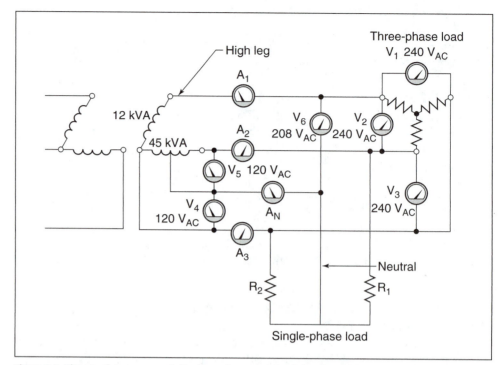

Figure 9-2 Three-phase open-delta transformer supplying both three-phase and single-phase loads.

been connected between the two lines of the larger transformer and its center tap. These two voltmeters will indicate a voltage of 120 volts each. Notice that it is these two lines and the center tap that are used to supply the single-phase power needed. The center tap of the larger transformer is used as a neutral conductor for the single-phase loads. Voltmeter V_6 has been connected between the center tap of the larger transformer and the line of the **high leg**   smaller transformer. This line is known as a **high leg**, because the voltage between this line and the neutral conductor will be higher than the voltage between the neutral and either of the other two conductors. The high-leg voltage can be computed by increasing the single-phase center-tapped voltage value by 1.732 (square root of 3). In this case, the high-leg voltage will be 208 V (120 × 1.732 = 208). When this type of connection is employed, the *National Electrical Code®* requires that the high leg be identified by connect-**orange wire**   ing it to an **orange wire** or by **tagging** it at any point where it enters an **tagging**   enclosure with the neutral conductor.

Load Conditions

In the first load condition, it will be assumed that only the three-phase load is in operation and none of the single-phase load is operating. If the three-phase load is operating at maximum capacity, ammeters A_1, A_2, and A_3 will indicate a current flow of 48.1 amps each (20 kVA/240 volts × 1.732 = 48.1 amps). Notice that when only the three-phase load is in operation, the current on each line is balanced.

Now assume that none of the three-phase load is in operation and only the single-phase load is operating. If all the single-phase load is operating at maximum capacity, ammeters A_2 and A_3 will each indicate a value of 125 amps (30 kVA/240 volts = 125 amps). Ammeter A_1 will indicate a current flow of 0 amp because all the load is connected between the other two lines of the transformer connection. Ammeter A_N will also indicate a value of 0 amp. Ammeter A_N is connected in the neutral conductor, and the neutral conductor carries the sum of the unbalanced load between the two phase conductors. Another way of stating this is to say that the neutral conductor carries the difference between the two line currents. Since both of these conductors are now carrying the same amount of current, the difference between them is 0 amp.

Now assume that one side of the single-phase load, resistor R_2, has been opened and no current flows through it. If the other line maintains a current flow of 125 A, the neutral conductor will have a current flow of 125 amps also (125 − 0 = 125).

Now assume that resistor R_2 has a value that will permit a current flow of 50 amps on that phase. The neutral current will now be 75 amps (125 − 50 = 75). Since the neutral conductor carries the sum of the unbalanced load, the size of the neutral conductor never needs to be larger than the largest line conductor.

It will now be assumed that both three-phase and single-phase loads are operating at the same time. If the three-phase load is operating at maximum capacity and the single-phase load is operating in such a manner that 125 amps flow through resistor R_1 and 50 amps flow through resistor R_2, the ammeters will indicate the following values:

A_1 = 48.1 amps

A_2 = 173.1 amps (48.1 + 125 = 173.1)

A_3 = 98.1 amps (48.1 + 50 = 98.1)

A_N = 75 amps (125 − 50 = 75)

Notice that the smaller of the two transformers is supplying current to only the three-phase load, but the larger transformer must supply current for both the single-phase and three-phase loads.

Although the circuit shown in *Figure 9-2* is the most common method of connecting both three-phase and single-phase loads to an open-delta transformer bank, it is possible to use the high leg to supply power to a single-phase load also. The circuit shown in *Figure 9-3* is a circuit of this type. Resistors R_1 and R_2 are connected to the lines of the transformer that has been center-tapped, and resistor R_3 is connected to the line of the other

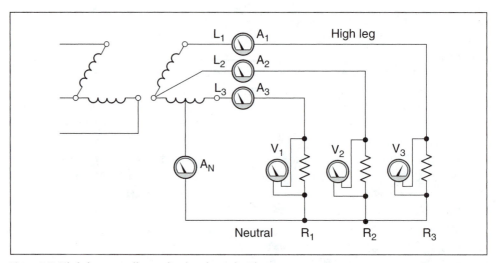

Figure 9-3 High leg supplies a single-phase load.

transformer. If the line-to-line voltage is 240 volts, voltmeters V_1 and V_2 will each indicate a value of 120 volts across resistors R_1 and R_2. Voltmeter V_3, however, will indicate that a voltage of 208 volts is applied across resistor R_3.

Calculating Neutral Current

The amount of current flow in the neutral conductor will still be the sum of the unbalanced load between lines L_2 and L_3, with the addition of the current flow in the high leg, L_1. To determine the amount of neutral current, use the formula

$$A_N = A_1 + (A_2 - A_3)$$

For example, assume line L_1 has a current flow of 100 amps, line L_2 has a current flow of 75 amps, and line L_3 has a current flow of 50 amps. The amount of current flow in the neutral conductor would be:

$$A_N = A_1 + (A_2 - A_3)$$
$$A_N = 100 + (75 - 50)$$
$$A_N = 100 + 25$$
$$A_N = 125 \text{ amps}$$

In this circuit, it is possible for the neutral conductor to carry more current than any of the three-phase lines. This circuit is more of an example of why the National Electrical Code® requires a high leg to be identified than it is a practical working circuit. It is seldom that the high-leg side of this type of connection will be connected to the neutral conductor.

CLOSED DELTA WITH CENTER TAP

Another three-phase transformer configuration used to supply power to single-phase and three-phase loads is shown in *Figure 9-4*. This circuit is virtually identical to the circuit shown in *Figure 9-2* with the exception that a third transformer has been added to close the delta. Closing the delta permits more power to be supplied for the operation of three-phase loads. In this circuit, it is assumed that the three-phase load has a power requirement of 75 kVA, and the single-phase load requires an additional 50 kVA. Three 25-kVA transformers could be used to supply the three-phase power needed (25 kVA × 3 = 75 kVA). The addition of the single-phase load, however, requires one of the transformers to be larger. This transformer must supply both the three-phase and single-phase load, which requires it to have a rating of 75 kVA (25 kVA + 50 kVA = 75 kVA). In this circuit, the primary is connected in a delta configuration. Since the secondary side of the transformer bank is a delta connection, either a wye or a delta primary could have been used. This, however, will not be true of all three-phase transformer connections supplying single-phase loads.

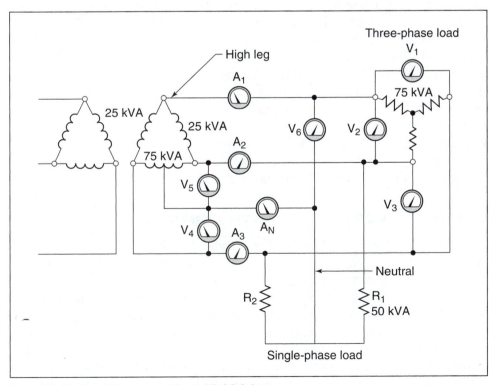

Figure 9-4 Closed-delta connection with high leg.

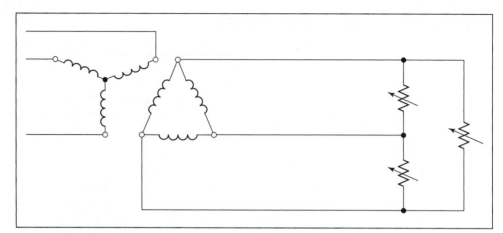

Figure 9-5 Single-phase loads supplied by a wye-delta transformer connection.

CLOSED DELTA WITHOUT CENTER TAP

In the circuit shown in *Figure 9-5,* the transformer bank has been connected in a wye-delta configuration. Notice that there is no transformer secondary with a center-tapped winding. In this circuit, there is no neutral conductor. The three loads have been connected directly across the three-phase lines. Since these three loads are connected directly across the lines, they form a delta-connected load. If these three loads are intended to be used as single-phase loads, they will in all likelihood have changing resistance values. The result of this connection is a three-phase delta-connected load that can be unbalanced in different ways. The amount of current flow in each phase is determined by the impedance of the load and the vectorial relationships of each phase. Each time one of the single-phase loads is altered, the vector relationship changes also. No one phase will become overloaded, however, if the transformer bank has been properly sized for the maximum connected load.

DELTA-WYE CONNECTION WITH NEUTRAL

The circuit shown in *Figure 9-6* is a three-phase transformer connection with a delta-connected primary and wye-connected secondary. The secondary has been center-tapped to form a neutral conductor. This is one of the most common connections used to provide power for single-phase loads. Typical voltages for this type of connection are 208/120 and 480/277. The neutral conductor will carry the vector sum of the unbalanced current. However, it should be noted that in this circuit, the sum of the unbalanced current is not the difference between two phases. In the delta connection where one transformer was center-tapped to form a neutral conductor, the

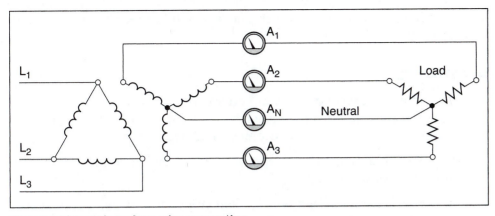

Figure 9-6 Three-phase four-wire connection.

two lines were 180° out of phase when compared with the center tap. In the wye connection, the lines will be 120° out of phase. When all three lines are carrying the same amount of amperage, the neutral current will be zero.

It should be noted that a wye-connected secondary with center tap can, under the right conditions, experience extreme unbalance problems. *If this transformer connection is powered by a three-phase three-wire system, the primary winding must be connected in a delta configuration.* If the primary is connected as a wye connection, the circuit will become exceedingly unbalanced when load is added to the circuit. Connecting the center tap of the primary to the center tap of the secondary will not solve the unbalance problem if a wye primary is used on a three-wire system.

If the incoming power is a three-phase four-wire system as shown in *Figure 9-7,* however, a wye-connected primary can be used without problems. The neutral conductor connected to the center tap of the primary prevents the unbalance problems. It is a common practice with this type of connection to tie the neutral conductor of both primary and secondary together as

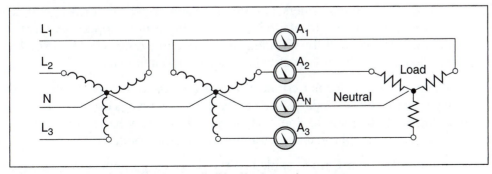

Figure 9-7 Neutral conductor is supplied by the incoming power.

shown. When this is done, however, line isolation between the primary and secondary windings is lost.

Summary

1. It is common practice to center-tap one of the transformers in a delta connection to provide power for single-phase loads. When this is done, the remaining phase connection becomes a high leg.

2. The National Electrical Code® requires that a high leg be identified by an orange wire or by tagging.

3. The center connection of a wye is often tapped to provide a neutral conductor for three-phase loads. This produces a three-phase four-wire system. Common voltages produced by this type of connection are 208/120 and 480/277.

4. Transformers should not be connected as a wye-wye unless the incoming power line contains a neutral conductor.

Review Questions

1. How does the National Electrical Code® specify that the high leg of a four-wire delta connection be marked?

2. An open-delta three-phase transformer system has one transformer center-tapped to provide a neutral for single-phase voltages. If the voltage from line to center tap is 277 volts, what is the high-leg voltage?

3. If a single-phase load is connected across the two line conductors and neutral of the transformer in question #2, and one line has a current of 80 amps and the other line has a current of 68 amps, how much current is flowing in the neutral conductor?

4. A three-phase transformer connection has a delta-connected secondary and one of the transformers has been center-tapped to form a neutral conductor. The phase-to-neutral value of the center-tapped secondary winding is 120 volts. If the high leg should become connected to a single-phase load, how much voltage will be applied to that load?

5. A three-phase transformer connection has a delta-connected primary and a wye-connected secondary. The center tap of the wye is used as a neutral conductor. If the line-to-line voltage is 480 volts, what is the voltage between any one phase conductor and the neutral conductor?

6. A three-phase transformer bank has the secondary connected in a wye configuration. The center tap is used as a neutral conductor. If the voltage

across any phase conductor and neutral is 120 volts, how much voltage would be applied to a three-phase load connected to the secondary of this transformer bank?

7. A three-phase transformer bank has the primary and secondary windings connected in a wye configuration. The secondary center tap is being used as a neutral to supply single-phase loads. Will connecting the center-tap connection of the secondary to the center-tap connection of the primary permit the secondary voltage to stay in balance when a single-phase load is added to the secondary?

8. Referring to the transformer connection in question 7, if the center tap of the primary is connected to a neutral conductor on the incoming power, will it permit the secondary voltages to be balanced when single-phase loads are added?

UNIT 10

Transformer Installation

Objectives:

After studying this unit the student should be able to:

- Determine the correct conductor size for the installation of a transformer in accord with the National Electrical Code®.

- Determine the proper size of the branch circuit protective device for transformers of different ratings.

This unit will cover the National Electrical Code® requirements for the installation of transformers. The two main factors to be discussed will be the selection of the size of the branch circuit protective device, and the selection of the conductor size.

TRANSFORMER PROTECTION

Fuses or circuit breakers can be used as the branch circuit protective devices in circuits supplying power to transformers. In general, fuses are selected for protection of high voltage circuits because they are less expensive than other types of protection, are extremely reliable, and do not require as much maintenance as circuit breakers. Regardless of which type of protection is employed, the ampere ratings will be the same and either must be capable of interrupting the maximum available fault current of the circuit. This ability of the fuse or circuit breaker to open under the maximum fault current is expressed as the **interrupting rating** of the device. According to NEC 240.60 (C)(3), fuses must be plainly marked with the interrupting current or rating if it is other than 10,000 amperes. NEC 240.83 (C) states that circuit breakers must be plainly marked with the interrupting rating if it is other than 5,000 amperes. The National Electrical Code® requires these markings because the minimum interrupting rating of a fuse is 10,000 amperes and the minimum interrupting rating of a circuit breaker is 5,000 amperes.

interrupting rating

The minimum interrupting rating permitted for a fuse or circuit breaker in a specific installation is the maximum symmetrical fault current available at the location of the protective device. Power companies will provide this information when requested and will recommend an interrupting rating in excess of this value.

Both fuses and circuit breakers used for protection of the primary are generally of the time delay type because of the high in-rush current associated with transformers. Fast acting fuses or circuit breakers will often open when power is applied to the primary winding.

Determining Interrupting Rating

The maximum rating of overcurrent devices for 600 volt and higher transformers is set forth in NEC Table 450.3(A), *Figure 10-1*. To use this table, the percent impedance (%Z) of the transformer must be known. This value is commonly stamped on the nameplate of transformers rated 25 kVA and higher. The actual impedance of a transformer is determined by its physical construction, such as the gage of the wire in the winding, the number of turns, the type of core material, and the magnetic efficiency of the core construction. **Percent impedance** is an empirical value that can be used to predict transformer performance. It is common practice to use the symbol %Z to represent the percent impedance. Percentages must be converted to a decimal form before they can be used in a mathematical formula. When this conversion has been made, the symbol ($\cdot$Z) will be used to represent the percent impedance in decimal form. This will be called the **decimal impedance**. The percent value is converted to the numeric value by moving the decimal point two places to the left, thus, 5.75% becomes 0.0575. This value has no units as it represents a ratio.

percent impedance

decimal impedance

When working with any transformer it is important to keep in mind the full meaning of the terms primary and secondary and high-voltage and low-voltage. The **primary** is the winding that is connected to a voltage source; the **secondary** is the winding that is connected to an electrical load. The source may be connected to either the low-voltage or the high-voltage terminals of the transformer. If a person should inadvertently connect a high-voltage source to the low-voltage terminals, the transformer

Table 450.3(A) Maximum Rating or Setting of Overcurrent Protection for Transformers Over 600 Volts (as a Percentage of Transformer-Rated Current)

| Location Limitations | Transformer Rated Impedance | Primary Protection Over 600 Volts | | Secondary Protection (See Note 2.) | | |
| | | | | Over 600 Volts | | 600 Volts or Less |
		Circuit Breaker (See Note 4.)	Fuse Rating	Circuit Breaker (See Note 4.)	Fuse Rating	Circuit Breaker or Fuse Rating
Any location	Not more than 6%	600% (See Note 1.)	300% (See Note 1.)	300% (See Note 1.)	250% (See Note 1.)	125% (See Note 1.)
	More than 6% and not more than 10%	400% (See Note 1.)	300% (See Note 1.)	250% (See Note 1.)	225% (See Note 1.)	125% (See Note 1.)
Supervised locations only (See Note 3.)	Any	300% (See Note 1.)	250% (See Note 1.)	Not required	Not required	Not required
	Not more than 6%	600%	300%	300% (See Note 5.)	250% (See Note 5.)	250% (See Note 5.)
	More than 6% and not more than 10%	400%	300%	250% (See Note 5.)	225% (See Note 5.)	250% (See Note 5.)

Notes:

1. Where the required fuse rating or circuit breaker setting does not correspond to a standard rating or setting, a higher rating or setting that does not exceed the next higher standard rating or setting shall be permitted.

2. Where secondary overcurrent protection is required, the secondary overcurrent device shall be permitted to consist of not more than six circuit breakers or six sets of fuses grouped in one location. Where multiple overcurrent devices are utilized, the total of all the device ratings shall not exceed the allowed value of a single overcurrent device. If both circuit breakers and fuses are used as the overcurrent device, the total of the device ratings shall not exceed that allowed for fuses.

3. A supervised location is a location where conditions of maintenance and supervision ensure that only qualified persons monitor and service the transformer installation.

4. Electronically actuated fuses that may be set to open at a specific current shall be set in accordance with settings for circuit breakers.

5. A transformer equipped with a coordinated thermal overload protection by the manufacturer shall be permitted to have separate secondary protection omitted.

Figure 10-1 *NEC*® Table 450.3(A).

would increase the voltage by the ratio of the turns. A 600V to 200V transformer would become a 600V to 1800V transformer if the connections were reversed. This will not only create a very dangerous situation, but could also result in permanent damage to the transformer because of excessive current flow in the winding. *Always be careful when working with transformers and never touch a terminal unless the power source has been disconnected.*

The percent impedance is measured by connecting an ammeter across the low-voltage terminals and a variable voltage source across the high-voltage terminals. This arrangement is shown in *Figure 10-2*. The connection of the ammeter is short-circuiting the secondary of the transformer. An

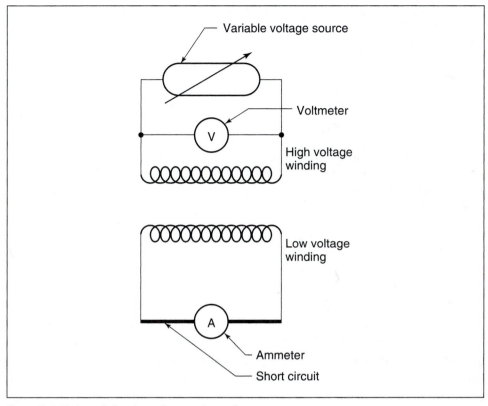

Figure 10-2 Determining transformer impedance.

ammeter should be chosen that has a scale that has about twice the range of the value to be measured so that the reading will be taken in the middle of the range. If the current to be measured is expected to be about 30 amperes, a meter with a 0–60 ampere range would be ideal. Using a meter with a range under 40 amperes or over 100 amperes may not permit an accurate reading.

After the connections have been made, the voltage is increased until the ammeter indicates the rated full load current of the secondary (low-voltage winding). The value of the source voltage is then used to calculate the decimal impedance (·Z). The ·Z is found by determining the ratio of the source voltage as compared to the rated voltage of the high-voltage winding.

Example 1

Assume that the transformer shown in *Figure 10-2* is a 2400/480 volt 15 kVA transformer. To determine the impedance of the transformer, first compute the full load current rating of the secondary winding.

$$I = \frac{VA}{E}$$

$$I = \frac{15{,}000}{480}$$

$$I = 31.25A$$

Next, increase the source voltage connected to the high-voltage winding until a current of 31.25 amperes flows in the low-voltage winding. For the purpose of this example assume that voltage value to 138 volts. Finally, determine the ratio of applied voltage as compared to the rated voltage.

$$\cdot Z = \frac{SourceVoltage}{RatedVoltage}$$

$$\cdot Z = \frac{138}{2400}$$

$$\cdot Z = 0.0575$$

To change the decimal value to %Z, move the decimal point two places to the right and add a % sign. This is the same as multiplying the decimal value by 100.

$$\%Z = 5.75\%$$

Transformer impedance is a major factor in determining the amount of voltage drop a transformer will exhibit between no load and full load and in determining the amount of current flow in a short-circuit condition. When transformer impedance is known, it is possible to calculate the maximum possible short-circuit current. This would be a worse case scenario and the available short-circuit current would decrease as the length of the connecting wires increased the impedance. The upcoming formulas can be used to calculate the short-circuit current value when the transformer impedance is known.

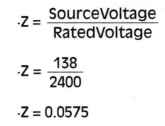

$$\text{(Single Phase) } I_{sc} = \frac{VA}{E \times \cdot Z}$$

$$\text{(Three Phase) } I_{sc} = \frac{VA}{E \times \sqrt{3} \times \cdot Z}$$

Since the formula for finding the rated current for a single phase transformer is

$$I = \frac{VA}{E}$$

and the formula for finding the rated current for a three phase transformer is

$$I = \frac{VA}{E \times \sqrt{3}}$$

the short circuit current can be determined by dividing the rated secondary current by the decimal impedance of the transformer.

$$I_{sc} = \frac{I_{secondary}}{\cdot Z}$$

The short-circuit current for the transformer in the previous example would be

$$I_{sx} = \frac{31.25}{0.0575}$$

$$I_{sc} = 543.5 \text{ amperes}$$

Determining Transformer Fuse or Circuit Breaker Size

The transformer impedance value is also used to determine the fuse or circuit breaker size for the primary and secondary windings. It will be assumed that the transformer shown in *Figure 10-2* is a step-down transformer and the 2400 volt winding is used as the primary and the 480 volt winding is used as the secondary. NEC Table 450.3(A) indicates that the size of the protective device for a primary over 600 volts and having an impedance of 6% or less is 300% of the rated current. The rated current for the primary winding in this example is

$$I = \frac{15,000}{2400}$$

$$I = 6.25 \text{ A}$$

The fuse or circuit breaker size will be 6.25 × 3.00 = 18.75 amperes. Note No.1 of NEC Table 450.3(A) permits the next higher rating to be used if the computed value does not correspond to one of the standard fuse sizes listed in NEC 240.6. The next higher standard size is 20 amperes.

NEC Table 450.3(A) indicates that if the secondary voltage is 600 volts or less, the fuse size will be set at 125% of the rated secondary current. In this example, the fuse size will be

$$31.25 \times 1.25 = 39 \text{ amperes}$$

A 40-ampere fuse will be used as the secondary short circuit protective device, *Figure 10-3*.

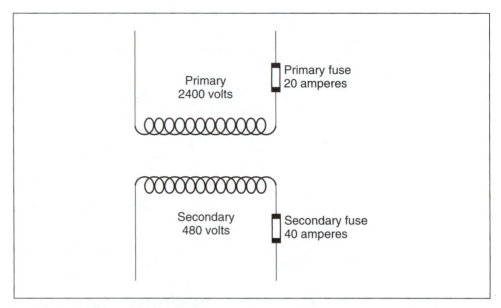

Figure 10-3 Transformer fusing.

Transformers Rated 600 Volts or Less

Fuse protection for transformers rated 600 volts or less is determined by NEC Table 450.3(B), *Figure 10-4.* The upper part of NEC Table 450.3(B) is concerned with transformers that have primary overcurrent protection only (no secondary protection). The percentage value of the fuse is determined by the amount of rated primary current. If the primary current is 9 amperes or more, the fuse or circuit breaker size is set at 125% of the primary current. If this current does not correspond to one of the standard fuse sizes listed in NEC Section 240.6, *Figure 10-5,* Note 1 of Table 450.3(B) states that the next higher standard rating can be used.

If the rated primary current is less than 9 amperes, the overcurrent device can be set at *not more than* 167% of that value. If the primary current is less than 2 amperes, the short circuit protective device can be set at *not more than* 300% of that value.

Notice that if the primary current is 9 amperes or more, it is permissible to increase the fuse size to the next highest standard rating. If the primary current is less than 9 amperes, the next lowest fuse size must be used.

The lower half of NEC Table 450.3(B) deals with transformers that have short circuit protection in both the primary and secondary windings. The percentage rating for determining fuse or circuit breaker size is again determined by rated current. If the secondary current is 9 amperes or more, the fuse size is 125%. Note 1 permits the fuse size to be increased to the next

Table 450.3(B) Maximum Rating or Setting of Overcurrent Protection for Transformers 600 Volts and Less (as a Percentage of Transformer-Rated Current)

Protection Method	Primary Protection			Secondary Protection (See Note 2.)	
	Currents of 9 Amperes or More	Currents Less Than 9 Amperes	Currents Less Than 2 Amperes	Currents of 9 Amperes or More	Currents Less Than 9 Amperes
Primary only protection	125% (See Note 1.)	167%	300%	Not required	Not required
Primary and secondary protection	250% (See Note 3.)	250% (See Note 3.)	250% (See Note 3.)	125% (See Note 1.)	167%

Notes:

1. Where 125 percent of this current does not correspond to a standard rating of a fuse or nonadjustable circuit breaker, a higher rating that does not exceed the next higher standard rating shall be permitted.

2. Where secondary overcurrent protection is required, the secondary overcurrent device shall be permitted to consist of not more than six circuit breakers or six sets of fuses grouped in one location. Where multiple overcurrent devices are utilized, the total of all the device ratings shall not exceed the allowed value of a single overcurrent device. If both breakers and fuses are utilized as the overcurrent device, the total of the device ratings shall not exceed that allowed for fuses.

3. A transformer equipped with coordinated thermal overload protection by the manufacturer and arranged to interrupt the primary current shall be permitted to have primary overcurrent protection rated or set at a current value that is not more than six times the rated current of the transformer for transformers having not more than 6 percent impedance and not more than four times the rated current of the transformer for transformers having more than 6 percent but not more than 10 percent impedance.

Figure 10-4 *NEC*® Table 450.3(B).

240.6 Standard Ampere Ratings.

(A) Fuses and Fixed-Trip Circuit Breakers. The standard ampere ratings for fuses and inverse time circuit breakers shall be considered 15, 20, 25, 30, 35, 40, 45, 50, 60, 70, 80, 90, 100, 110, 125, 150, 175, 200, 225, 250, 300, 350, 400, 450, 500, 600, 700, 800, 1000, 1200, 1600, 2000, 2500, 3000, 4000, 5000, and 6000 amperes. Additional standard ampere ratings for fuses shall be 1, 3, 6, 10, and 601. The use of fuses and inverse time circuit breakers with nonstandard ampere ratings shall be permitted.

Figure 10-5 NEC 240.6.

higher standard value. If the secondary current is less than 9 amperes, the fuse or circuit breaker cannot be rated greater than 167% of the secondary current.

Example 2

Assume that a 4 kVA transformer is rated 480/120 volts. Also assume that the transformer is mounted in a piece of equipment that is protected by a 30 ampere circuit breaker. Note 3 of NEC Table 450.3(B) states that if the transformer is provided with a thermal overload device in the primary winding by the manufacturer, no further primary protection is required if the feeder overcurrent protective device is not greater than 6 times the primary current for a transformer with a rated impedance of not more than 6%, and not more than four times the primary rated current for a transformer with a rated impedance greater than 6% but not more than 10%. The rated primary current for this transformer is

$$I = \frac{VA}{E}$$

$$I = \frac{4000}{480}$$

$$I = 8.33 \text{ amperes}$$

Six times the rated primary current would be 49.98 amperes and four times the rated current would be 33.32 amperes. The 30-ampere circuit breaker can serve as the primary overcurrent protection for this transformer.

Example 3

Assume that a transformer has a primary winding rated at 240 volts and is provided with a thermal overload device by the manufacturer. Also assume that the primary has a rated current of 3 amperes and an impedance of 4%. To determine if separate overcurrent protection is needed for the primary, multiply the rated primary current by 6 (3 × 6 = 18 amperes). If the branch circuit protective device supplying power to the transformer primary has an overcurrent protection device rated at 18 amperes or less, no additional protection is required. If the branch circuit overcurrent protection device is greater than 18 amperes, a separate overcurrent protection device for the primary will be required.

If separate overcurrent protection is required, it will be computed at 167% of the primary current rating, because the primary current is less than 9 amperes but greater than 2 amperes. In this example the primary overcurrent protective device should be rated at

$$3 \times 1.67 = 5.01 \text{ amperes}$$

A 5-ampere fuse should be used to provide primary overcurrent protection.

Overcurrent Protection For Autotransformers

NEC 450.4 (A) states that an autotransformer rated at 600 volts or less must be protected by an overcurrent device in each of the input ungrounded conductors. This section also states that the overcurrent device is to be rated at not more than 125% of the rated full-load input current of the autotransformer. If the rated primary current is 9 amperes or more, and the computed fuse size does not correspond to a standard fuse or

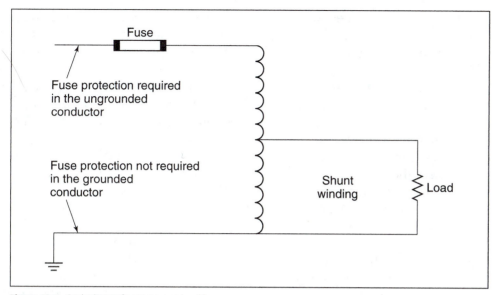

Figure 10-6 Autotransformer protection.

circuit breaker size listed in NEC 240.6, the next standard size may be used. The code further states that an overcurrent device shall not be installed in series with the shunt winding, *Figure 10-6*. The shunt winding is the section of the autotransformer winding that is common to both the primary and secondary.

The exception to NEC 450.4(A) states that if the input current is rated less than 9 amperes, however, the rating of the fuse or circuit breaker cannot be greater than 167% of the full-load input current. Autotransformers with rated voltages over 600 volts are subject to the provisions set forth in NEC 450.3(A).

DETERMINING CONDUCTOR SIZE FOR A TRANSFORMER

There is more than one method that can be employed to determine the conductor size needed to supply power to a transformer installation. Electrical engineers often determine the conductor size by computing the resistance of the wire when the current, amount of voltage drop, and length of conductor are known. Electricians generally determine the conductor size by referring to the NEC. This is the reason the conductor size recommended by the manufacturer will often be different than the size determined by using the NEC. This text will use the NEC to determine the conductor size.

NEC 215.2(A) states that the rating of a branch-circuit shall not be less than noncontinuous load plus 125% of the continuous load. Most transformer installations are intended for continuous operation (more than 3 hours). About the only exception is transformers intended for the operation of welding machines. Article 630 of the National Electrical Code® should be consulted for these special provisions.

There are several steps that should be followed when selecting a conductor. The first step is to determine the amount of current the conductor must carry. This can be done by calculating the rated current of the winding and then increasing this value by 25%.

The next step is to select an insulation type that corresponds to the conditions where the installation is to be made. Factors such as the ambient temperature and whether the conductors will be in a wet or dry location are important. Table 310.13 of the NEC lists different types of insulation and characteristics of each.

Another factor that must be taken into consideration when determining the conductor size is the temperature rating of the devices and terminals as specified in NEC 110.14(C). This section states that the conductor is to be so selected and coordinated as to not exceed the lowest temperature rating of any connected termination, any connected conductor, or any connected device. This means that regardless of the temperature rating of the conductor, **ampacity** the **ampacity** (current carrying capacity) must be selected from a column that does not exceed the temperature rating of the termination or any device connected to the circuit. The conductors listed in the first column of NEC Table 310.16, *Figure 10-7*, have a temperature rating of 60°C, the conductors in the second column have a rating of 75°C, and the conductors in the third column have a rating of 90°C. The temperature ratings of devices such as circuit breakers, fuses, and terminals are found in Underwriters Laboratories' (UL's) product directories. Occasionally, the temperature rating may be found on the piece of equipment, but this is the exception and not the rule. As a general rule the temperature rating of most devices will not exceed 75°C.

When the termination temperature rating is not listed or known, NEC 110.14(C)(1)(a) states that for circuits rated at 100 amperes or less, or for #14 AWG through #1 AWG conductors, the ampacity of the wire, regardless of its temperature rating, will be selected from the 60°C column. This does not mean that only those insulation types listed in the 60°C column can be used, but that the ampacity must be selected from that column. For example, assume that a copper conductor with type THWN-2 insulation is to be connected to a 50 ampere circuit breaker that does not have a listed temperature rating. According to NEC Table 310.16, a #8 AWG copper conductor with THWN-2 insulation is rated to carry 55 amperes of current. Type THWN-2 insulation is located in the 90°C column, but the temperature rating of the circuit breaker is not known. Therefore, the wire size must be selected from the ampacity ratings in the 60°C column. A #8 AWG copper conductor has a current rating of

Table 310.16   Allowable Ampacities of Insulated Conductors Rated 0 Through 2000 Volts, 60°C Through 90°C (140°F Through 194°F), Not More Than Three Current-Carrying Conductors in Raceway, Cable, or Earth (Directly Buried), Based on Ambient Temperature of 30°C (86°F)

Size AWG or kcmil	Temperature Rating of Conductor (See Table 310.13.)						Size AWG or kcmil
	60°C (140°F)	75°C (167°F)	90°C (194°F)	60°C (140°F)	75°C (167°F)	90°C (194°F)	
	Types TW, UF	Types RHW, THHW, THW, THWN, XHHW, USE, ZW	Types TBS, SA, SIS, FEP, FEPB, MI, RHH, RHW-2, THHN, THHW, THW-2 THWN-2, USE-2, XHH, XHHW, XHHW-2, ZW-2	Types TW, UF	Types RHW, THHW, THW, THWN, XHHW, USE	Types TBS, SA, SIS, THHN, THHW, THW-2, THWN-2, RHH, RHW-2, USE-2, XHH XHHW, XHHW-2, ZW-2	
	COPPER			ALUMINUM OR COPPER-CLAD ALUMINUM			
18	—	—	14	—	—	—	—
16	—	—	18	—	—	—	—
14*	20	20	25	—	—	—	—
12*	25	25	30	20	20	25	12*
10*	30	35	40	25	30	35	10*
8	40	50	55	30	40	45	8
6	55	65	75	40	50	60	6
4	70	85	95	55	65	75	4
3	85	100	110	65	75	85	3
2	95	115	130	75	90	100	2
1	110	130	150	85	100	115	1
1/0	125	150	170	100	120	135	1/0
2/0	145	175	195	115	135	150	2/0
3/0	165	200	225	130	155	175	3/0
4/0	195	230	260	150	180	205	4/0
250	215	255	290	170	205	230	250
300	240	285	320	190	230	255	300
350	260	310	350	210	250	280	350
400	280	335	380	225	270	305	400
500	320	380	430	260	310	350	500
600	355	420	475	285	340	385	600
700	385	460	520	310	375	420	700
750	400	475	535	320	385	435	750
800	410	490	555	330	395	450	800
900	435	520	585	355	425	480	900
1000	455	545	615	375	445	500	1000
1250	495	590	665	405	485	545	1250
1500	520	625	705	435	520	585	1500
1750	545	650	735	455	545	615	1750
2000	560	665	750	470	560	630	2000

CORRECTION FACTORS

Ambient Temp. (°C)	For ambient temperatures other than 30°C (86°F), multiply the allowable ampacities shown above by the appropriate factor shown below.						Ambient Temp. (°F)
21–25	1.08	1.05	1.04	1.08	1.05	1.04	70–77
26–30	1.00	1.00	1.00	1.00	1.00	1.00	78–86
31–35	0.91	0.94	0.96	0.91	0.94	0.96	87–95
36–40	0.82	0.88	0.91	0.82	0.88	0.91	96–104
41–45	0.71	0.82	0.87	0.71	0.82	0.87	105–113
46–50	0.58	0.75	0.82	0.58	0.75	0.82	114–122
51–55	0.41	0.67	0.76	0.41	0.67	0.76	123–131
56–60	—	0.58	0.71	—	0.58	0.71	132–140
61–70	—	0.33	0.58	—	0.33	0.58	141–158
71–80	—	—	0.41	—	—	0.41	159–176

* See 240.4(D).

Figure 10-7 *NEC®* Table 310.16.

only 40 amperes in the 60°C column. Therefore, a #6 AWG conductor, which has a current rating of 55 amperes in the 60°C column, will be used.

If the termination temperature of the device is known, a conductor of that rating or higher may be used. Assume that the 50 ampere circuit breaker in the above example has a known temperature rating of 75°C. The ampacities listed in the 75°C column will be used to select the wire size. A #8 AWG THWN-2 conductor could now be used in this circuit because NEC Table 310.16 lists a current rating of 50 amperes for a #8 AWG conductor in the 75°C column.

For circuits rated over 100 amperes, or for conductor sizes larger than #1 AWG, NEC Section 110.14(C)(1)(b) states that the ampacity ratings listed in the 75°C column may be used to select wire sizes unless a conductor with a 60°C temperature rating has been selected for use. For example, types TW and UF insulation are listed in the 60°C column. If one of these two insulation types has been specified, the wire size must be chosen from the 60°C column regardless of the ampere rating of the circuit.

Example 4

A 25 kVA single-phase transformer has a primary voltage of 480 volts. Copper conductors with type THW insulation are to be used. Determine the conductor size for this installation.

The first step will be to determine the full load rated current of the winding and increase this value by 125%.

$$I = \frac{VA}{E}$$

$$I = \frac{25,000}{480}$$

$$I = 52.08 \text{ amperes}$$

$$I_{(Total)} = 52.08 \times 1.25$$

$$I_{(Total)} = 65.1 \text{ amperes}$$

Although type THW insulation is located in the 75°C column, the current is less than 100 amperes. Therefore, in accord with NEC Section 110-14(C), the conductor must be selected from the 60°C column of NEC Table 310.16. A #4 AWG conductor will be used.

Summary

1. Fuses or circuit breakers can be used as branch circuit protective devices.

2. Fuses are generally used for protection in high voltage circuits.

3. Fuses or circuit breakers must be capable of interrupting the maximum fault current of the circuit.

4. The ability of a fuse or circuit breaker to open under fault current is expressed as the interrupting rating.

5. The interrupting rating of fuses must be plainly marked if it is other than 10,000 amperes.

6. The interrupting rating of circuit breakers must be plainly marked if it is other than 5,000 amperes.

7. The impedance of a transformer is determined by its physical construction.

8. The percent impedance of a transformer can be determined by connecting an ammeter across the low-voltage terminals. Power is applied to the high-voltage terminals until rated current flows in the secondary. The percent impedance is the ratio of the applied voltage as compared to the rated voltage.

9. The short circuit current of a transformer can be determined by dividing the secondary current by the decimal impedance.

10. For transformers rated at 600 volts or less and having a primary current of 9 amperes or more, the protective device is determined by multiplying the maximum current rating by 125%.

11. For transformers rated at 600 volts or less and having a primary current less than 9 amperes, the overcurrent protective device shall be not more than 167% of the rated current of the secondary.

12. When determining the conductor size for a transformer, the 60°C column is generally used for circuits rated less than 100 amperes and the 75°C column is generally used for circuits rated more than 100 amperes.

Review Questions

1. A 20 kVA transformer has secondary voltage of 240 volts, and a listed impedance of 4.2%. What is the short circuit current?

2. What size overcurrent device should be used to protect the primary of a 480/120 volt 15 kVA transformer?

3. A 150 kVA transformer has a primary voltage of 13,800 volts, and a listed impedance of 2.5%. What size fuse should be used to provide primary overcurrent protection?

4. The secondary of the transformer in question #3 has a rated voltage of 2400 volts. What size fuse should be used to provide branch circuit protection?

5. A 10 kVA transformer has a primary voltage of 208 volts and a secondary voltage of 120 volts. What size circuit breaker should be used to protect the primary winding?

6. A 10 kVA transformer has a primary voltage rating of 240 volts and a secondary voltage rating of 480 volts. The transformer is connected to a circuit protected by a 100 ampere circuit breaker. The secondary is protected by a 25-ampere fuse. Does the primary require separate overcurrent protection?

7. What size conductor should be used to connect the primary of the transformer described in question #6? Assume the conductor to be copper with type THW insulation.

UNIT 11

Transformer Cooling

Objectives

After studying this unit, you should be able to:

- Discuss the sources of heat in transformers.
- Identify various methods of transformer cooling.
- Identify different types of transformer cooling methods.
- Describe external devices that aid in transformer cooling.

Transformers are very efficient machines with 90% to 99% efficiency. The 1% to 10% losses are mostly from I^2R loss, eddy current loss, and hysteresis loss. These losses are primarily manifested as heat gain in the transformer. If heat is not removed from the transformer it will build up to an extremely high temperature and eventually destroy the insulation in the transformer.

AIR-COOLED TRANSFORMERS

A control transformer is a transformer that supplies voltage to energize motor starters and similar electrical equipment *(Figure 11-1)*. Many small control transformers are shell-type transformers. This design places some of the core material outside the windings, exposing the core material to free air and allows heat to be radiated away from the transformer. The sizes of these

Figure 11-1 Control transformer with fuse protection added to the secondary winding (courtesy of Hevi-Duty Electric Co.).

transformers limit the heat produced to a few watts, so heat is not a major problem. For example, a typical 500 VA single-phase control transformer with 95% efficiency would produce only about 25 VA (watts) of heat at maximum rated load. This amount of heat is radiated from the transformer without the need for special cooling provisions.

Larger Control Transformers Supplying a Motor Control Center

Larger control transformers *(Figure 11-2)* feed controls for many starters in a motor control center and might also furnish lighting and convenience outlets for the control center area. These transformers are usually placed inside the MCC (Motor Control Center) cubicle and have no external case. The core and windings are exposed to free air that is allowed to circulate by convection around the transformer. These cubicles have ventilation openings in the area near the transformer, allowing the heated air to escape by convection and ambient air to be admitted. These transformers must be inspected periodically for accumulations of dust and dirt that would impede the flow of cooling air. A vacuum cleaner is used to remove dust and dirt from the **deenergized** transformer. Make sure the ventilation openings in the cubicle are not blocked with dust or debris.

deenergized

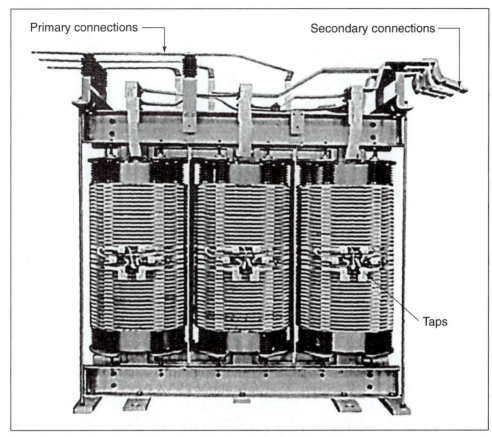

Figure 11-2 Dry-type transformer used in a motor control center.

Small Dry-Type Power Transformers for Commercial and Industrial Use

Small industrial and commercial dry-type transformers are used to provide power for lighting and convenience outlets and other small loads. They are typically installed in utility rooms and are enclosed in a case with ventilation openings. Transformers are subject to accumulations of dust and debris and must be inspected and cleaned periodically. One hazard with this type of transformer is that because they are usually installed in utility rooms they often have material stored on or around them, blocking the ventilation. Sizes from 25 kVA to 100 kVA are typical.

Large Industrial Dry-Type Transformers for Use in Unit Substations

Large industrial dry-type transformers are used in unit substations *(Figure 11-3)* to reduce distribution voltages, typically 13,800 volts, to voltages

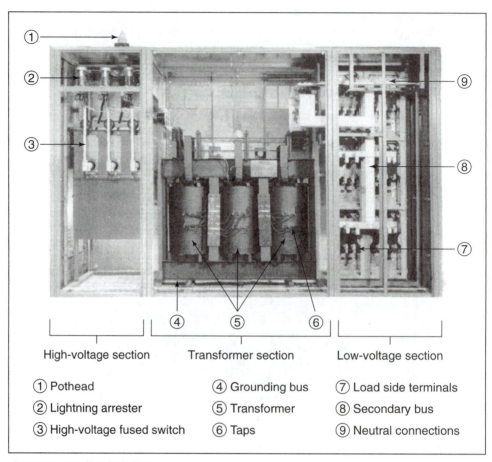

High-voltage section   Transformer section   Low-voltage section

① Pothead   ④ Grounding bus   ⑦ Load side terminals

② Lightning arrester   ⑤ Transformer   ⑧ Secondary bus

③ High-voltage fused switch ⑥ Taps   ⑨ Neutral connections

Figure 11-3 A unit substation.

used in the industrial unit. These transformers are installed in large cubicles with many square feet of ventilation openings at the top and bottom of the cubicle. Dust and dirt are a problem with these installations, depending on the environment in which the substation is installed. Because of their large size and loads, they produce more heat, necessitating more ventilating air circulated through the cubicle, and bringing with it more dirt. Although air filters are often installed to help prevent the entrance of dust and dirt, frequent inspection and cleaning are usually indicated for these transformers. Sizes from 100 kVA through 1 MVA are typical.

Convection and Fan Cooling for Dry-Type Transformers

While most dry-type air-cooled transformers are cooled by natural convection, it is sometimes necessary to design a transformer with forced-air fans for additional cooling in order to reach maximum load rating. In cases where fans are used for cooling of dry-type transformers, it is common practice to install filters in the air inlets to the fans.

LIQUID-COOLED TRANSFORMERS

Some transformers require increased cooling provided by a material that will absorb heat from the windings and core material. This cooling medium then carries the heat to the outer case or to special cooling devices. Cooling media can be circulated by natural convection or by pumps. In some cases special gasses are used as the cooling medium. More often, a special transformer oil is used that has dual functions of providing insulation for the energized transformer parts and cooling the parts. The oil must have high dielectric strength and be capable of absorbing heat well. One common oil used for this purpose is known as **Askarel**. A special type of Askarel oil, called polychlorinated **Askarel** biphenyl (PCB), is a highly toxic substance now banned in the United States. Before working with older transformers, they must be tested to determine whether the cooling and insulating medium is PCB. If PCB is found, it must be removed and disposed of in a manner approved by the U.S. Environmental Protection Agency (EPA). The transformer must then be cleaned and retested for the presence of trace amounts of PCB.

Pole-Mounted Distribution Transformers for Residential and Commercial Use

Transformers mounted on poles and used to supply power to residential and small commercial locations are filled with special oil. This oil provides insulation between the metal parts of the transformer and the energized windings. It also carries heat away from the windings and core to the housing (tank) of the unit, where it is radiated to the surrounding air. As the oil is heated by the core and windings it becomes less dense and rises in the tank. The oil nearest the outside gives up some of its heat to the housing and outside air and is cooled *(Figure 11-4)*. The cooler oil is more dense and falls to the bottom of the tank. In this way the oil circulates within the tank by natural convection, moving heat away from the heat-producing parts.

Fins or Tubes on Pole-Mounted Transformers

Some larger pole-mounted transformers have fins welded to the outside of the tank to allow more surface area for the transfer of heat to the surrounding air *(Figure 11-5)*. Other transformers have hollow tubes connected to the tank at the top and bottom. Oil circulates through these tubes by natural convection and is cooled by air around the tubes *(Figure 11-6)*. Other designs employ hollow plates that increase the surface area and permit a greater degree of cooling *(Figure 11-7)*.

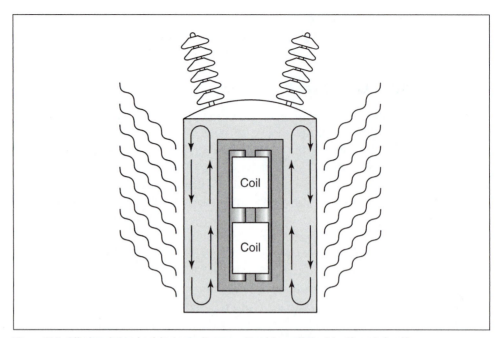

Figure 11-4 Oil circulates inside transformer. Heat is radiated to the air by the case.

Figure 11-5 Fins increase the surface area and promote cooling.

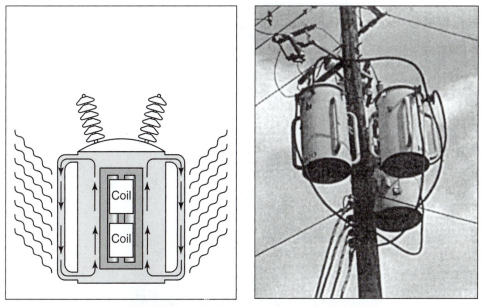

Figure 11-6 A & B Oil circulates through hollow tubes to increase cooling.

Figure 11-7 Flat plates increase surface area and permit a faster rate of cooling.

Pad-Mounted Oil-Cooled Transformers

Pad-mounted oil-cooled power transformers range in size from a few hundred kVAs to several hundred MVAs *(Figure 8-8)*. The design and construction of the transformer and its housing permit the insulating and cooling oil to flow around the heat-producing core and windings, removing heat. The

Figure 11-8 Pad-mounted transformer.

heated oil rises by convection and loses some of its heat to the housing of the transformer. The cooled oil goes to the bottom and displaces the hot oil. This circulation moves the heat generated in the transformer to the outside air.

Cooling Fins or Radiators on Pad-Mounted Transformers

As in other transformers, some pad-mounted transformers have cooling fins welded to the outside of the housing to aid in the dissipation of heat. Others have radiators mounted to the outside of the housing through which the heated oil flows by convection *(Figure 11-9)*. Radiators are constructed to give the maximum possible surface area to transfer heat to the surrounding air. Some larger transformers also include thermostatically controlled fans that circulate air through the radiators to further assist in heat removal. The maximum power ratings of large transformers equipped with radiators, fans, or other heat-removing equipment is partially based on the proper functioning of the heat-removing devices.

Water Cooling of Transformer Oil

Some large transformers, especially those enclosed in vaults where air circulation is not practical, may have their cooling oil circulated by pumps through water-cooled heat exchangers *(Figure 11-10)*. Heat removed from

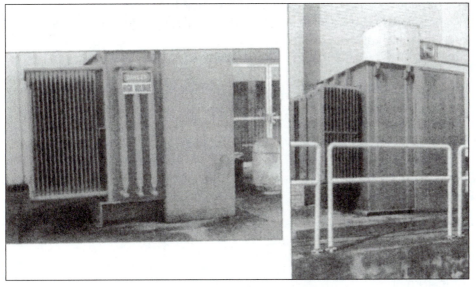

Figure 11-9 A & B Pad-mounted transformers with radiators.

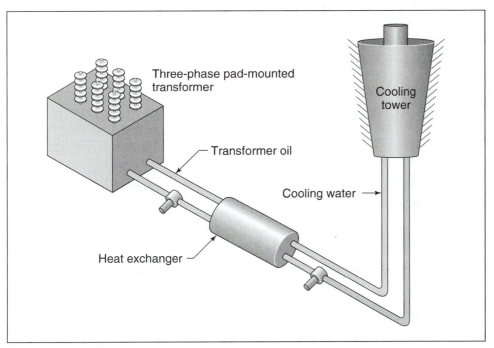

Figure 11-10 Water is used to cool the transformer oil.

the transformer by oil is transferred to the water to be removed by a cooling tower or other means. It is very important in this type of cooling that internal pressure of the transformer and the oil be maintained at a higher pressure than the water pressure in the heat exchanger to prevent water entering the transformer in case of a leak. If water is allowed to enter the transformer it could destroy the dielectric strength of the insulating liquid and cause a disastrous short circuit between energized parts.

Summary

1. Transformers develop heat from I^2R loss, eddy current loss, and hysteresis loss.

2. If heat is not removed from the transformer, it will destroy the insulation in the transformer.

3. Small transformers depend on natural convention of the surrounding air to remove heat.

4. Larger dry-heat transformers sometimes use fans to circulate air through the transformer to assist cooling.

5. Oil-cooled transformers use special insulating oils to insulate and cool the transformer.

6. Some oil-cooled transformers have external radiators through which the oil circulates by convection or pumps.

7. Some oil-cooled transformers use external water-cooled heat exchangers to remove heat from the cooling medium.

Review Questions

1. What are the three main sources of heat in a transformer?

2. What is one advantage of shell-type construction in small control transformers?

3. How are larger control transformers supplying a motor control center usually cooled?

4. What is a way to increase cooling on a large dry-type transformer used in a unit substation?

5. What is one special problem with dry-type transformers used in unit substations?

6. What are two types of cooling medium used in transformers?

7. Name two functions of transformer oil.

8. What type of coolant/insulator is banned in the USA?

9. Name two ways of increasing the cooling of oil-cooled transformers.

10. What hazard is involved in water-cooled transformers?

UNIT 12

Transformer Maintenance

Objectives

After studying this unit, you should be able to:

- List safety procedures used when maintaining transformers.

- Discuss the necessity for regular preventive maintenance for transformers.

- Describe the procedures to be used for inspection, maintenance, cleaning, and testing of various types of transformers.

- Perform maintenance procedures on transformers.

Transformers are usually thought of as stationary objects with no moving parts. Because of this misconception, transformers are often neglected and left out of routine preventive maintenance schedules. This could prove to be a very expensive omission. Transformers must be inspected and maintained on a regular schedule to get maximum performance and life from them. This applies to all transformers, no matter how large or small they are. Environmental conditions such as changing temperatures caused by varying loads and changing ambient temperature affect the operation and life of the transformer. Dust, moisture, and corrosive chemicals in the air surrounding the transformer will greatly affect its operation and life. The type of maintenance procedures and intervals between procedures are governed by the type, size, location, and application of the transformer.

SAFETY PROCEDURES

As with any electrical equipment, the primary consideration when working on or near transformers must be the safety of personnel. Before working on any transformer establish whether it is energized and whether the work can be done safely with power on the transformer. Most maintenance procedures will require that power be disconnected and locked or tagged out. On larger transformers with high-voltage connections it is usually advisable to prepare a written switching procedure detailing each step of the process of deenergizing the equipment. By following a written procedure and initialing each step as it is taken, errors in switching can be avoided.

In many larger installations, grounds are placed on each side of the transformer after it is deenergized to protect workers. If these grounds are not removed before the transformer is energized the windings could be severely damaged. A written switching procedure will include the placement and removal of these grounding connections. This will help avoid energizing a transformer with the grounds still in place.

After the power to the transformer has been disconnected and before doing any work, it is advisable to test all exposed connections for voltage. Use proper test instruments with a voltage rating at least as high as the voltage rating of the connection. This is especially important when there is more than one source of power, as in a double-ended substation *(Figure 12-1)*. Double-ended substations permit power to be supplied from another source in the event of equipment failure. Although the circuit kVA capacity is reduced, power can be maintained until the defective equipment is repaired or replaced. This can, however, cause a backfeed to the secondary side of a transformer that has the primary disconnected. Extreme care must be taken when working with double-ended systems to ensure that power is not being applied to either the primary or secondary windings.

ENTERING A TRANSFORMER TANK

In some of the maintenance procedures, it is necessary to enter a transformer tank. When this is part of the maintenance procedure, the atmosphere in the tank must be tested for the presence of combustible and/or toxic gases and also for the presence of sufficient oxygen. Oxygen is normally present in

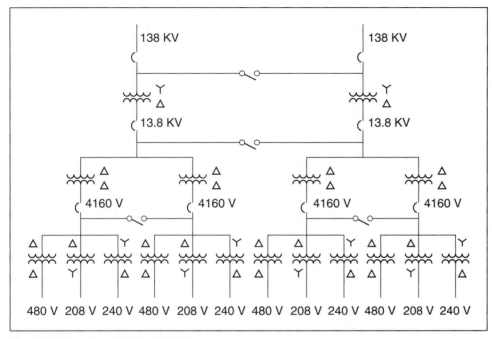

Figure 12-1 Double-ended substations.

the atmosphere at about 21.2%. If transformer tank concentration is less than 20%, it could be a health threat to the worker. If there are dangerous gases present or if there is insufficient oxygen in the tank, fresh air ventilation is necessary until safe conditions are met. When anyone is inside the tank, there must be a person outside the entrance to observe the worker in the tank and be alert for any problems.

MAINTENANCE OF SMALL CONTROL TRANSFORMERS

Inspection

The first step in any preventive maintenance procedure is inspection of equipment. Transformer inspection will reveal the presence of rust, corrosion, dirt, or dust buildup that should be noted at this time.

Cleaning

The outside of the transformer should be cleaned with an approved solvent or cleaner. Rust and corrosion should be removed and the housing painted if necessary.

Tightening

All connections and mounting bolts should be tightened. Any corroded connections should be replaced.

Testing

Small transformers should be tested annually for short circuits and grounds. A megger or megohmmeter test between the primary and secondary windings will test the insulation between windings *(Figure 12-2)*, and a megger test from each winding to the housing or core will show any insulation weakness in this area *(Figure 12-3)*. Use a megger with voltage ratings close to transformer winding rating. For example, a 500-volt megger would be used to test the insulation on a transformer with a 480-volt rated winding.

After testing the windings with a megger, each winding should be tested for continuity with an ohmmeter *(Figure 12-4)* by connecting the ohmmeter leads across the terminals or leads connected to the ends of each winding. This test will determine if any of the windings are open, but will probably not determine if the windings are shorted. In some instances, the insulation of the wire breaks down and permits the turns to short together. When this occurs, it has the effect of reducing the number of turns for that winding. If these shorted windings do not make contact with the case or core of the transformer, a megger test will not reveal the problem. This type of problem is generally found by connecting the transformer to power and measuring

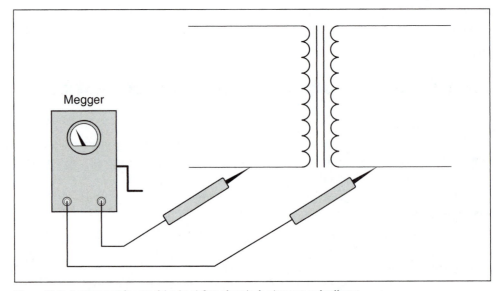

Figure 12-2 A megger is used to test for shorts between windings.

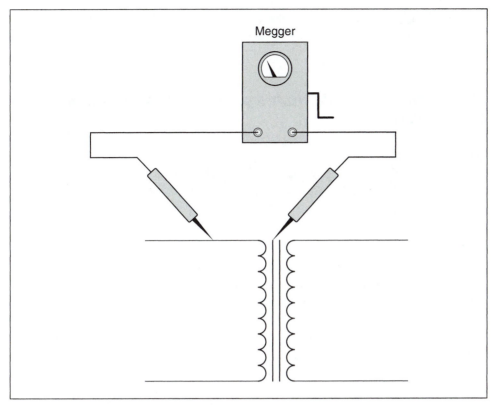

Figure 12-3 Testing for grounds with a megger.

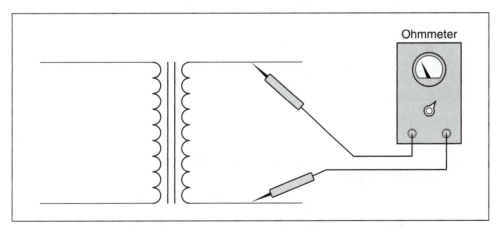

Figure 12-4 An ohmmeter is used to test windings for continuity.

current and voltage values. Excessive current draw or a large deviation from the rated voltage of a winding are good indicators of a shorted winding. When making this test be aware that it is not uncommon for the secondary voltage to be higher than the rated value under a no-load condition. Voltage ratings are listed for full load, not no load. It would not be uncommon for a transformer winding rated at 24 volts to measure 28 or 30 volts with no load connected.

LARGE CONTROL TRANSFORMERS SUPPLYING A MOTOR CONTROL CENTER

Inspection

Check the housing and windings for dust and dirt accumulation. Check the windings and connecting wires for insulation deterioration, dryness, or evidence of charring. Examine the connections for corrosion or looseness.

Cleaning

Because these transformers are usually located near the floor in the cabinet of the motor control center, they are subject to dust and dirt accumulations. Use a vacuum cleaner to remove loose dust from the deenergized transformer. Make sure all air passages are open and free of debris. Wipe any dirt off with a clean dry cloth. Clean dirt and debris from ventilation openings in the cabinet.

Tightening

Tighten all mounting bolts and electrical connections. Any corroded or burned connections must be replaced.

Testing

As with smaller control transformers, these units should be tested annually with a megger to determine the condition of the insulation. Results should be logged for comparison with previous tests in order to detect any trends that might indicate insulation deterioration.

SMALL DRY-TYPE TRANSFORMERS FOR COMMERCIAL AND INDUSTRIAL USE

Inspection

These transformers are enclosed in a housing with ventilation openings. They are typically installed in locations that are shared with custodial or other supplies and often have extraneous material stacked around and on them. Inspection requires removing surrounding material and opening the inspection covers. Examine housing and core material for rust or corrosion.

Check windings and wiring for dry, cracked insulation. Inspect connections for corrosion or looseness.

Cleaning

Remove all debris from around the transformer housing. Make sure the ventilation openings are not blocked. Use a vacuum cleaner to remove loose dust and dirt from the windings, housing, and area around the transformer. Use a clean, dry cloth to wipe away any remaining dirt. If the housing is rusty or corroded or if the paint is in poor condition, the housing should be repainted.

Testing

Annual testing should include a megger insulation test between the primary and secondary windings and between the windings and the core material or ground. Record the readings for future reference and comparison.

LARGE INDUSTRIAL DRY-TYPE TRANSFORMERS FOR USE IN UNIT SUBSTATIONS

Special Safety Precautions

These transformers are frequently installed in double-ended substations, and, therefore, have more than one source of power. Opening the source of power to the primary windings may not deenergize the transformer if the secondary is connected to another transformer secondary through a tie breaker or switch *(Figure 12-1)*. It is possible for voltage to be stepped up from secondary to primary voltage in this configuration. Be sure all connections to primary and secondary windings are locked or tagged open and checked for voltage before working on any transformer. Be aware that there might be dangerous voltages present in the cubicle for control circuits or space heaters that might not be deenergized when the transformer is deenergized.

Inspection

If possible, remove enough panels from the cubicle enclosing the transformer to allow both sides to be seen and inspected *(Figure 11-3)*. Check for dirt and debris on windings and in air passages. Inspect the support structure for rust and corrosion. Check all coils for distortion of windings and evidence of overheating. Make sure all spacers are in place. Check all bolts in structural members and electrical connections for looseness. Inspect electrical connections and buswork for corrosion or evidence of overheating. If equipment is available, an infrared thermographic inspection of the transformer, connections, and busses, or wiring under normal load conditions, will indicate any trouble spots.

Cleaning

Cleaning schedules for substation transformers will vary according to the environment in which the transformer is installed. Most installations will require annual inspection and cleaning. In dusty, dirty, or corrosive environments, the maintenance schedule will be determined by experience. Monthly inspections when the transformer is first installed will indicate how often the unit will have to be taken out of service for maintenance. Cleaning of transformer coils and core and the structure and surrounding areas is done with a vacuum cleaner. Use clean, dry cloths to wipe off any dirt left by the vacuum cleaner. Make sure all air passages are open and clean the ventilation openings in the cubicle panels in the vicinity of the transformer.

Tightening

Tighten all bolts in the support structure. Make sure all electrical connections are tight. If any connections are corroded or show evidence of overheating, they must be cleaned and reconnected. All bus connections must be tightened to the manufacturer's recommended torque.

Testing

Annual testing with a megger will indicate the general condition of the insulation. On transformers with higher voltages, special testing equipment and procedures might be required. Follow the manufacturer's recommendations for these transformers. Check the operation of any auxiliary cooling fans and associated temperature switches. Clean or replace ventilation air filters if used. Check calibration of any winding temperature gages, switches, or alarms. Check operation of space-heating equipment and controls.

MEDIA-COOLED TRANSFORMERS

Pole-Mounted Distribution Transformers

Pole-mounted distribution transformers are usually oil-filled and require little maintenance. They should be inspected annually for leaks, rust, and corrosion of the housing. If equipped with cooling fins or radiators, make sure they are clean and that air flow is not blocked by debris. Inspect the condition of the paint and repaint if necessary. Inspect bushings for dirt and damage; clean or replace if necessary. Tighten all connections, making sure all mounting bolts are tight and that any arms or brackets are in good condition. Test the condition of insulation with a megger and/or high-voltage test equipment.

PAD-MOUNTED OIL-COOLED TRANSFORMERS

External Inspection

The first step in maintaining transformers is a thorough external inspection. Look for evidence of leaks in the housing or cooling radiators. Inspect the housing for rust, corrosion, or damage and note the general condition of the paint. Inspect bushings for cracks or chips. Look for loose, corroded, or discolored connections. Inspect the housing ground connection to make sure it is tight and corrosion free. Check external gages that indicate temperature and level of the cooling oil and internal pressure *(Figure 12-5)*.

Cooling Equipment

If the transformer is equipped with auxiliary cooling equipment, it should be checked for proper operation. Check radiator connections to the tank for leaks and make necessary repairs. Cooling fans should be operated manually to be sure they work. Temperature and pressure switches and gages should be removed and calibrated yearly to ensure proper operation. Transformers with a gas blanket (usually dry nitrogen) over the oil should have the gas pressure checked at least once a week.

Figure 12-5 External gages indicate level and temperature of the transformer oil and internal pressure of the transformer housing.

Transformer Protective Relaying

Transformers that have a gas blanket on top of the insulating oil have pressure switches that actuate an alarm system if the gas pressure on the blanket drops below a certain point. These switches should be tested frequently along with any temperature or pressure alarm devices on the transformer windings or tank. Protective relaying usually includes overcurrent relays, sudden pressure relays, reverse current relays, and winding and oil over-temperature relays of various types. These devices should be tested and calibrated by qualified technicians on a regularly scheduled basis, but at least once a year.

INTERNAL INSPECTION AND MAINTENANCE

On larger transformers it will be necessary to open manholes or inspection covers to determine the condition of the windings, connections, and other parts inside the housing. Before any covers are removed, it is advisable to have new gaskets available for replacement when reclosing. Relieve any internal pressure in the transformer before loosening flange bolts. It is very important that no tools or equipment be left inside the housing. Inventory all tools, parts, and equipment brought to the work area before opening the transformer and before closing it. Anything left in the transformer could cause a short circuit or interfere with normal circulation of the cooling medium and destroy the transformer. Make sure all safety precautions are followed and atmosphere is tested before entering the transformer.

Look for loose, corroded, or discolored connections, distorted or damaged windings, and broken or missing spacers between windings. Check the general condition of the insulation for deterioration. Clean and tighten connections where necessary. Be sure to follow manufacturer's recommendations for torque when tightening connections. Check and tighten mounting bolts. Look for deposits of sludge on windings, core material, or other structures. Sludge deposits indicate contamination of the oil and will reduce the dielectric strength of the insulation. Sludge can also act as thermal insulation and decrease the transfer of heat from the internal parts to the cooling oil. While inspecting the internal parts of the transformer, observe any evidence of rust on the inside of the housing or covers. This might indicate condensation on these parts, which could be caused by a leaky gasket that allows ambient air to be admitted into the housing.

INSULATION TESTING

As with any other transformer, the dielectric strength of the insulation must be tested at least once a year. Megger testing can be done on the lower-voltage transformers. Hand crank and battery-operated meggers are shown

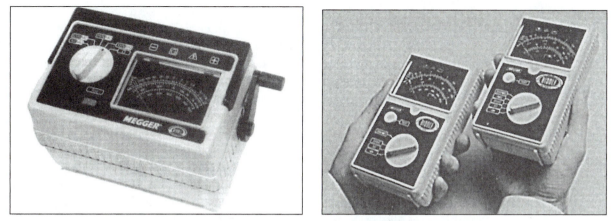

Figure 12-6 Meggers used to test transformer windings (courtesy of Biddle Instruments).

Figure 12-7 High-voltage tester (courtesy of Biddle Instruments).

in *Figure 12-6*. Special high-voltage equipment is necessary to test insulation on higher voltage units. A high-voltage tester, generally referred to as a "HiPot" is shown in *Figure 12-7*. This unit develops a high voltage and measures any current leakage caused by weak or defective insulation.

For voltages above 13,800 volts, it is usually advisable to contract out high-voltage insulation tests to a company that specializes in this type of testing and has trained technicians and the proper equipment available. As with any insulation testing, a record should be kept of test results to establish trends in insulation dielectric strength.

OIL TESTING

Transformer oil testing should be conducted at least once a year and more frequently in cases of frequent overloads or if there is a history of marginal oil test results. Oil samples should be in clean, dry containers labeled with the identity of the transformer. After the sample is drawn, it should be allowed to stand for a while to permit any free water to settle to the bottom of the sample. Glass containers make it easier to see any free water in the sample.

Testing for dielectric strength is done in a special device that has a cup for the sample and electrodes placed 0.1 inch apart. Thoroughly clean and dry the sample cup and then rinse it with a portion of the sample. Fill the cup and allow it to settle for at least three minutes to eliminate air bubbles. Turn the device on and gradually increase the voltage until it arcs across the sample. Record the voltage and repeat the test five times on each of three samples from each transformer. Calculate the average of the fifteen tests done in this manner to get the representative dielectric strength of the oil. An average dielectric strength of 26 kV to 29 kV is considered usable, 29 kV to 30 kV is good, and under 26 kV is poor and the oil should be replaced or filtered to increase dielectric strength. Special equipment is required to filter transformer oil, and the transformer must be deenergized. This process is usually contracted to companies specializing in transformer maintenance.

Other tests conducted on transformer oil include water content, gas content, and color. A water content of less than 25 parts-per-million is usually acceptable for units operating at voltages up to 228 kV. Excess water can come from condensation or leaks in the housing or cooling system and will reduce the dielectric strength of the insulation and oil. Filtering is necessary to remove excess water from the oil.

Arcing or overheating can cause combustible gases such as acetylene, hydrogen, methane, and ethane to be formed in the oil. The presence of these gases can be detected only by specialized test equipment and should be done by qualified technicians. Samples should be sent to laboratories specializing in this type of testing. Most transformer consulting firms prefer to have their technicians collect the samples in order to ensure uniform sampling procedures. In most cases, the companies doing this testing will submit a report listing the conditions found, probable causes, suggested remedies, and suggested frequency of retesting.

Summary

1. Regular preventive maintenance is vital to the performance and life of any transformer.

2. Environmental conditions such as dust, moisture, excessive heat, and corrosive atmospheres affect the life of a transformer.

3. Personnel safety is the primary consideration when working on or around a transformer.

4. It is advisable to use written switching sheets when working on substation transformers.

5. Before entering a transformer tank, the atmosphere must be tested.

6. The basic steps in transformer maintenance are inspection, cleaning, repair, and testing.

7. Accurate records of test results will indicate any trends in transformer condition.

8. Protective relaying and instrumentation should be included in any transformer maintenance schedule.

9. Dielectric testing of transformer insulation and testing of transformer oil should be done by qualified technicians or companies specializing in this type of testing.

Review Questions

1. What is the primary consideration for personnel performing transformer maintenance?

2. What are the basic steps of transformer maintenance?

3. Why is it important to keep accurate records of inspections and tests?

4. What environmental conditions affect the life of a transformer?

5. What factors govern the type of maintenance procedures and the maintenance intervals for transformers?

6. What tests must be made before entering a transformer tank?

7. What devices and procedures should be used to clean a dry-type transformer?

8. What is a special hazard encountered in large industrial transformers used in unit substations?

9. List some of the protective relays found with large transformers.

10. Before opening manholes on large transformers, what material should be available?

11. Why is it important to inventory tools and equipment before and after working inside a transformer?

12. What effect will sludge deposits have on the internal parts of a transformer?

13. What voltages indicate good oil in the standard dielectric test of oil?

14. What amount of water in transformer oil is considered acceptable for units operating at less than 288 kV?

UNIT **13**

Harmonics

Objectives

After studying this unit, you should be able to:

- Define a harmonic.
- Discuss the problems concerning harmonics.
- Identify the characteristics of different harmonics.
- Perform a test to determine if harmonic problems exist.
- Discuss methods of dealing with harmonic problems.

Harmonics are voltages or currents that operate at a frequency that is a multiple of the fundamental power frequency. If the fundamental power frequency is 60 Hz, for example, the second harmonic would be 120 Hz, and the third harmonic would be 180 Hz, and so on. Harmonics are produced by nonlinear loads that draw current in pulses rather than in a continuous manner. Harmonics on single-phase power lines are generally caused by devices such as computer power supplies, electronic ballasts in fluorescent lights, triac light dimmers, and so on. Three-phase harmonics are generally produced by variable frequency drives for AC motors and electronic drives for DC motors. A good example of a pulsating load is one which converts AC current into DC and then regulates the DC voltage by pulse width modulation *(Figure 13-1)*. Many regulated power supplies operate in this manner. The bridge rectifier in *Figure 13-1* changes the alternating current into pulsating direct current. A filter capacitor is used to smooth the pulsations. The transistor turns on and off to supply power to the load. The amount of time the transistor is turned on as compared to the time it is turned off determines the output DC voltage. Each time the transistor turns

harmonics

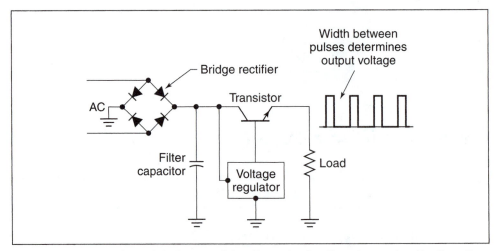

Figure 13-1 Pulse width modulation regulates the output voltage by varying the time the transistor conducts as compared to the time it is turned off.

on it causes the capacitor to begin discharging. When the transistor turns off, the capacitor begins to charge again. Current is drawn from the AC line each time the capacitor charges. These pulsations of current produced by the charging capacitor can cause the AC sine wave to become distorted. These distorted current and voltage waveforms flow back into the other parts of the power system *(Figure 13-2)*.

HARMONIC EFFECTS

Harmonics can have very detrimental effects on electrical equipment. Some common symptoms of harmonics are overheated conductors and transformers and circuit breakers that seem to trip when they should not. Harmonics are classified by name, frequency, and sequence. The name refers to whether the harmonic is the second, third, fourth, and so on of the fundamental frequency. The frequency refers to the operating frequency of the harmonic. The second harmonic operates at 120 Hz, the third at 180 Hz, the fourth at 240 Hz, and so on. The sequence refers to the phasor rotation with respect to the fundamental waveform. In an induction motor, a positive sequence harmonic rotates in the same direction as the fundamental frequency. A negative sequence harmonic rotates in the opposite direction of the fundamental frequency. A particular set of harmonics called "triplens" has a zero sequence. **Triplens** are the odd multiples of the third harmonic (3rd, 9th, 15th, 21st, etc.). A chart showing the sequence of the first nine harmonics is shown in *Figure 13-3*.

triplens

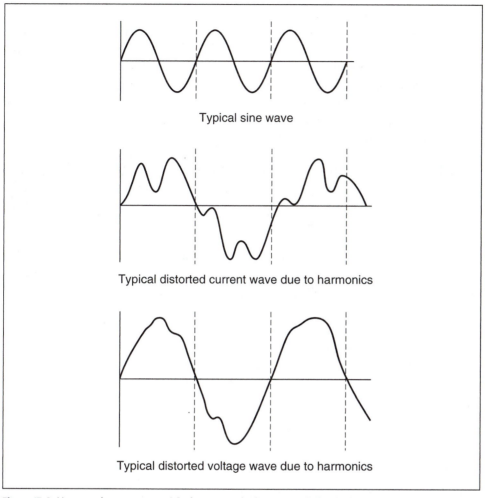

Typical sine wave

Typical distorted current wave due to harmonics

Typical distorted voltage wave due to harmonics

Figure 13-2 Harmonics cause an AC sine wave to become distorted.

Name	Fund.	2nd	3rd	4th	5th	6th	7th	8th	9th
Frequency	60	120	180	240	300	360	420	480	540
Sequence	+	−	0	+	−	0	+	−	0

Figure 13-3 Name, frequency, and sequence of the first nine harmonics.

Harmonics with a positive sequence generally cause overheating of conductors and transformers and circuit breakers. Negative sequence harmonics
can cause the same heating problems as positive harmonics, in addition to
problems with motors. Since the phasor rotation of a negative harmonic is

opposite that of the fundamental frequency, it tends to weaken the rotating magnetic field of an induction motor, causing it to produce less torque. The reduction of torque causes the motor to operate below normal speed. The reduction in speed results in excessive motor current and overheating.

Although triplens do not have a phasor rotation, they can cause a great deal of trouble in a three-phase four-wire system, such as a 208/120 volt or 480/277 volt system. In a common 208/120 volt wye-connected system, the primary is generally connected in delta, and the secondary is connected in wye *(Figure 13-4)*.

Single-phase loads that operate on 120 volts are connected between any phase conductor and the neutral conductor. The neutral current is the vector sum of the phase currents. In a balanced three-phase circuit (all phases having equal current), the neutral current is zero. Although single-phase loads tend to cause an unbalanced condition, the vector sum of the currents generally causes the neutral conductor to carry less current than any of the phase conductors. This is true for loads that are linear and draw a continuous sine wave current. When pulsating (nonlinear) currents are connected to a three-phase four-wire system, triplen harmonic frequencies disrupt the normal phasor relationship of the phase currents and can cause the phase currents to add in the neutral conductor instead of canceling it. Since the neutral conductor is not protected by a fuse or circuit breaker, there is real danger of excessive heating in the neutral conductor.

Harmonic currents are also reflected in the delta primary winding where they circulate and cause overheating. Other heating problems are caused by eddy current and hysteresis losses. Transformers are typically designed for 60 Hz operation. Higher harmonic frequencies produce greater core losses than the transformer is designed to handle. Transformers that are connected to

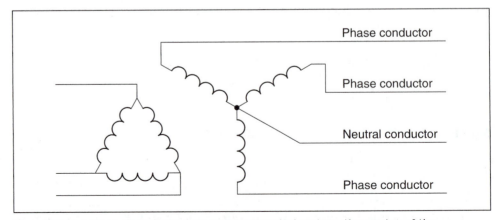

Figure 13-4 In a three-phase four-wire wye-connected system, the center of the wye-connected secondary is tapped to form a neutral conductor.

circuits that produce harmonics must sometimes be derated or replaced with transformers that are specially designed to operate with harmonic frequencies.

Transformers are not the only electrical component to be affected by harmonic currents. Emergency and standby generators can be affected in the same way as transformers. This is especially true for standby generators used to power data processing equipment in the event of a power failure. Some harmonic frequencies can distort the zero crossing of the waveform produced by the generator.

CIRCUIT BREAKER PROBLEMS

Thermal-magnetic circuit breakers use a bimetallic trip mechanism, which is sensitive to the heat produced by the circuit current. These circuit breakers are designed to respond to the heating effect of the true RMS current value. If the current becomes too great, the bimetallic mechanism trips the breaker open. Harmonic currents cause a distortion of the RMS value which can cause the breaker to trip when it should not, or not to trip when it should. Thermal-magnetic circuit breakers, however, are generally better protection against harmonic currents than electronic circuit breakers. Electronic breakers sense the peak value of current. The peaks of harmonic currents are generally higher than the fundamental sine wave *(Figure 13-5)*. Although the peaks of harmonic currents are generally higher than the fundamental frequency, they

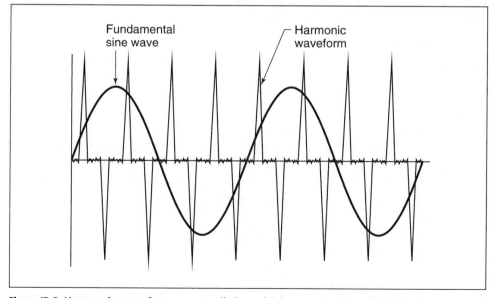

Figure 13-5 Harmonic waveforms generally have higher peak values than the fundamental waveform.

can be lower. In some cases electronic breakers may trip at low currents, and in other cases they may not trip at all.

BUSS DUCTS AND PANEL PROBLEMS

Triplen harmonic currents can also cause problems with neutral buss ducts and connecting lugs. A neutral buss is sized to carry the rated phase current. Since triplen harmonics can cause the neutral current to be higher than the phase current, it is possible for the neutral buss to become overloaded.

Electrical panels and buss ducts are designed to carry currents that operate at 60 Hz. Harmonic currents produce magnetic fields that operate at higher frequencies. If these fields should become mechanically resonant with the panel or buss duct enclosures, the panels and buss ducts can vibrate and produce buzzing sounds at the harmonic frequency.

Telecommunications equipment is often affected by harmonic currents. Telecommunication cable is often run close to power lines. To minimize interference, communication cables are run as far from phase conductors as possible and as close to the neutral conductor as possible. Harmonic currents in the neutral conductor induce high frequency currents into the communication cable. These high frequency currents can be heard as a high-pitch buzzing sound on telephone lines.

DETERMINING HARMONIC PROBLEMS ON SINGLE-PHASE SYSTEMS

There are several steps that can be followed in determining if there is a problem with harmonics. One step is to do a survey of the equipment. This is especially important in determining if there is a problem with harmonics in a single-phase system.

1. Make an equipment check. Equipment such as personal computers, printers, and fluorescent lights with an electronic ballast are known to produce harmonics. Any piece of equipment that draws current in pulses can produce harmonics.
2. Review maintenance records to see if there have been problems with circuit breakers tripping for no apparent reason.
3. Check transformers for overheating. If the cooling vents are unobstructed and the transformer is operating excessively hot, harmonics could be the problem. Check transformer currents with an ammeter capable of indicating a true RMS current value. Make sure that the voltage and current ratings of the transformer have not been exceeded.

It is necessary to use an ammeter that responds to true RMS current when making this check. Some ammeters respond to the average value, not the

RMS value. Meters that respond to the true RMS value generally state this on the meter. Meters that respond to the average value are generally less expensive and do not state that they are RMS meters.

Meters that respond to the average value use a rectifier to convert the alternating current into direct current. This value must be increased by a factor of 1.111 to change the average reading into the RMS value for a sine wave current. True RMS responding meters calculate the heating effect of the current. The chart in *Figure 13-6* shows some of the differences between average indicating meters and true RMS meters. In a distorted waveform, the true RMS value of current will no longer be average × 1.111 *(Figure 13-7)*.

Ammeter type	Sine wave response	Square wave response	Distorted wave response
Average responding	Correct	Approximately 10% high	As much as 50% low
True RMS responding	Correct	Correct	Correct

Figure 13-6 Comparison of average-responding and true RMS-responding ammeters.

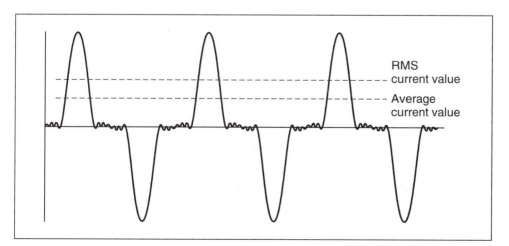

Figure 13-7 Average current values are generally less than the true RMS value in a distorted waveform.

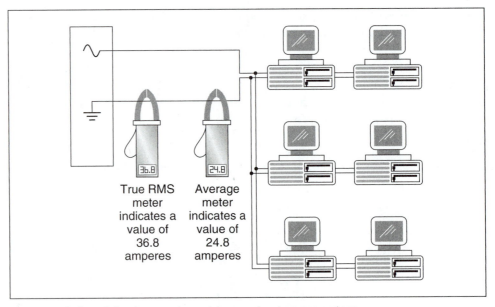

Figure 13-8 Determining harmonic problems using two ammeters.

The distorted waveform generally causes the average value to be as much as 50% less than the RMS value.

Another method of determining if a harmonic problem exists in a single-phase system is to make two separate current checks. One check is made using an ammeter that indicates the true RMS value, and the other is made using a meter that indicates the average value *(Figure 13-8)*. In this example, it is assumed that the true RMS ammeter indicates a value of 36.8 amperes and the average ammeter indicates a value of 24.8 amperes. Determine the ratio of the two measurements by dividing the average value by the true RMS value.

$$\text{Ratio} = \frac{\text{Average}}{\text{RMS}}$$

$$\text{Ratio} = \frac{24.8}{36.8}$$

$$\text{Ratio} = 0.674$$

A ratio of 1 would indicate no harmonic distortion. A ratio of 0.5 would indicate extreme harmonic distortion. This method does not reveal the name or sequence of the harmonic distortion, but it does give an indication that there is a problem with harmonics. To determine the name, sequence, and amount of harmonic distortion present, employ use of a harmonic analyzer.

DETERMINING HARMONIC PROBLEMS ON THREE-PHASE SYSTEMS

Determining if a problem with harmonics exists in a three-phase system is similar to determining the problem in a single-phase system. Since harmonic problems in a three-phase system generally occur in a wye-connected four-wire system, this example will assume a delta-connected primary and wye-connected secondary with a center-tapped neutral as shown in *Figure 13-4*. To test for harmonic distortion in a three-phase four-wire system, measure all phase currents and the neutral current with both a true RMS-indicating ammeter and an average-indicating ammeter. Assume that the three-phase system being tested is supplied by a 200 kVA transformer and the current values shown in *Figure 13-9* were recorded. The current values indicate that a problem with harmonics does exist in the system. Note the higher current measurements made with the true RMS-indicating ammeter and the fact that the neutral current is higher than any phase current.

DEALING WITH HARMONIC PROBLEMS

After it has been determined that harmonic problems exist, something must be done to deal with the problem. It is generally not practical to remove the equipment causing the harmonic distortion, so other methods must be employed. It is a good idea to consult a power quality expert to determine the exact nature and amount of harmonic distortion present. Some general procedures for dealing with harmonics are as follows:

1. In a three-phase four-wire system, the 60 Hz part of the neutral current can be reduced by balancing the current on the phase conductors. If all phases have equal current flow, the neutral current will be zero.
2. If triplen harmonics are present on the neutral conductor, harmonic filters can be added at the load. These filters help reduce the amount of harmonics on the line.

Conductor	True RMS-responding ammeter	Average-responding ammeter
Phase 1	365	292
Phase 2	396	308
Phase 3	387	316
Neutral	488	478

Figure 13-9 Measuring phase and neutral currents in a three-phase four-wire wye-connected system.

3. Pull extra neutral conductors. The ideal situation is to use a separate neutral for each phase, instead of using a shared neutral.

4. Install a larger neutral conductor. If it is impractical to supply a separate neutral conductor for each phase, increase the size of the common neutral.

5. Derate or reduce the amount of load on the transformer. Harmonic problems generally involve overheating of the transformer. In many instances, it is necessary to derate the transformer to a point where it can handle the extra current caused by the harmonic distortion. When this is done, it is generally necessary to add a second transformer and divide the load between the two transformers.

DETERMINING TRANSFORMER HARMONIC DERATING FACTOR

The most practical and straightforward method for determining the derating factor for a transformer is recommended by the Computer & Business Equipment Manufacturers Association. To use this method, two ampere measurements must be made. One is the true RMS current of the phases, and the second is the instantaneous peak phase current. The instantaneous peak current can be determined with an oscilloscope connected to a current probe or with an ammeter capable of measuring the peak value. Many of the digital clamp-on ammeters are capable of measuring the average, true RMS, and peak values of current. For this example, assume that peak current values are measured for the 200 kVA transformer discussed previously. These values are added to the previous data obtained with the true RMS- and average-indicating ammeters *(Figure 13-10)*. The formula for determining the transformer harmonic derating factor is:

$$THDF = \frac{(1.414)(RMS\ phase\ current)}{Instantaneous\ peak\ phase\ current}$$

Conductor	True RMS-responding ammeter	Average-responding ammeter	Instantaneous peak current
Phase 1	365	292	716
Phase 2	396	308	794
Phase 3	387	316	737
Neutral	488	478	957

Figure 13-10 Peak currents are added to the chart.

This formula produces a derating factor somewhere between 0 and 1.0. Since the instantaneous peak value of current is equal to the RMS value × 1.414, if the current waveforms are sinusoidal (no harmonic distortion), the formula produces a derating factor of 1.0. Once the derating factor is determined, multiply the derating factor by the kVA capacity of the transformer. The product is the maximum load that should be placed on the transformer.

If the phase currents are unequal, find an average value by adding the currents together and dividing by three.

$$\text{Phase (RMS)} = \frac{365 + 396 + 387}{3}$$

$$\text{Phase (RMS)} = 382.7$$

$$\text{Phase (Peak)} = \frac{716 + 794 + 737}{3}$$

$$\text{Phase (Peak)} = 749$$

$$\text{THDF} = \frac{(1.414)(382.7)}{749}$$

$$\text{THDF} = 0.722$$

The 200 kVA transformer in this example should be derated to 144.4 kVA (200 kVA × 0.722).

Summary

1. Harmonics are generally caused by loads that pulse the power line.

2. Harmonic distortion on single-phase lines is often caused by computer power supplies, copy machines, fax machines, and light dimmers.

3. Harmonic distortion on three-phase power lines is generally caused by variable frequency drives and electronic DC drives.

4. Harmonics can have a positive rotation, negative rotation, or no rotation.

5. Positive rotating harmonics rotate in the same direction as the fundamental frequency.

6. Negative rotating harmonics rotate in the opposite direction of the fundamental frequency.

7. Triplen harmonics are the odd multiples of the third harmonic.

8. Harmonic problems can generally be determined by using a true RMS ammeter and an average-indicating ammeter, or by using a true RMS ammeter and an ammeter that indicates the peak value.

9. Triplen harmonics generally cause overheating of the neutral conductor on three-phase four-wire systems.

Review Questions

1. What is the frequency of the second harmonic?

2. Which of the following are considered triplen harmonics—3rd, 6th, 9th, 12th, 15th, or 18th?

3. Would a positive rotating harmonic or a negative rotating harmonic be more harmful to an induction motor? Explain your answer.

4. What instrument should be used to determine what harmonics are present in a power system?

5. A 22.5 kVA single-phase transformer is tested with a true RMS ammeter and an ammeter that indicates the peak value. The true RMS reading is 94 amperes. The peak reading is 204 amperes. Should this transformer be derated, and if so, by how much?

UNIT 14

Direct Current Generators

Objectives

After studying this unit, you should be able to:

- Discuss the theory of operation of direct current generators.
- List the factors that determine the amount of output voltage produced by a generator.
- List the three major types of direct current generators.
- List different types of armature windings.
- Describe the differences between series and shunt field windings.
- Discuss the operating differences between different types of generators.
- Draw schematic diagrams for different types of direct current generators.
- Set the brushes to the neutral plane position on the commutator of a DC machine.

Although most of the electric power generated throughout the world is alternating current, direct current is used for some applications. Many industrial plants use direct current generators to produce the power needed to operate large direct current motors. Direct current motors have characteristics that make them superior to alternating current motors for certain applications. Direct current generators and motors are also used in diesel locomotives. The diesel engine in most locomotives is used to operate a large direct current generator. The generator is used to provide power to direct current motors connected to the wheels.

WHAT IS A GENERATOR?

generator

A **generator** is a device that converts mechanical energy into electrical energy. Direct current generators operate on the principle of magnetic induction *(Figure 14-1)*. In this example, the ends of the wire loop have been connected to two sliprings mounted on the shaft. Brushes are used to carry the current from the loop to the outside circuit.

In *Figure 14-2*, an end view of the shaft and wire loop is shown. At this particular instant, the loop of wire is parallel to the magnetic lines of flux, and no cutting action is taking place. Since the lines of flux are not being cut by the loop, no voltage is induced in the loop.

In *Figure 14-3*, the shaft has been turned 90° clockwise. The loop of wire cuts through the magnetic lines of flux, and a voltage is induced in the loop. When the loop is rotated 90°, it is cutting the maximum number of lines of flux per second and the voltage reaches its maximum, or peak, value.

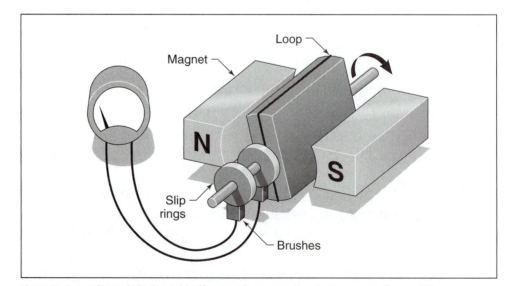

Figure 14-1 A voltage is induced in the conductor as it cuts magnetic lines of flux.

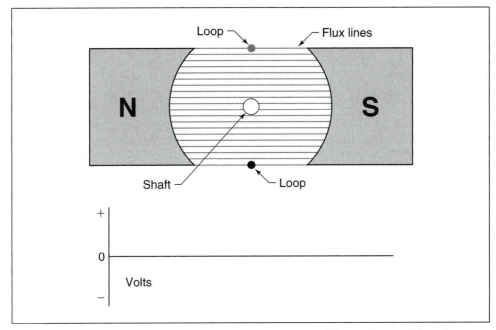

Figure 14-2 The loop is parallel to the lines of flux and no cutting action is taking place.

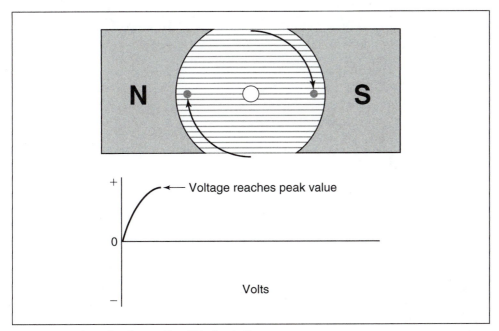

Figure 14-3 Induced voltage after 90° of rotation.

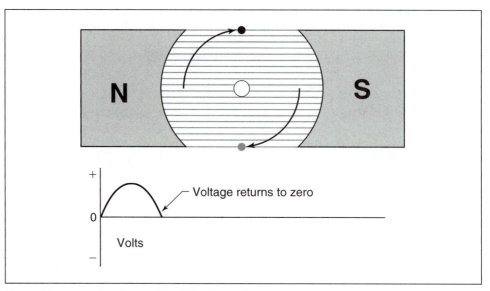

Figure 14-4 Induced voltage after 180° of rotation.

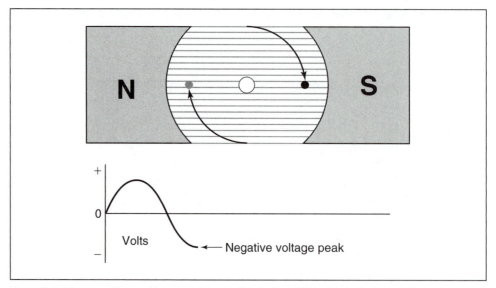

Figure 14-5 The negative voltage peak is reached after 270° of rotation.

After another 90° of rotation *(Figure 14-4)*, the loop has completed 180° of rotation and is again parallel to the lines of flux. As the loop was turned, the voltage decreased until it again reached zero.

As the loop continues to turn, the conductors again cut the lines of magnetic flux *(Figure 14-5)*. This time, however, the conductor that previously cut

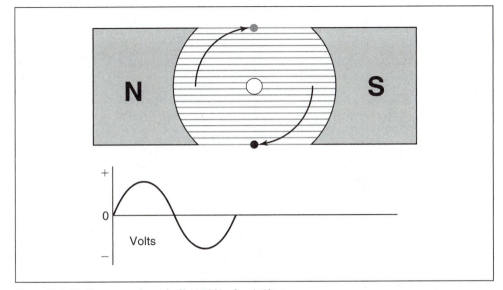

Figure 14-6 Voltage produced after 360° of rotation.

through the flux lines of the south magnetic field is cutting the lines of the north magnetic field, and the conductor that previously cut the lines of the north magnetic field is cutting the lines of the south field. Since the conductors are cutting the flux lines of opposite magnetic polarity, the polarity of voltage reverses. After 270° of rotation, the loop has rotated to the position shown, and the maximum amount of voltage in the negative direction is being produced.

After another 90° of rotation, the loop has completed one rotation of 360° and returned to its starting position *(Figure 14-6)*. The voltage decreased from its negative peak back to zero. Notice that the voltage produced in the **armature** (the rotating member of the machine), alternates polarity. *The voltage produced in all rotating armatures is alternating voltage.*   **armature**

Since direct current generators must produce DC current instead of AC current, some device must be used to change the alternating voltage produced in the armature windings into direct voltage before it leaves the generator. This job is performed by the **commutator**. The commutator is   **commutator** constructed from a copper ring split into segments with insulating material between the segments *(Figure 14-7)*. Brushes riding against the commutator segments carry the power to the outside circuit.

In *Figure 14-8*, the loop has been placed between the poles of two magnets. At this point in time, the loop is parallel to the magnetic lines of flux, and no voltage is induced in the loop. Note that the brushes make contact with both of the commutator segments at this time. The position at which the windings are parallel to the lines of flux and there is no induced voltage is

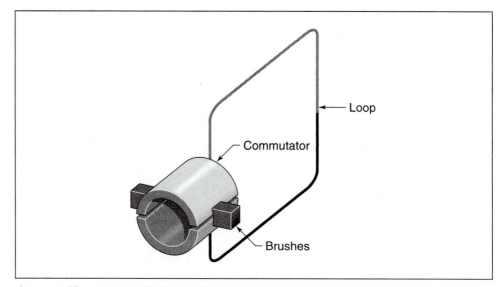

Figure 14-7 The commutator is used to convert the AC voltage produced in the armature into DC voltage.

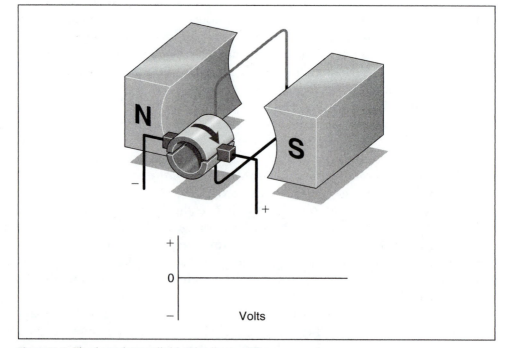

Figure 14-8 The loop is parallel to the lines of flux.

called the **neutral plane**. The brushes should be set to make contact between commutator segments when the armature windings are in the neutral plane position.

As the loop rotates, the conductors begin to cut through the magnetic lines of flux. The conductor cutting through the south magnetic field is connected to the positive brush, and the conductor cutting through the north magnetic field is connected to the negative brush *(Figure 14-9)*. Since the loop is cutting lines of flux, a voltage is induced into the loop. After 90° of rotation, the voltage reaches its most positive point.

As the loop continues to rotate, the voltage decreases to zero. After 180° of rotation, the conductors are again parallel to the lines of flux, and no voltage is induced in the loop. Note that the brushes again make contact with both segments of the commutator at the time when there is no induced voltage in the conductors *(Figure 14-10)*.

During the next 90° of rotation, the conductors again cut through the magnetic lines of flux. This time, however, the conductor that previously cut through the south magnetic field is now cutting the flux lines of the north field, and the conductor that previously cut the lines of flux of the north magnetic field is cutting the lines of flux of the south magnetic field *(Figure 14-11)*. Since

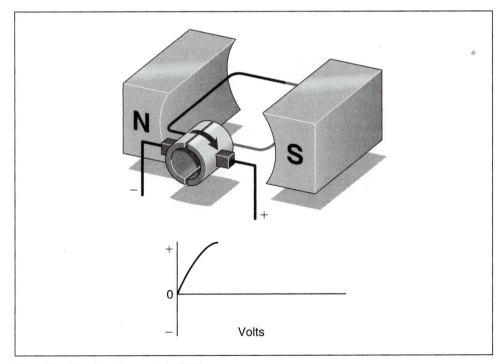

Figure 14-9 The loop has rotated 90°.

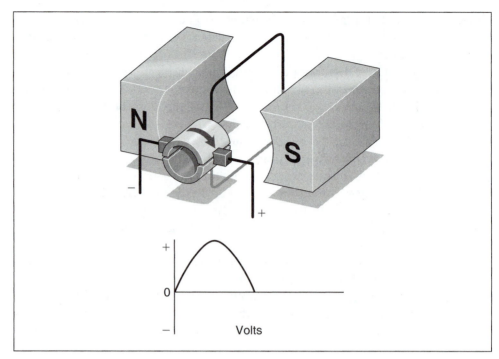

Figure 14-10 The loop has rotated 180°.

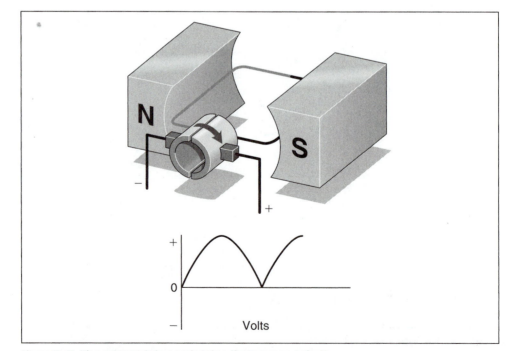

Figure 14-11 The commutator maintains the proper polarity.

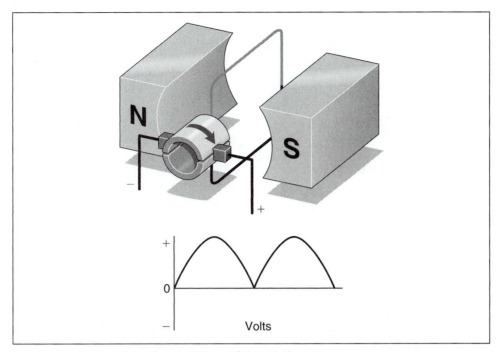

Figure 14-12 The loop completes one complete rotation.

these conductors are cutting the lines of flux of opposite magnetic polarities, the polarity of induced voltage is different for each of the conductors. The commutator, however, maintains the correct polarity to each brush. The conductor cutting through the north magnetic field will always be connected to the negative brush, and the conductor cutting through the south field will always be connected to the positive brush. Since the polarity at the brushes has remained constant, the voltage will increase to its peak value in the same direction.

As the loop continues to rotate *(Figure 14-12)*, the induced voltage again decreases to zero when the conductors become parallel to the magnetic lines of flux. Notice that during this 360° rotation of the loop the polarity of voltage remained the same for both halves of the waveform. This is called rectified DC voltage. The voltage is pulsating. It does turn on and off, but it never reverses polarity. Since the polarity for each brush remains constant, the output voltage is DC.

To increase the amount of output voltage, it is common practice to increase the number of turns of wire for each loop *(Figure 14-13)*. If a loop contains 20 turns of wire, the induced voltage will be 20 times greater than that for a single-loop conductor. The reason for this is that each loop is connected in series with the other loops. Since the loops form a series path, the voltage induced in the loops will add. In this example, if each loop

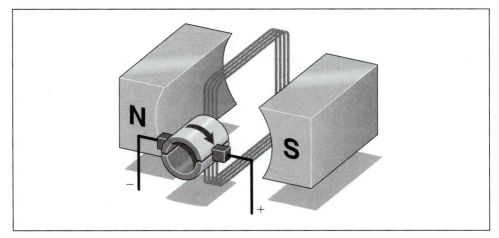

Figure 14-13 Increasing the number of turns increases the output voltage.

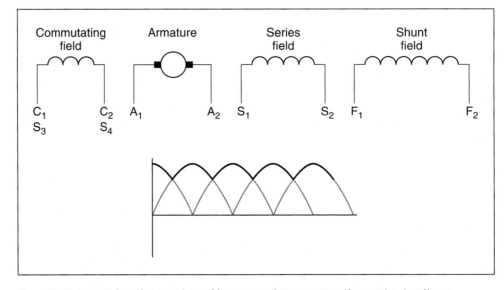

Figure 14-14 Increasing the number of loops produces a smoother output voltage.

has an induced voltage of 2 V, the total voltage for this winding would be 40 V (2 V × 20 loops = 40 V).

It is also common practice to use more than one loop of wire *(Figure 14-14)*. When more than one loop is used, the average output voltage is higher and there is less pulsation of the rectified voltage. This pulsation is called *ripple*.

The loops are generally placed in slots of an iron core *(Figure 14-15)*. The iron acts as a magnetic conductor by providing a low-reluctance path for

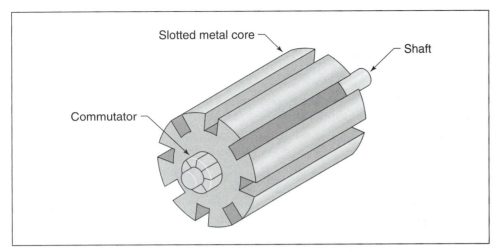

Slotted metal core

Shaft

Commutator

Figure 14-15 The loops of wire are wound around slots in a metal core.

Figure 14-16 DC machine armature.

magnetic lines of flux to increase the inductance of the loops and provide a higher induced voltage. The commutator is connected to the slotted iron core. The entire assembly of iron core, commutator, and windings is called the armature *(Figure 14-16)*. The windings of armatures are connected in different ways depending on the requirements of the machine. The three basic types of armature windings are the *lap, wave,* and *frogleg.*

ARMATURE WINDINGS

Lap Wound Armatures

**lap wound
armatures**

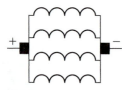

Figure 14-17 Lap wound armatures have their windings connected in parallel. They are used in machines intended for high-current and low-voltage operation.

**wave wound
armatures**

**frogleg wound
armatures**

Lap wound armatures are used in machines designed for low voltage and high current. These armatures are generally constructed with large wire because of high current. A good example of where lap wound armatures are used is in the starter motor of almost all automobiles. One characteristic of machines that use a lap wound armature is that they will have as many pairs of brushes as there are pairs of poles. The windings of a lap wound armature are connected in parallel *(Figure 14-17)*. This permits the current capacity of each winding to be added and provides a higher operating current. Lap wound armatures have as many parallel paths through the armature as there are pole pieces.

Wave Wound Armatures

Wave wound armatures are used in machines designed for high voltage and low current. These armatures have their windings connected in series as shown in *Figure 14-18*. When the windings are connected in series, the voltage of each winding adds, but the current capacity remains the same. A good example of where wave wound armatures are used is in the small generator in hand-cranked megohmmeters. Wave wound armatures never contain more than two parallel paths for current flow regardless of the number of pole pieces, and they never contain more than one set of brushes (a set being one brush or group of brushes for positive and one brush or group of brushes for negative).

Frogleg Wound Armatures

Frogleg wound armatures are probably the most used. These armatures are used in machines designed for use with moderate current and moderate voltage. The windings of a frogleg wound armature are connected in series-parallel as shown in *Figure 14-19*. Most large DC machines use frogleg wound armatures.

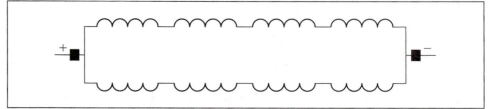

Figure 14-18 Wave wound armatures have their windings connected in series. Wave windings are used in machines intended for high-voltage, low-current operation.

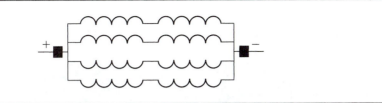

Figure 14-19 Frogleg wound armatures are connected in series-parallel. These windings are generally used in machines intended for medium voltage and current operation.

BRUSHES

The **brushes** ride against the commutator segments and are used to connect the armature to the external circuit of the DC machine. Brushes are made from a material that is softer than the copper bars of the commutator. This permits the brushes, which are easy to replace, to wear instead of the commutator. The brush leads are generally marked A_1 and A_2 and are referred to as the armature leads.

brushes

POLE PIECES

The **pole pieces** are located inside the housing of the DC machine *(Figure 14-20)*. The pole pieces provide the magnetic field necessary for the operation of the machine. They are constructed of some type of good magnetic conductive material such as soft iron or silicon steel. Some DC generators use permanent magnets to provide the magnetic field instead of electromagnets. These machines are generally small and rated about one horsepower or less. A DC generator that uses permanent magnets as its field is referred to as a ***magneto***.

pole pieces

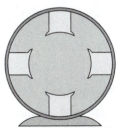

Figure 14-20 Pole pieces are constructed of soft iron and placed on the inside of the housing.

magneto

FIELD WINDINGS

Most DC machines use wound electromagnets to provide the magnetic field. Two types of field windings are used. One is the series field, and the other is the shunt field. **Series field windings** are made with relatively few turns of very large wire and have a very low resistance. They are so named because they are connected in series with the armature. The terminal leads of the series field are labeled S_1 and S_2. It is not uncommon to find the series field of large horsepower machines wound with square or rectangular wire *(Figure 14-21)*. The use of square wire permits the windings to be laid closer together, which increases the number of turns that can be wound in a

series field windings

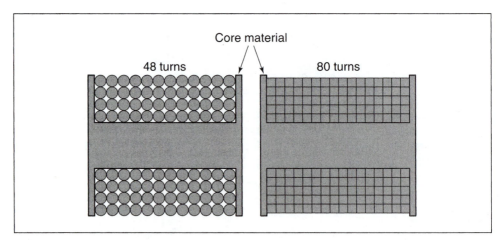

Figure 14-21 Square wire permits more turns than round wire in the same area.

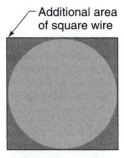

Figure 14-22 A square wire of equal size contains more surface area than round wire.

particular space. Square and rectangular wire can also be made physically smaller than round wire and still contain the same surface area *(Figure 14-22)*.

Shunt field windings are made with many turns of small wire. Since the shunt field is constructed with relatively small wire, it has a much higher resistance than the series field. The shunt field is intended to be connected in parallel with, or to shunt, the armature. The resistance of the shunt field must be high because its resistance is used to limit current flow through the field. The shunt field is often referred to as the "field," and its terminal leads are labeled F_1 and F_2.

When a DC machine uses both series and shunt fields, each pole piece will contain both windings *(Figure 14-23)*. The windings are wound on the pole pieces in such a manner that when current flows through the winding it will produce alternate magnetic polarities. In the illustration shown in *Figure 14-23*, two pole pieces will form north magnetic polarities, and two will form south magnetic polarities. A DC machine with two field poles and one interpole is shown in *Figure 14-24*. Interpoles will be discussed later in this unit.

SERIES GENERATORS

There are three basic types of direct current generators: the series, shunt, and compound. The type is determined by the arrangement and connection of field coils. The **series generator** contains only a series field connected in series with the armature *(Figure 14-25)*. A schematic diagram used to represent a series-connected DC machine is shown in *Figure 14-26*. The series generator must be *self-excited,* which means that the pole pieces contain some amount of residual magnetism. This residual magnetism produces an initial output voltage that permits current to flow through the field if a load is

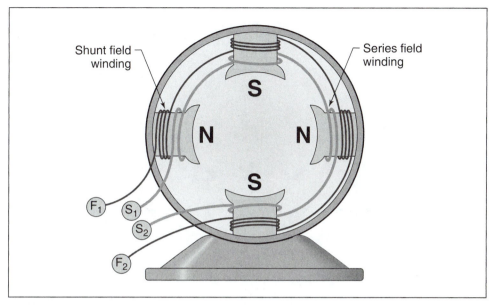

Shunt field winding

Series field winding

Figure 14-23 Both series and shunt field windings are contained on each pole piece.

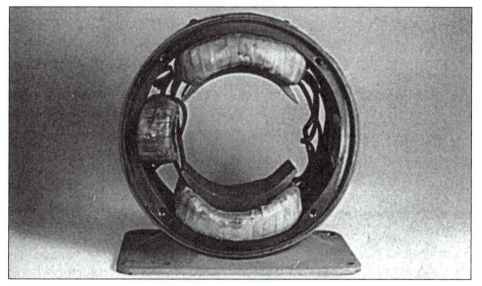

Figure 14-24 A two-pole DC machine with one interpole.

connected to the generator. The amount of output voltage produced by the generator is proportional to three factors:

1. The number of turns of wire in the armature
2. The strength of the magnetic field of the pole pieces
3. The speed of the cutting action (speed of rotation)

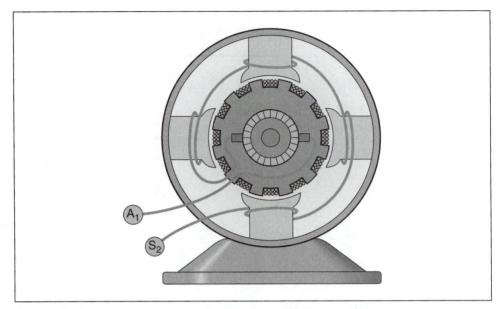

Figure 14-25 The series field is connected in series with the armature.

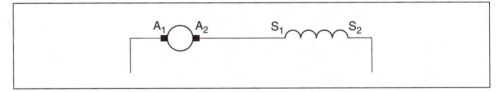

Figure 14-26 Schematic drawing of a series generator.

To understand why these three factors determine the output voltage of a generator, recall that 1 V is induced in a conductor when it cuts magnetic lines of flux at a rate of 1 weber per second (1 weber = 100,000,000 lines of flux). When conductors are wound into a loop, each turn acts as a separate conductor. Since the turns are connected in series, the voltage induced into each conductor will add. If one conductor has an induced voltage of 0.5 V and there are 20 turns, the total induced voltage would be 10 V.

The second factor is the strength of the magnetic field. Flux density is a measure of the strength of a magnetic field. If the number of turns of wire in the armature remains constant and the speed remains constant, the output voltage can be controlled by the number of flux lines produced by the field poles. Increasing the lines of flux will increase the number of flux lines cut per second and, therefore, the output voltage. The magnetic field strength can be increased until the iron of the pole pieces reaches saturation.

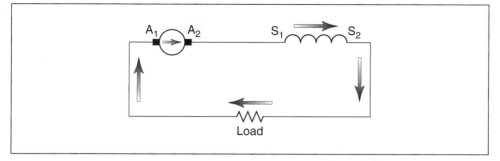

Figure 14-27 Residual magnetism produces an initial voltage used to provide a current flow through the load.

Induced voltage is proportional to the number of flux lines cut per second. If the strength of the magnetic field remains constant and the number of turns of wire in the armature remains constant, the output voltage will be determined by the speed at which the conductors cut the flux lines. Increasing the speed of the armature will increase the speed of the cutting action, which will increase the output voltage. Likewise, decreasing the speed of the armature will decrease the output voltage.

Connecting Load to the Series Generator

When a load is connected to the output of a series generator, the initial voltage produced by the residual magnetism of the pole pieces produces a current flow through the load *(Figure 14-27)*. Since the series field is connected in series with the armature, the current flowing through the armature and load must also flow through the series field. This causes the magnetism of the pole pieces to become stronger and produce more magnetic lines of flux. When the strength of the magnetic pole pieces increases, the output voltage increases also.

If another load is added *(Figure 14-28)*, more current flows, and the pole pieces produce more magnetic lines of flux, which again increases the output voltage. Each time a load is added to the series generator, its output voltage increases. This increase of voltage will continue until the iron in the pole pieces and armature becomes saturated. At that point, an increase of load will result in a decrease of output voltage *(Figure 14-29)*.

SHUNT GENERATORS

Shunt generators contain only a shunt field winding connected in parallel with the armature *(Figure 14-30)*. A schematic diagram used to represent a shunt-connected DC machine is shown in *Figure 14-31*. Shunt generators can be either self-excited or separately excited. Self-excited shunt generators are

shunt generators

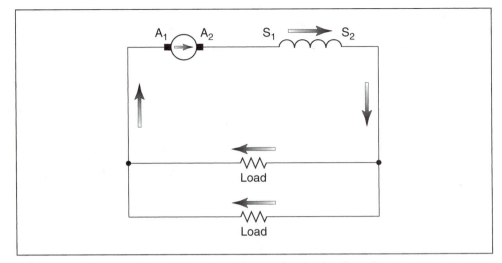

Figure 14-28 If more load is added, current flow and output voltage increase.

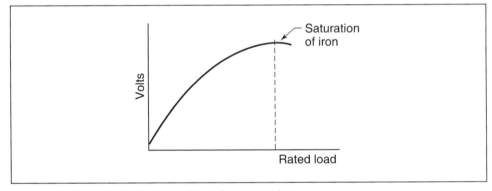

Figure 14-29 Characteristic curve of a series generator.

similar to self-excited series generators in that residual magnetism in the pole pieces is used to produce an initial output voltage. In the case of a shunt generator, however, the initial voltage is used to produce a current flow through the shunt field. This current increases the magnetic field strength of the pole pieces, which produces a higher output voltage *(Figure 14-32)*. This buildup of voltage continues until a maximum value, determined by the speed of rotation, the turns of wire in the armature, and the turns of wire on the pole pieces, is reached.

Another difference between the series generator and the self-excited shunt generator is that the series generator must be connected to a load before voltage can increase. The load is required to form a complete path for current to flow through the armature and series field *(Figure 14-27)*. In a self-excited

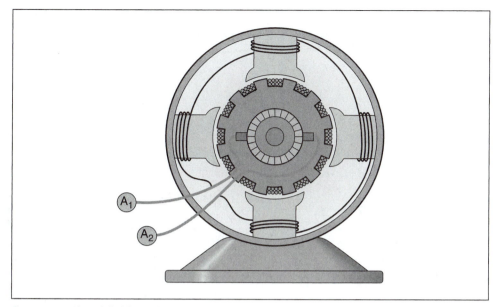

Figure 14-30 Shunt field windings are connected in parallel with the armature.

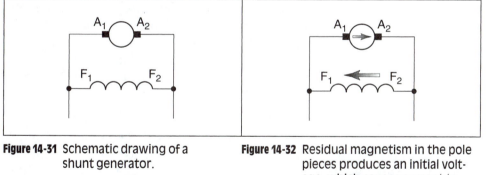

Figure 14-31 Schematic drawing of a shunt generator.

Figure 14-32 Residual magnetism in the pole pieces produces an initial voltage, which causes current to flow through the shunt field, increasing field flux.

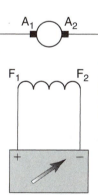

DC power supply

Figure 14-33 Separately excited shunt generators must have an external power source to provide excitation current for the shunt field.

shunt generator, the shunt field winding provides a complete circuit across the armature, permitting the full output voltage to be obtained before a load is connected to the generator.

Separately excited generators have their fields connected to an external source of direct current *(Figure 14-33)*. The advantages of the separately excited machine are that it gives better control of the output voltage and that its voltage drop is less when load is added. The characteristic curves of both self-excited and separately excited shunt generators are shown in *Figure 14-34*.

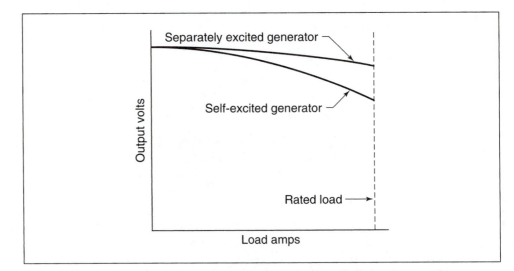

Figure 14-34 Characteristic curves of self- and separately excited shunt generators.

The self-excited generator exhibits a greater drop in voltage when load is added because the armature voltage is used to produce the current flow in the shunt field. Each time the voltage decreases, the current flow through the field decreases, causing a decrease in the amount of magnetic flux lines in the pole pieces. This decrease of flux in the pole pieces causes a further decrease of output voltage. The separately excited machine does not have this problem because the field flux is held constant by the external power source.

Field Excitation Current

Regardless of which type of shunt generator is used, the amount of output voltage is generally controlled by the amount of field excitation current. **Field excitation current** is the DC current that flows through the shunt field winding. This current is used to turn the iron pole pieces into electromagnets. Since one of the factors that determines the output voltage of a DC generator is the strength of the magnetic field, the output voltage can be controlled by the amount of current flow through the field coils. A simple method of controlling the output voltage is by the use of a shunt field rheostat. The shunt field rheostat is connected in series with the shunt field winding *(Figure 14-35)*. By adding or removing resistance connected in series with the shunt field winding, the amount of current flow through the field can be controlled. This in turn controls the strength of the magnetic field of the pole pieces.

When it is important that the output voltage remain constant regardless of load, an electronic voltage regulator can be used to adjust the shunt field

field excitation current

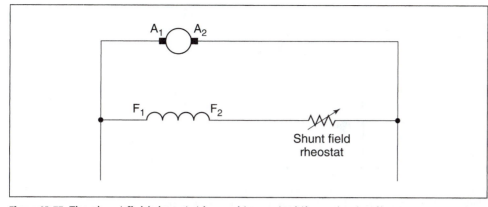

Figure 14-35 The shunt field rheostat is used to control the output voltage.

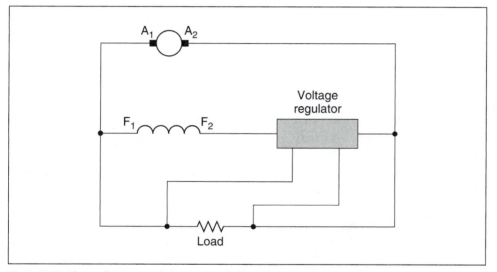

Figure 14-36 The voltage regulator controls the amount of shunt field current.

current *(Figure 14-36)*. The voltage regulator connects in series with the shunt field in a similar manner as the shunt field rheostat. The regulator, however, senses the amount of voltage across the load. If the output voltage should drop, the regulator will permit more current to flow through the shunt field. If the output voltage should become too high, the regulator will decrease the current flow through the shunt field.

Generator Losses

When load is added to the shunt generator, the output voltage will drop. This voltage drop is due to losses that are inherent to the generator. The

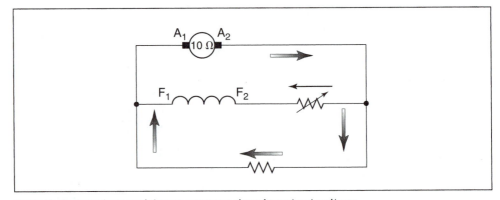

Figure 14-37 Armature resistance causes a drop in output voltage.

largest of these losses is generally caused by the resistance of the armature. In *Figure 14-37*, it is assumed that the armature has a wire resistance of 10 Ω. When a load is connected to the output of the generator, current will flow from the armature, through the load, and back to the armature. As current flows through the armature, the resistance of the wire causes a voltage drop. Assume that the armature has a current flow of 2 A. If the resistance of the armature is 10 Ω, it will require 20 V to push the current through the resistance of the armature.

Now assume that the armature has a resistance of 2 Ω. The same 2 A of current flow requires only 4 V to push the current through the armature resistance. A low-resistance armature is generally a very desirable characteristic for DC machines. *In the case of a generator, the voltage regulation is determined by the resistance of the armature.* Voltage regulation is measured by the amount that output voltage will drop as load is added. A generator with good voltage regulation has a small amount of voltage drop as load is added.

Some other losses are I^2R losses, eddy current losses, and hysteresis losses. Recall that I^2R is one of the formulas for finding power, or watts. In the case of a DC machine, it describes the power loss associated with heat due to the resistance of the wire in both the armature and field windings.

Eddy currents are currents that are induced into the metal core material by the changing magnetic field as the armature spins through the flux lines of the pole pieces. Eddy currents are so named because they circulate around inside the metal in a manner similar to the swirling eddies in a river *(Figure 14-38)*. These swirling currents produce heat, which is a power loss. Many machines are constructed with laminated pole pieces and armature cores to help reduce eddy currents. The surface of each lamination forms a layer of iron oxide, which acts as an insulator to help prevent the formation of eddy currents.

Hysteresis losses are losses due to molecular friction. As discussed previously, alternating current is produced inside the armature. This reversal of

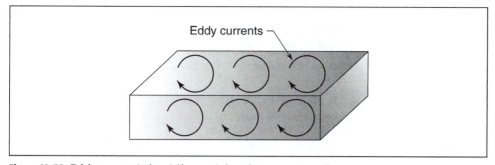

Figure 14-38 Eddy currents heat the metal and cause power loss.

the direction of current flow causes the molecules of iron in the core to realign themselves each time the current changes direction. The molecules of iron are continually rubbing against each other as they realign magnetically. The friction of the molecules rubbing together causes heat, which is a power loss. Hysteresis loss is proportional to the speed of rotation of the armature. The faster the armature rotates, the more current reversals there are per second, and the more heat is produced because of friction.

COMPOUND GENERATORS

Compound generators contain both series and shunt fields. Most large DC machines are compound wound. The series and shunt fields can be connected in two ways. One connection is called **long shunt** *(Figure 14-39).* The long shunt connection has the shunt field connected in parallel with both the armature and series field. This is the more used of the two connections.

compound generators

long shunt

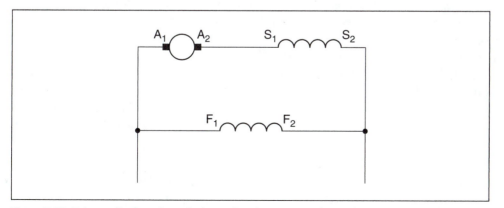

Figure 14-39 Schematic drawing of a long shunt compound generator.

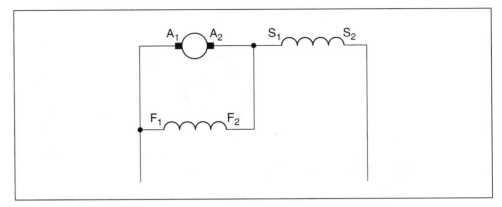

Figure 14-40 Schematic drawing of a short shunt compound generator.

short shunt

The second connection is called **short shunt** *(Figure 14-40)*. The short shunt connection has the shunt field connected in parallel with the armature. The series field is connected in series with the armature. This is a very common connection for DC generators that must be operated in parallel with each other.

COMPOUNDING

compounding

The relationship of the strengths of the two fields in a generator determines the amount of **compounding** for the machine. A machine is over-compounded when the series field has too much control and the output voltage increases each time a load is added to the generator. Basically, the generator begins to take on the characteristics of a series generator. **Over-compounding** is characterized by the fact that the output voltage at full load will be greater than the output voltage at no load *(Figure 14-41)*.

over-compounding

When the generator is flat-compounded, the output voltage will be the same at full load as it is at no load. **Flat compounding** is accomplished by permitting the series field to increase the output voltage by an amount that is equal to the losses of the generator.

flat compounding

If the series field is too weak, however, the generator will become **under-compounded**. This condition is characterized by the fact that the output voltage will be less at full load than it is at no load. When a generator is undercompounded it has characteristics similar to those of a shunt generator.

under-compounded

Controlling Compounding

Most DC machines are constructed in such a manner that they are over-compounded if no control is used. This permits the series field strength to be weakened and thereby permits control of the amount of compounding.

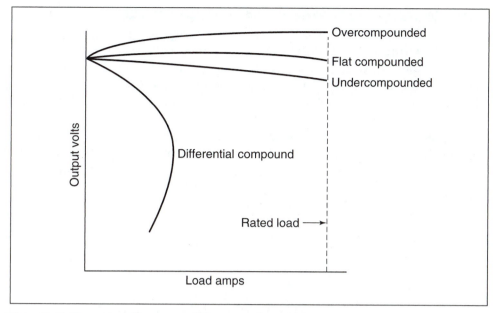

Figure 14-41 Characteristic curves of compound generators.

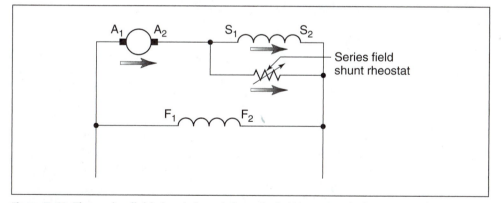

Figure 14-42 The series field shunt rheostat controls the amount of compounding.

The amount of compounding is controlled by connecting a low-value variable resistor in parallel with the series field *(Figure 14-42)*. This resistor
is known as the series field shunt rheostat, or the **series field diverter**.
The rheostat permits part of the current that normally flows through the
series field to flow through the resistor. This reduces the amount of
magnetic flux produced by the series field, which reduces the amount of
compounding.

**series field
diverter**

Cumulative and Differential Compounding

Direct current generators are generally connected in such a manner that they are a **cumulative compound**. This means that the shunt and series fields are connected in such a manner that when current flows through them, they aid each other in the production of magnetism *(Figure 14-43)*. In the example shown, each of the field windings would produce the same magnetic polarity for the pole piece.

A **differential compound** generator has its fields connected in such a manner that they oppose each other in the production of magnetism *(Figure 14-44)*. In this example, the shunt and series fields are attempting to

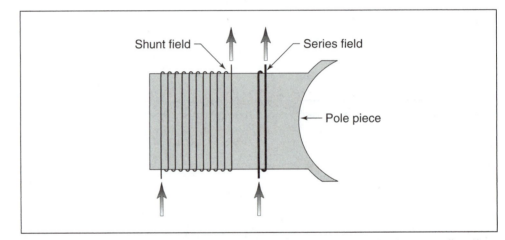

Figure 14-43 In a cumulative-compound machine, the current flows in the same direction through both the series and shunt field.

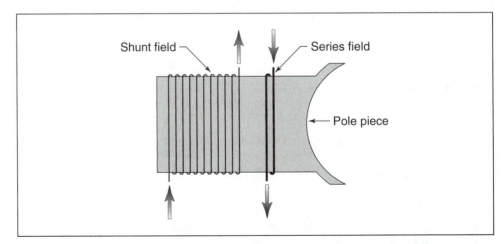

Figure 14-44 In a differential-compound machine, the current flows through the shunt field in a direction opposite that of the current flow through the series field.

produce opposite magnetic polarities for the same pole piece. This results in the magnetic field becoming weaker as current flow through the series field increases. Although there are some applications for a differential-compound machine, they are very limited.

COUNTERTORQUE

When a load is connected to the output of a generator, current flows from the armature, through the load, and back to the armature. As current flows through the armature, a magnetic field is produced around the armature *(Figure 14-45)*. In accord with Lenz's law, the magnetic field of the armature will be opposite in polarity to that of the pole pieces. Since these two magnetic fields are opposite in polarity, they are attracted to each other. This magnetic attraction causes the armature to become hard to turn. This turning resistance is called **countertorque**, and it must be overcome by the device used to drive **countertorque** the generator. This is the reason that as load is added to the generator, more power is required to turn the armature. Since countertorque is produced by the attraction of the two magnetic fields, it is proportional to the output or armature current if the field excitation current remains constant. *Countertorque is a measure of the useful electrical energy produced by the generator.*

Countertorque is often used to provide a braking action in DC motors. If the field excitation current remains turned on, the motor can be converted into a generator very quickly by disconnecting the armature from its source of power and reconnecting it to a load resistance. The armature now supplies current to the load resistance. The countertorque developed by the generator action

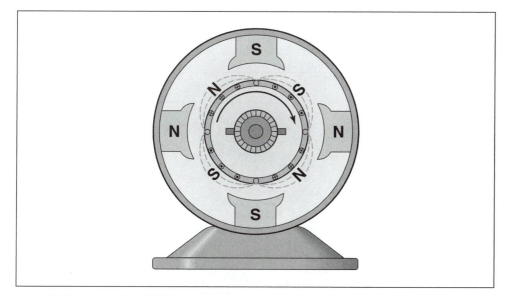

Figure 14-45 A magnetic field is produced around the armature.

dynamic braking

regenerative braking

causes the armature to decrease in speed. When this type of braking action is used, it is referred to as **dynamic braking** or **regenerative braking**.

ARMATURE REACTION

armature reaction

Armature reaction is the twisting or bending of the magnetic lines of flux of the pole pieces. It is caused by the magnetic field produced around the armature as it supplies current to the load *(Figure 14-46)*. This distortion of the main magnetic field causes the position of the neutral plane to change position. When the neutral plane changes, the brushes no longer make contact between commutator segments at the point when no voltage is induced in the armature. This results in power loss and arcing and sparking at the brushes, which can cause overheating and damage to both the commutator and brushes. The amount of armature reaction is proportional to armature current.

Correcting Armature Reaction

Armature reaction can be corrected in several ways. One method is to rotate the brushes an amount equal to the shift of the neutral plane *(Figure 14-47)*. This method would be satisfactory, however, only if the generator delivered a constant current. Since the distortion of the main magnetic field is proportional to armature current, the brushes would have to be adjusted each time the load current changed. In the case of a generator, the brushes would be rotated in the direction of rotation of the armature. In the case of a motor, the brushes would be rotated in a direction opposite that of armature rotation.

interpole

Another method that is used often is to insert small pole pieces, called **interpoles** or commutating poles, between the main field poles *(Figure 14-24)*.

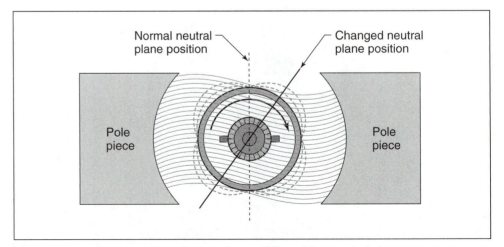

Figure 14-46 Armature reaction changes the position of the neutral plane.

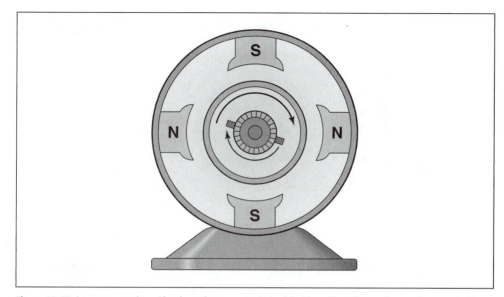

Figure 14-47 In a generator, the brushes are rotated in the direction of armature rotation to correct armature reaction.

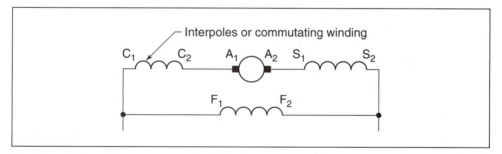

Figure 14-48 Interpoles are connected in series with the armature.

The interpoles are sometimes referred to as the commutating winding because they are wound with a few turns of large wire similar to the series field winding. The interpoles are connected in series with the armature, which permits their strength to increase with an increase of armature current *(Figure 14-48)*. Interpole connections are often made inside the housing of the machine. When the interpole connection is made internally, the A_2 lead is actually connected to one end of the interpole winding. When the interpole windings are brought out of the machine separately, they are generally labeled C_1 and C_2, which stands for commutating field. It is not unusual, however, to find them labeled S_3 and S_4.

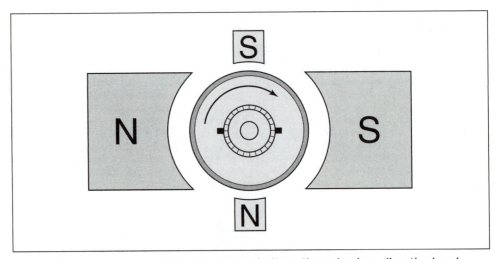

Figure 14-49 Interpoles must have the same polarity as the pole piece directly ahead of them.

In a generator, the magnetic field of the armature tends to bend the main magnetic field upward as shown in *Figure 14-46*. In a motor, the armature field bends the main field downward. The function of the interpoles is to restore the field to its normal condition. When a DC machine is used as a generator, the interpoles will have the same polarity as the main field pole directly ahead of them (ahead in the sense of the direction of rotation of the armature) *(Figure 14-49)*. When a DC machine is used as a motor, the interpoles will have the same polarity as the pole piece behind them in the sense of direction of rotation of the armature.

Interpoles do have one disadvantage. They restore the field only in their immediate area and are not able to overcome all the field distortion. Large DC generators use another set of windings called **compensating windings** to help restore the main magnetic field. Compensating windings are made by placing a few large wires in the face of the pole piece parallel to the armature windings *(Figure 14-50)*. The compensating winding is connected in series with the armature so that its strength increases with an increase of output current.

compensating windings

SETTING THE NEUTRAL PLANE

Most DC machines are designed in such a manner that the position of the brushes on the commutator can be set or adjusted. An exposed view of the brushes and brush yoke of a direct current machine is shown in *Figure 14-51*. The simplest method of setting the brushes to the neutral plane

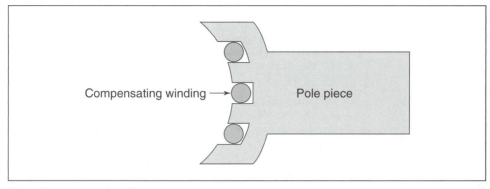

Figure 14-50 Compensating winding helps correct armature reaction.

Figure 14-51 Exposed view of a direct current machine (courtesy of GE Motors – DM&G).

position is to connect an AC voltmeter across the shunt field leads. Low-voltage alternating current is then applied to the armature *(Figure 14-52)*. The armature acts as the primary of a transformer, and the shunt field acts as the secondary. If the brushes are not set at the neutral plane position, the changing magnetic field of the armature will induce a voltage into the shunt

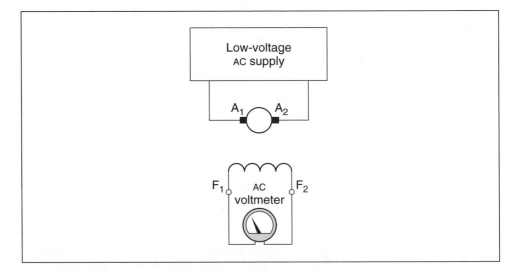

Figure 14-52 Setting the brushes at the neutral plane.

field. The brush position can be set by observing the action of the AC voltmeter. If the brush yoke is loosened to permit the brushes to be moved back and forth on the commutator, the voltmeter pointer will move up and down the scale. The brushes are set to the neutral plane position when the voltmeter is at its lowest possible reading.

FLEMING'S LEFT-HAND GENERATOR RULE

left-hand generator rule

Fleming's **left-hand generator rule** can be used to determine the relationship of the motion of the conductor in a magnetic field to the direction of the induced current. To use the left-hand rule, place the thumb, forefinger, and center finger at right angles to each other as shown in *Figure 14-53. The*

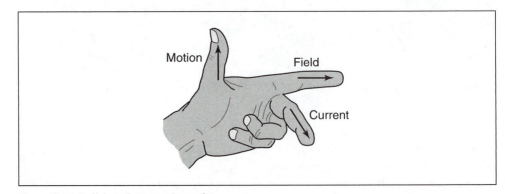

Figure 14-53 Left-hand generator rule.

forefinger points in the direction of the field flux, assuming that magnetic lines of force are in a direction of north to south. *The thumb points in the direction of thrust,* or movement of the conductor, *and the center finger shows the direction of the current induced into the armature.* An easy method of remembering which finger represents which quantity is shown here:

THumb = THrust
Forefinger = Flux
Center finger = Current

The left-hand rule can be used to clearly illustrate that if the polarity of the magnetic field is changed or if the direction of armature rotation is changed, the direction of induced current will change also.

PARALLELING GENERATORS

There may be occasions when one direct current generator cannot supply enough current to operate the connected load. In such a case, another generator is connected in parallel with the first. Direct current generators should never be connected in parallel without an equalizer connection *(Figure 14-54).* The equalizing connection is used to connect the series fields of the two machines in parallel with each other. This arrangement prevents one machine from taking the other over as a motor.

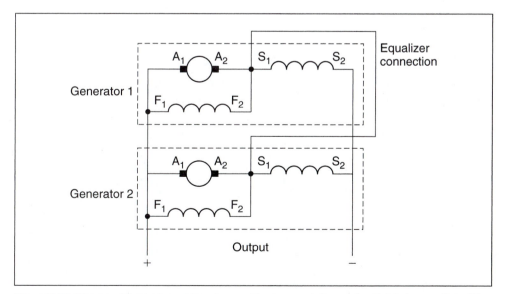

Figure 14-54 The equalizer connection is used to connect the series fields in parallel with each other.

Assume that two generators are to be connected in parallel, and the equalizing connection has not been made. Unless both machines are operating with identical field excitation when they are connected in parallel, the machine with the greatest excitation will take the entire load and begin operating the other machine as a motor. The series field of the machine that accepted the load would be strengthened, and the series field of the machine that gave up the load would be weakened. The machine with the stronger series field will take even greater load, and the machine with the weaker series field will reduce load even further.

The generator that begins motoring will have the current flow through its series field reversed, which will cause it to operate as a differential-compound motor *(Figure 14-55)*. If the motoring generator is not removed from the line, the magnetic field strength of the series field will become greater than the field strength of the shunt field. This will cause the polarity of the residual magnetism in the pole pieces to reverse. This is often referred to as flashing the field. Flashing the field results in the polarity of the output voltage being reversed when the machine is restarted as a generator. The equalizer connection prevents field reversal even if the generator becomes a motor.

The resistance of the equalizer cable should not exceed 20% of the resistance of the series field winding of the smallest paralleled generator. This will ensure that the current flow provided to the series fields will divide in the approximate inverse ratio of the respective series field winding.

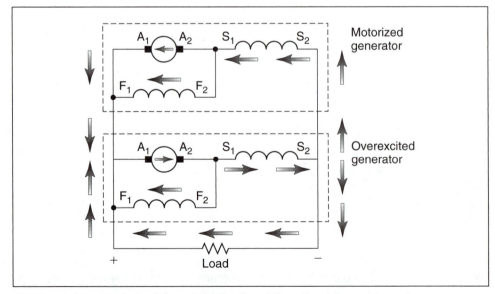

Figure 14-55 One generator takes all the load, and the other becomes a motor.

Summary

1. A generator is a machine that converts mechanical energy into electrical energy.

2. Generators operate on the principle of magnetic induction.

3. Alternating current is produced in all rotating armatures.

4. The commutator changes the alternating current produced in the armature into direct current.

5. The brushes are used to make contact with the commutator and to carry the output current to the outside circuit.

6. The position at which there is no induced voltage in the armature is called the neutral plane.

7. Lap wound armatures are used in machines designed for low-voltage and high-current operation.

8. Wave wound armatures are used in machines designed for high-voltage and low-current operation.

9. Frogleg wound armatures are the most used and are intended for machines designed for moderate voltages and current.

10. The loops of wire, iron core, and commutator are made as one unit and are referred to as the armature.

11. The armature connection is marked A_1 and A_2.

12. Series field windings are made with a few turns of large wire and have a very low resistance.

13. Series field windings are connected in series with the armature.

14. Series field windings are marked S_1 and S_2.

15. Shunt field windings are made with many turns of small wire and have a high resistance.

16. Shunt field windings are connected in parallel with the armature.

17. The shunt field windings are marked F_1 and F_2.

18. Three factors that determine the voltage produced by a generator are:
    A. The number of turns of wire in the armature
    B. The strength of the magnetic field of the pole pieces
    C. The speed of the armature

19. Series generators increase their output voltage as load is added.

20. Shunt generators decrease their output voltage as load is added.

21. The voltage regulation of a DC generator is proportional to the resistance of the armature.

22. Compound generators contain both series and shunt field windings.

23. A long shunt compound generator has the shunt field connected in parallel with both the armature and series field.

24. A short shunt compound generator has the shunt field connected in parallel with the armature, but in series with the series field.

25. When a generator is overcompounded, the output voltage will be higher at full load than it is at no load.

26. When a generator is flat-compounded, the output voltage will be the same at full load and no load.

27. When a generator is undercompounded, the output voltage will be less at full load than it is at no load.

28. Cumulative-compound generators have their series and shunt fields connected in such a manner that they aid each other in the production of magnetism.

29. Differential-compound generators have their series and shunt field winding connected in such a manner that they oppose each other in the production of magnetism.

30. Armature reaction is the twisting or bending of the main magnetic field.

31. Armature reaction is caused by the interaction of the magnetic field produced in the armature.

32. Armature reaction is proportional to armature current.

33. Interpoles are small pole pieces connected between the main field poles used to help correct armature reaction.

34. Interpoles are connected in series with the armature.

35. Interpoles used in a generator must have the same polarity as the main field pole directly ahead of them in the sense of rotation of the armature.

36. Interpoles used in a motor must have the same polarity as the main field pole directly behind them in the sense of rotation of the armature.

37. Interpole leads are sometimes marked C_1 and C_2 or S_3 and S_4.

38. Interpole leads are not always brought out of the machine.

39. The neutral plane can be set by connecting an AC voltmeter to the shunt field and a source of low-voltage AC to the armature. The brushes are then adjusted until the voltmeter indicates the lowest possible voltage.

40. When a generator supplies current to a load, countertorque is produced, which makes the armature harder to turn.

41. Countertorque is proportional to the armature current if the field excitation current remains constant.

42. Countertorque is a measure of the useful electrical energy produced by the generator.

Review Questions

1. What is a generator?

2. What type of voltage is produced in all rotating armatures?

3. What are the three types of armature windings?

4. What type of armature winding would be used for a machine intended for high-voltage, low-current operation?

5. What are interpoles, and what is their purpose?

6. How are interpoles connected in relation to the armature?

7. What type of field winding is made with many turns of small wire?

8. How is the series field connected in relation to the armature?

9. How is the shunt field connected in relation to the armature?

10. What is armature reaction?

11. To what is armature reaction proportional?

12. What are eddy currents?

13. What condition characterizes overcompounding?

14. What is the function of the shunt field rheostat?

15. What is used to control the amount of compounding for a generator?

16. Explain the difference between cumulative- and differential-compounded connections.

17. What three factors determine the amount of output voltage for a DC generator?

18. To what is the voltage regulation of a DC generator proportional?

19. To what is countertorque proportional?

20. Of what is countertorque a measure?

UNIT 15

Direct Current Motors

Objectives

After studying this unit, you should be able to:

- Discuss the principle of operation of direct current motors.
- Discuss different types of DC motors.
- Draw schematic diagrams of different types of DC motors.
- Be able to connect a DC motor for a particular direction of rotation.
- Discuss counter-EMF.
- Describe methods for controlling the speed of direct current motors.

Direct current motors are used throughout industry in applications where variable speed is desirable. The speed-torque characteristic of direct current motors makes them desirable for many uses. The automotive industry uses DC motors to start internal combustion engines and to operate blower fans, power seats, and other devices where a small motor is needed.

DC MOTOR PRINCIPLES

Direct current motors operate on the principle of repulsion and attraction of magnetism. Motors use the same types of armature windings and field windings as the generators discussed in Unit 14. A motor, however,

performs the opposite function of a generator. A **motor** is a device used to   **motor**
convert electrical energy into mechanical energy. Direct current motors
were, in fact, the first electric motors to be invented. For many years, it was
believed not possible to make a motor that could operate using alternating
current.

DC motors are the same basic machines as DC generators. In the case of a
generator, some device is used to turn the shaft of the armature, and the
power produced by the turning armature is supplied to a load. In the case of
a motor, power connected to the armature causes it to turn. To understand
why the armature turns when current is applied to it, refer to the simple one-
loop armature in *Figure 15-1*. Electrons enter the loop through the negative
brush, flow around the loop, and exit through the positive brush. As current
flows through the loop, a magnetic field is created around the loop. An end
view, illustrating the pole pieces and the two conductors of a single loop, is
shown in *Figure 15-2*. The X indicates electrons moving away from the
observer as the back of an arrow moving away. The dot represents electrons
moving toward the observer as the point of an approaching arrow. The
left-hand rule for magnetism can be used to check the direction of the magnetic field around the conductors.

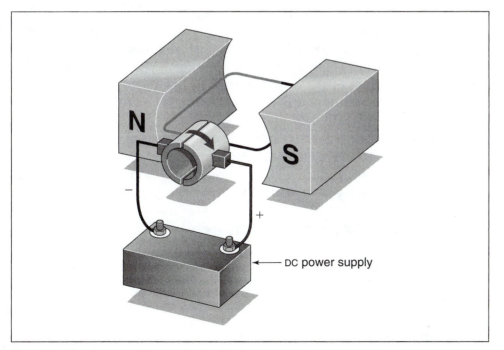

Figure 15-1 Direct current is supplied to the loop.

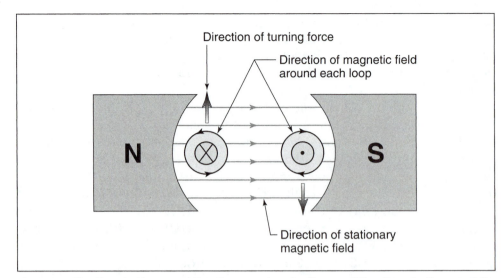

Direction of turning force

Direction of magnetic field around each loop

N   S

Direction of stationary magnetic field

Figure 15-2 Flux lines in the same direction repel each other, and flux lines in the opposite direction attract each other.

Torque

Magnetic lines of force flow in a direction of north to south between the poles of the stationary magnet (left to right in *Figure 15-2*). When magnetic lines of flux flow in the same direction, they repel each other. When they flow in opposite directions, they attract each other. The magnetic lines of flux around the conductors cause the loop to be pushed in the direction shown by the arrows. This pushing or turning force is called **torque** and is created by the magnetic field of the pole pieces and the magnetic field of the loop or armature. Two factors that determine the amount of torque produced by a direct current motor are:

- Strength of the magnetic field of the pole pieces
- Strength of the magnetic field of the armature

Notice that there is no mention of speed or cutting action. One characteristic of a direct current motor is that it can develop maximum torque at 0 RPM.

Increasing the Number of Loops

In the previous example, a single-loop armature was used to illustrate the operating principle of a DC motor. In actual practice, armatures are constructed with many turns of wire per loop and many loops. This provides a strong continuous turning force for the armature *(Figure 15-3)*.

The Commutator

When a DC machine is used as a generator, the commutator performs the function of a mechanical rectifier to change the alternating current produced

torque

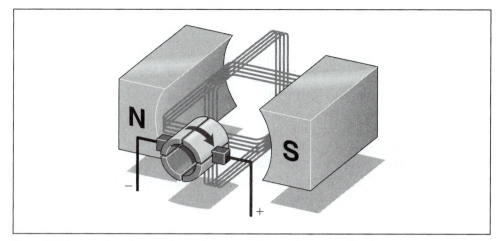

Figure 15-3 Increasing the number of loops and turns increases the torque.

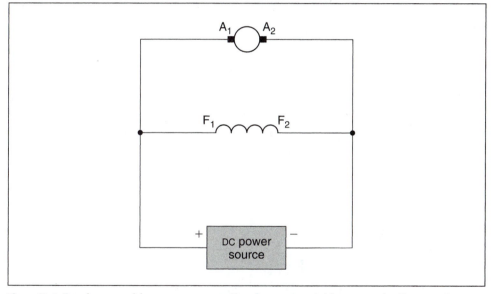

Figure 15-4 Brushes provide power connection from an outside source to the armature.

in the armature into direct current before it exits the machine through the brushes. When a DC machine is used as a motor, the commutator performs the function of a rotary switch and maintains the correct direction of current flow through the armature windings. In order for the motor to develop a turning force, the magnetic field polarity of the armature must remain constant in relation to the polarity of the pole pieces. The commutator forces the direction of current flow to remain constant through certain sections of the armature as it rotates. The brushes are used to provide power to the armature from an external power source *(Figure 15-4)*.

SHUNT MOTORS

shunt motor

There are three basic types of direct current motors: shunt, series, and compound. The direct current **shunt motor** has the shunt field connected in parallel with the armature *(Figure 15-5)*. This permits an external power source to supply current to the shunt field and maintain a constant magnetic field. The shunt motor has very good speed characteristics. The full-load speed will generally remain within 10% of the no-load speed. Shunt motors

constant speed motors

are often referred to as **constant speed motors**. Characteristic curves for shunt, series, and compound motors are shown in *Figure 15-6*. Note that the shunt motor maintains the most constant speed as load is added and armature current increases.

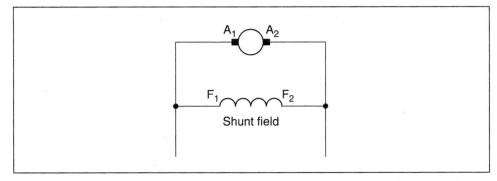

Figure 15-5 The shunt motor has the shunt field connected in parallel with the armature.

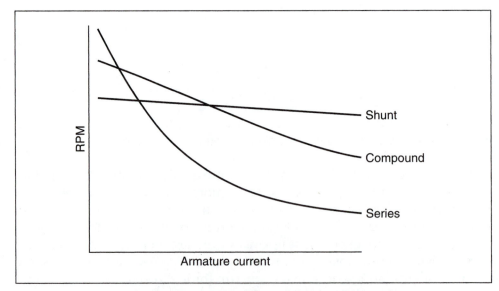

Figure 15-6 Characteristic speed curves for different direct current motors.

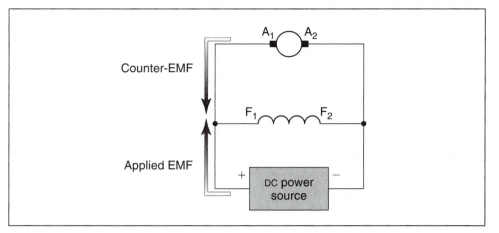

Figure 15-7 Counter-EMF limits the flow of current through the armature.

Counter-EMF

When the windings of the armature spin through the magnetic field produced by the pole pieces, a voltage is induced into the armature. This induced voltage is opposite in polarity to the applied voltage and is known as **counter-EMF (CEMF)** or **back-EMF** *(Figure 15-7)*. It is the CEMF that limits the flow of current through the armature when the motor is in operation. The amount of counter-EMF produced in the armature is proportional to three factors:

1. Number of turns of wire in the armature
2. Strength of the magnetic field of the pole pieces
3. Speed of the armature

counter-EMF (CEMF)

back-EMF

Speed-Torque Characteristics

When a DC motor is first started, the inrush of current can be high because no CEMF is being produced by the armature. The only current-limiting factor is the amount of wire resistance in the windings of the armature. When current flows through the armature, a magnetic field is produced and the armature begins to turn. As the armature windings cut through the magnetic field of the pole pieces, counter-EMF is induced in the armature. The counter-EMF opposes the applied voltage, causing current flow to decrease. If the motor is not connected to a load, the armature will continue to increase in speed until the counter-EMF is almost the same value as the applied voltage. At this point, the motor produces enough torque to overcome its own losses. Some of these losses are:

1. I^2R loss in the armature windings
2. Windage loss
3. Bearing friction
4. Brush friction

When a load is added to the motor, the torque will not be sufficient to support the load at the speed at which the armature is turning. The armature will, therefore, slow down. When the armature slows down, counter-EMF is reduced and more current flows through the armature windings. This produces an increase in magnetic field strength and an increase in torque. This is the reason that armature current increases when load is added to the motor.

Speed Regulation

speed regulation

The amount by which the speed decreases as load is added is called the **speed regulation**. *The speed regulation of a direct current motor is proportional to the resistance of the armature.* The lower the armature resistance, the better the speed regulation. The reason for this is that armature current determines the torque produced by the motor if the field excitation current is held constant. In order to produce more torque, more current must flow though the armature, which increases the magnetic field strength of the armature. The amount of current flowing through the armature will be determined by the counter-EMF and the armature resistance. If the field excitation current is constant, the amount of counter-EMF will be proportional to the speed of the armature. The faster the armature turns, the higher the counter-EMF. When the speed of the armature decreases, the counter-EMF decreases also.

Assume an armature has a resistance of 6 Ω. Now assume that when load is added to the motor, an additional 3 A of armature current will be required to produce the torque necessary to overcome the added load. In this example, a voltage of 18 V will be required to increase the armature current by 3 A (3 A × 6 Ω = 18 V). This means that the speed of the armature must drop enough so that the counter-EMF is 18 V less than it was before. The reduction in counter-EMF permits the applied voltage to push more current through the resistance of the armature.

Now assume that the armature has a resistance of 1 Ω. If a load is added that requires an additional 3 A of armature current, the speed of the armature must drop enough to permit a 3-V reduction in counter-EMF (3 A × 1 Ω = 3 V). The armature does not have to reduce speed as much to cause a 3-V reduction in counter-EMF as it does for a reduction of 18 V.

SERIES MOTORS

series motor

The operating characteristics of the direct current **series motor** are very different from those of the shunt motor. The reason is that the series motor has only a series field connected in series with the armature *(Figure 15-8)*. The armature current, therefore, flows through the series field. The speed of the series motor is controlled by the amount of load connected to the motor.

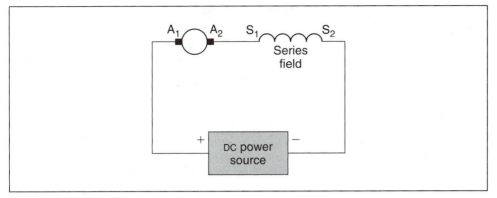

Figure 15-8 Series motor connection.

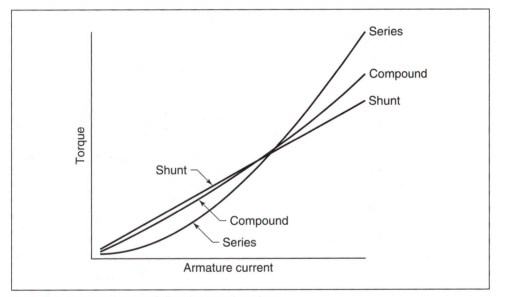

Figure 15-9 Torque curves of direct current motors.

When load is increased, the speed of the motor will decrease. This causes a reduction in the amount of counter-EMF produced in the armature and an increase in armature and series field current. Since the current increases in both the armature and series field, the torque will increase by the square of the current. In other words, if the current doubles, the torque will increase four times. Characteristic curves showing the relationship of torque and armature current for the three main types of DC motors are shown in *Figure 15-9.* Notice that the series motor produces the most torque of the three motors.

Series Motor Speed Characteristics

Series motors have no natural speed limit and should, therefore, never be operated in a no-load condition. Large series motors that suddenly lose their load will race to speeds that will destroy the motor. Series motors operating at no load have been known to develop such an extremely high RPM that centrifugal force slings both the windings out of the slots in the armature and the copper bars out of the commutator. For this reason, series motors should be coupled directly to a load. Belts or chains should never be used to connect a series motor to a load.

Series motors have the ability to develop extremely high starting torques. An average of about 450% of full torque is common. These motors are generally used for applications that require a high starting torque, such as the starter motor on an automobile, cranes and hoists, and electric buses and street cars.

COMPOUND MOTORS

compound motor

The **compound motor** uses both a series field and a shunt field. This motor is used to combine the operating characteristics of both the series and shunt motors. The series field of the compound motor permits the motor to develop high torque, and the shunt field permits speed control and regulation. The compound motor is used more than any other type of direct current motor in industry. The compound motor will not develop as much torque as the series motor, but it will develop more than the shunt motor. The speed regulation of a compound motor is not as good as a shunt motor, but it is much better than a series motor *(Figure 15-6)*.

Compound motors can be connected as short shunt or long shunt just as compound generators can *(Figure 15-10)*. The long shunt connection is more common because it has superior speed regulation. Compound motors can also be **cumulative-compounded motors** or **differential-compounded motors**. Although there are some applications for differential-compounded motors, it is generally a connection to be avoided. If a generator is inadvertently connected as a differential-compound machine, the greatest consequence will be that the output voltage drops rapidly as load is added. This is not the case with a motor, however. When a motor is connected as differential-compound, the shunt field will determine the direction of rotation of the motor at no load or light load. When load is added to the motor, the series field will become stronger. If enough load is added, the magnetic field of the pole pieces will reverse polarity, and the motor will suddenly stop, reverse direction, and begin operating as a series motor. This can damage the motor and the equipment to which the motor is connected.

cumulative-compounded motors

differential-compounded motors

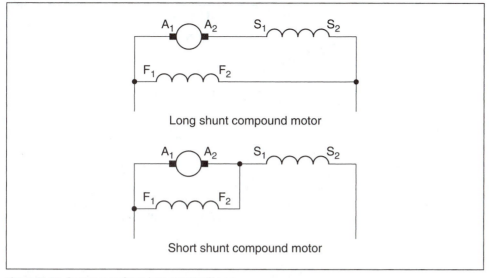

Figure 15-10 Compound motor connections.

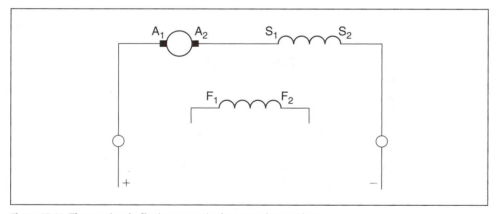

Figure 15-11 The motor is first connected as a series motor.

If a compound motor is to be connected, use the following steps to pre-vent the motor from being accidentally connected as differential-compound:

1. Disconnect the motor from the load.
2. Connect the series field and armature windings together to form a series motor connection, leaving the shunt field disconnected. Connect the motor to the power source *(Figure 15-11)*.

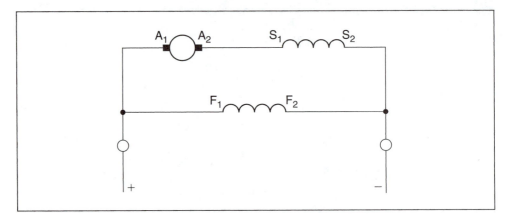

Figure 15-12 Connect the shunt field and again test for direction of rotation.

3. Turn the power on momentarily to determine the direction of rotation. This application of power must be of very short duration because the motor is now being operated as a series motor with no load. The idea is to "bump" the motor just to check for direction of rotation. If the motor turns in the opposite direction than desired, reverse the connection of the armature leads. This will reverse the direction of rotation of the motor.

4. Connect the shunt field leads to the incoming power *(Figure 15-12)*. Again, turn on the power and check the direction of rotation. If the motor operates in the desired direction, it is connected as cumulative-compound. If the motor turns in the opposite direction, it is connected as differential-compound, and the shunt field leads should be reversed.

This test can be used to check for a differential- or cumulative-compound connection because the shunt field controls the direction of rotation at no load. If the motor operates in the same direction as both a series and compound motor, the magnetic polarity of the pole pieces must be the same for both connections. This indicates that both the series and shunt field windings must be producing the same magnetic polarity and are, therefore, connected as cumulative-compound.

TERMINAL IDENTIFICATION FOR DC MOTORS

The terminal leads of DC machines are labeled so they can be identified when they are brought outside the motor housing to the terminal box. DC motors have the same terminal identification as that used for DC generators. *Figure 15-13* illustrates this standard identification. Terminals A_1 and

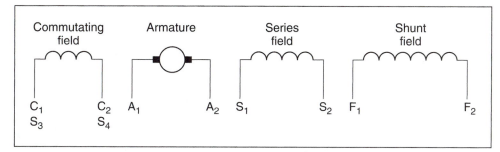

Figure 15-13 Terminal identification for DC machines.

A_2 are connected to the armature through the brushes. The ends of the series field are identified with S_1 and S_2, and the shunt field leads are marked F_1 and F_2. Some DC machines will provide access to another set of windings called the commutating field or interpoles. The ends of this winding will be labeled C_1 and C_2 or S_3 and S_4. It is common practice to provide access to the interpole winding on machines designed to be used as motors or generators.

DETERMINING THE DIRECTION OF ROTATION OF A DC MOTOR

The direction of rotation of a DC motor is determined by facing the commutator end of the motor. This is generally the back or rear of the motor. If the windings have been labeled in a standard manner, it is possible to determine the direction of rotation when the motor is connected. *Figure 15-14* illustrates the standard connections for a series motor. The standard connections for a shunt motor are illustrated in *Figure 15-15,* and the standard connections for a compound motor are shown in *Figure 15-16.*

The direction of rotation of a DC motor can be reversed by changing the connections of the armature leads or the field leads. It is common practice to change the connection of the armature leads. This is done to prevent changing a cumulative-compound motor into a differential-compound motor.

Although it is standard practice to change the connection of the armature leads to reverse the direction of rotation, it is not uncommon to reverse the rotation of small shunt motors by changing the connection of the field leads. If a motor contains only a shunt field, there is no danger of changing the motor from a cumulative- to a differential-compound motor. The shunt field leads are often changed on small motors because the amount of current flow through the field is much less than the current flow through the armature.

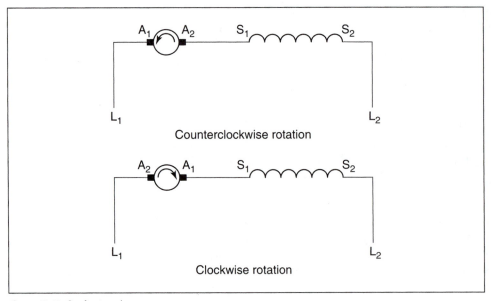

Figure 15-14 Series motors.

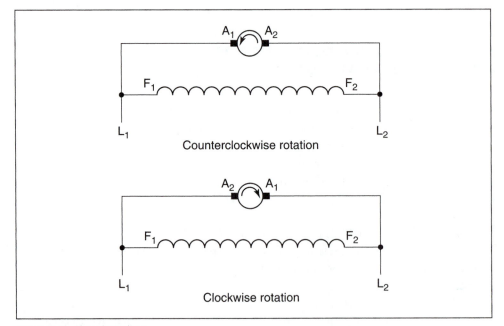

Figure 15-15 Shunt motors.

This permits a small double-pole double-throw switch to be used as a control for reversing the direction of rotation *(Figure 15-17)*.

Large compound motors often use a control circuit similar to the one shown in *Figure 15-18* for reversing the direction of rotation. This control circuit

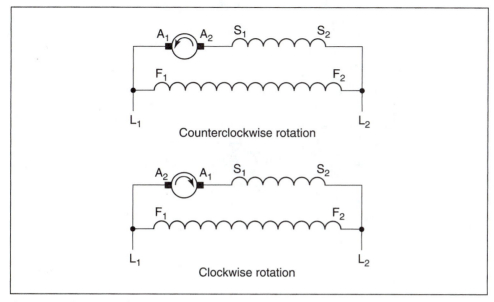

Figure 15-16 Compound motors.

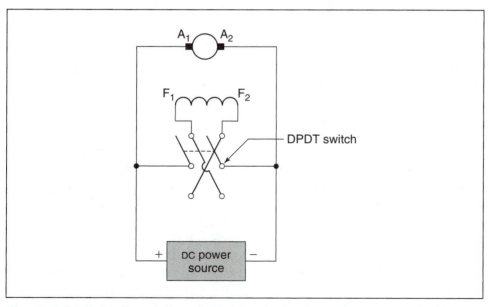

Figure 15-17 Reversing the rotation of a shunt motor by changing the shunt field leads.

uses magnetic contactors to reverse the flow of current through the armature. If the circuit is traced, it will be seen that when the forward or reverse direction is chosen, only the current through the armature changes direction. The current flow through the shunt and series fields remains the same.

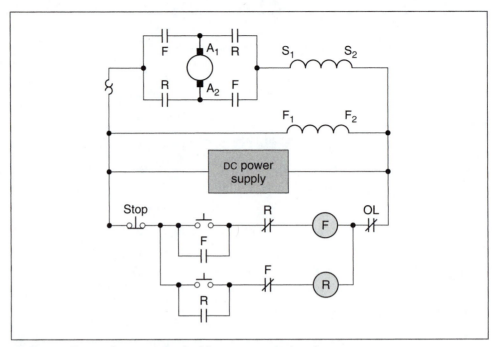

Figure 15-18 Forward-reverse control for a compound motor.

SPEED CONTROL

There are several ways of controlling the speed of direct current motors. The method employed is generally dictated by the requirements of the load. When full voltage is connected to both the armature and shunt field, the motor will operate at its base speed. If the motor is to be operated at below base speed or under speed, full voltage is maintained to the shunt field and the amount of armature current is reduced. The reduction of armature current causes the motor to produce less torque, and the speed decreases. One method of reducing armature current is to connect resistance in series with the armature *(Figure 15-19)*. In the example shown, three resistors are connected in series with the armature. Contacts S_1, S_2, and S_3 can be used to shunt out steps of resistance and permit more armature current to flow.

Although adding resistance in the armature circuit does permit the motor to be underspeeded, it has several disadvantages. As current flows through the resistors, they waste power in the form of heat. Also, the speed of the motor can be controlled only in steps. There is no smooth increase or decrease in speed. Most large direct current motors use an electronic controller to supply variable voltage to the armature circuit separately from the field *(Figure 15-20)*. This permits continuous adjustment of the speed from

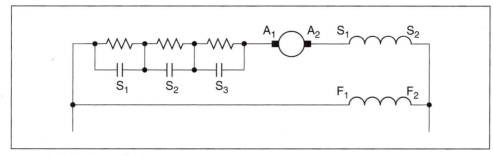

Figure 15-19 Resistors limit armature current.

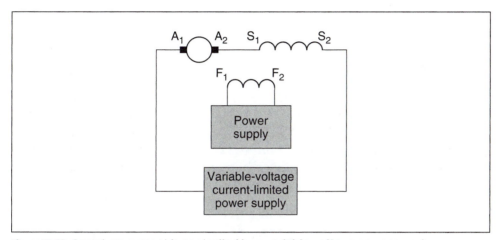

Figure 15-20 Armature current is controlled by a variable-voltage power supply.

zero to full RPM. Electronic power supplies can also sense the speed of the motor and maintain a constant speed as load is changed. Most of these power supplies are current-limited. The maximum current output can be set to a value that will not permit the motor to be harmed if it stalls or if the load becomes too great.

A direct current motor can be overspeeded by connecting full voltage to the armature and reducing the current flow through the shunt field. In *Figure 15-21*, a shunt field rheostat has been connected in series with the shunt field. As resistance is added to the shunt field, field current decreases, which causes a decrease in the flux density of the pole pieces. This decrease in flux density produces less counter-EMF in the armature, which permits more armature current to flow. The increased armature current causes an increase in the magnetic field strength of the armature. This increased magnetic field strength of the armature produces a net gain in torque, which causes the motor speed to increase.

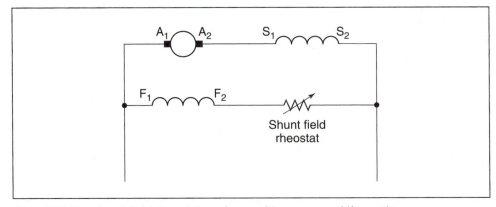

Figure 15-21 The shunt field rheostat can be used to overspeed the motor.

FIELD LOSS RELAY

Most large compound direct current motors have a protective device connected in series with the shunt field called the **field loss relay (FLR)** or shunt field relay *(Figure 15-22)*. The function of the field loss relay is to disconnect power to the armature if current flow through the shunt field should decrease below a certain level. If the shunt field current stopped completely,

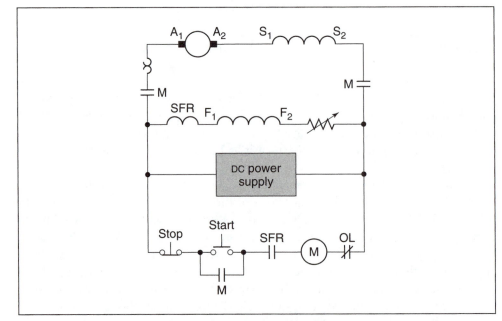

Figure 15-22 The shunt field relay disconnects power to the armature if shunt field current stops.

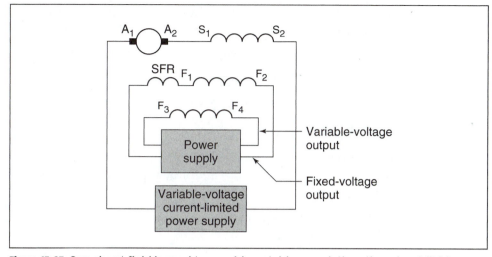

Figure 15-23 One shunt field is used to provide a stable speed; the other shunt field provides overspeed control.

the compound motor would become a series motor and would increase rapidly in speed. This could cause damage to both the motor and the load.

Many large DC compound motors intended to operate in an overspeed condition will actually contain two separate shunt fields *(Figure 15-23)*. One shunt field is connected to a fixed voltage and maintains a constant field to provide an upper limit to motor speed. This shunt field will be connected to the field loss, or shunt field relay. The second shunt field is connected to a source of variable voltage. This shunt field is used to increase speed above the base speed. For this type of motor, base speed is achieved by applying full voltage to the armature and both shunt fields. It should be noted that most large DC motors have voltage applied to the shunt field at all times, even when the motor is not in operation. The resistance of the winding produces heat, which is used to prevent any formation of moisture inside the motor.

HORSEPOWER

When James Watt first began to try to market steam engines, he found that he needed a way to compare them to the horses they were to replace. After conducting experiments, Watt found that the average horse could do work at a rate of 550 ft-lb/s. This became the basic horsepower measurement. Horsepower can also be expressed in the basic electrical unit for power, which is the watt.

1 horsepower = 746 W

Once horsepower has been converted to a basic unit of power, it can be converted to other power units, such as:

$$1 \text{ W} = 3.42 \text{ BTUs per hour}$$

$$1055 \text{ W} = 1 \text{ BTU per second}$$

$$4.19 \text{ W} = 1 \text{ calorie per second}$$

$$1.36 \text{ W} = 1 \text{ ft-lb per second}$$

In order to determine the horsepower output of a motor, the rate at which the motor is doing work must be known. The following formula can be used to determine the horsepower output of a motor.

$$hp = \frac{(1.59)\ (torque)\ (RPM)}{100{,}000}$$

where

hp = horsepower

1.59 = a constant

torque = torque in lb-in.

RPM = speed

100,000 = a constant

Example 1

How much horsepower is being produced by a motor turning a load of 350 lb-in. at a speed of 1375 RPM?

Solution

$$hp = \frac{1.59 \times 350 \times 1375}{100{,}000}$$

$$hp = 7.65$$

Once the output horsepower is known, it is possible to determine the efficiency of the motor by using the formula

$$Eff. = \frac{power\ out}{power\ in} \times 100$$

Example 2

A direct current motor is connected to a 120-V DC line and has a current draw of 1.3 A. The motor is operating a load that requires 8 lb-in. of torque and is turning at a speed of 1250 RPM. What is the efficiency of the motor?

Solution

The first step is to determine the horsepower output of the motor.

$$hp = \frac{1.59 \times 8 \times 1250}{100,000}$$

$$hp = 0.159$$

Now that the output horsepower is known, horsepower can be changed into watts using the formula

$$1\ hp = 746\ W$$

$$0.159 \times 746 = 118.6\ W$$

The amount of input power can be found by using the formula

$$watts = volts \times amps$$

$$watts = 120 \times 1.3$$

$$watts = 156$$

The efficiency of the motor can be found by using the formula

$$Eff. = \frac{power\ out}{power\ in} \times 100$$

The answer is multiplied by 100 to change it to a percent.

$$Eff. = \frac{118.6}{156} \times 100$$

$$Eff. = 76\%$$

Torque is often measured in pound-feet instead of pound-inches. Another formula often used to determine horsepower when the torque is measured in pound-feet is

$$hp = \frac{(2\pi)\ (torque)\ (RPM)}{33,000}$$

where

$$hp = horsepower$$

$$\pi = 3.1416$$

$$torque = lb\text{-}ft$$

$$RPM = speed\ in\ revolutions\ per\ minute$$

$$33,000 = a\ constant$$

BRUSHLESS DC MOTORS

brushless DC
motors

Brushless DC motors do not contain a wound armature, commutator, or brushes. The armature or rotor (rotating member) contains permanent magnets. The rotor is surrounded by fixed stator windings *(Figure 15-24)*. The stationary armature or stator winding is generally three-phase, but some motors are designed to operate on four-phase or two-phase power. Two-phase stator windings are commonly used for motors intended to operate small fans.

The phases are provided by a converter inside the motor housing that changes the direct current into alternating current *(Figure 15-25)*. Alternating

Figure 15-24 Brushless DC motor (courtesy: GE ECM™ Technologies).

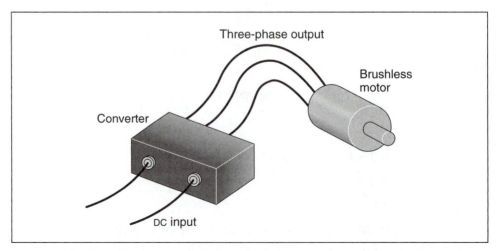

Figure 15-25 A converter changes direct current into three-phase alternating current.

current is used to create a rotating magnetic field inside the stator of the motor. This rotating magnetic field attracts the permanent magnets of the rotor and causes the rotor to turn in the same direction as the rotating field. Rotating magnetic fields will be discussed in Unit 17. The speed is determined by the number of stator poles per phase and the frequency of the AC voltage.

Some converters that supply power to brushless motors produce sine waves, but most produce a trapezoidal AC waveform *(Figure 15-26)*. Motors powered by the trapezoidal waveform produce about 10% more torque than those powered by sine waves. But motors powered by sine waves operate more smoothly and with less torque ripple at low speeds. Motors that require very smooth operation such as grinders, mills, or other machines that produce a finished surface are powered by sine waves. Applications where high torque and low speed are required often employ brushless motors that have a very high number of stator poles. Some of these motors have as many as 64 stator poles per phase. At 60 Hz this would produce a speed of 112.5 RPM. The following formula can be used to determine the speed of the rotating magnetic field when the frequency and number of stator poles per phase are known:

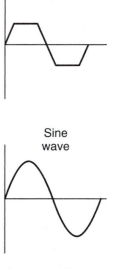

Trapezoidal wave

Sine wave

Figure 15-26 Most converters produce a trapezoidal wave, but some produce a sine wave.

$$S = \frac{120 \times f}{P}$$

where

S = speed in revolutions per minute (RPM)

120 = a constant

f = frequency in hertz

P = number of poles per phase

The speed of the rotating magnetic field for the motor described previously would be

$$S = \frac{120 \times 60}{64}$$

S = 112.5 RPM

The use of this many poles to provide low speed and high torque is often referred to as magnetic gearing, because it eliminates the need for mechanical gears and other speed-reducing equipment. This in turn eliminates friction and backlash associated with mechanical gears. Motors that use a high

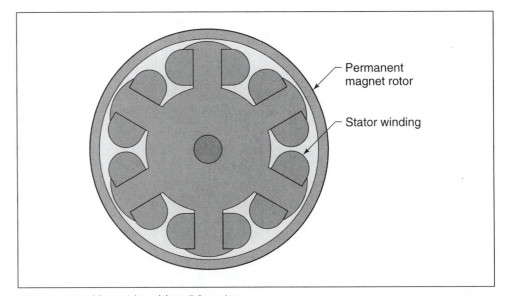

Figure 15-27 Inside-out brushless DC motor.

pole count for low-speed high-torque applications are often called ring motors or ring torquers.

Inside-Out Motors

Another type of brushless DC motor uses a rotor shaped like a hollow cylinder or cup. The stator windings are wound inside the rotor *(Figure 15-27)*. The permanent magnet rotor is mounted on the outside of the stator windings. The size and shape of the rotor cause it to act as a flywheel, which gives these motors a large amount of inertia. *Motors with a large amount of inertia exhibit superior speed regulation characteristics.* Inside-out motors are often used to drive the hard disks in computers, operate tape cartridge drives, provide power for robots, and operate fans and blowers in high-speed air conditioning systems.

Differences between Brush Type and Brushless Motors

Since brushless motors do not have a commutator or brushes, they are generally smaller and cost less than brush type motors. The added expense of the converter, however, makes the cost about the same as a brush type motor. Brushless motors dissipate heat more quickly because the stator windings can dissipate heat faster than a wound armature. They are more efficient than brush type motors, require less maintenance, and as a general rule have less down time.

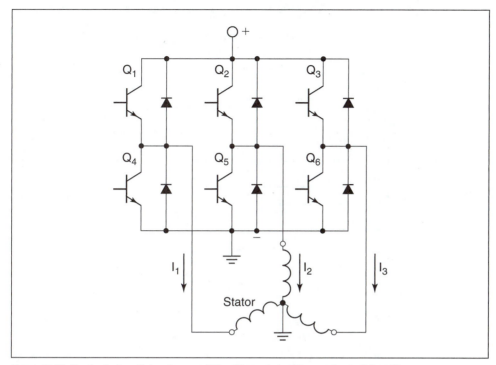

Figure 15-28 Typical circuit to change DC voltage into three-phase AC voltage.

CONVERTERS

As stated previously, converters are used to change the direct current into multiphase alternating current to produce a rotating magnetic field in the stationary armature or stator of the brushless DC motor. An example circuit for a DC-to-three-phase converter is shown in *Figure 15-28*. In this example, transistors Q_1 through Q_6 are used as switches. When the transistors are turned on or off in the proper sequence, current can be routed through the stator windings in such a manner that it produces three alternating currents 120° out of phase with each other. The commutation sequence provides an action similar to that of the commutator in a brush type motor. A chart illustrating the firing order of the transistors to produce this commutation sequence is shown in *Figure 15-29*. The current flow through the three stator windings is illustrated in the same chart.

PERMANENT MAGNET MOTORS

Permanent magnet motors contain a wound armature and brushes as does a conventional direct current motor. The pole pieces, however, are permanent magnets. This eliminates the need for shunt or series field windings

**permanent
magnet motors**

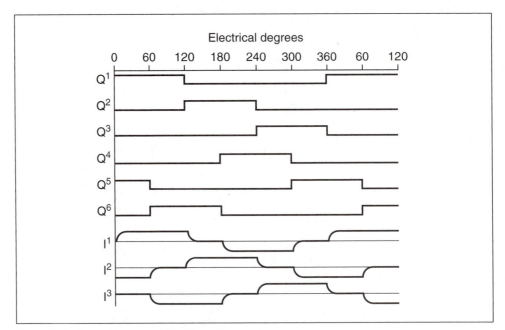

Figure 15-29 Converter commutation sequence.

(Figure 15-30). Permanent magnet motors have a higher efficiency than conventional field wound motors because power is supplied to the armature circuit only. These motors have been popular for many years in applications where batteries must be used to supply power to the motor, such as trolling motors on fishing boats and small electric vehicles.

The horsepower rating of PM (permanent magnet) motors has increased significantly since the introduction of rare-earth magnets such as samarium cobalt and neodymium in the mid-1970s. These materials have replaced Alnico and ferrite magnets in most PM motors. The increased strength of rare-earth magnets permits motors to produce torques that range from 7.0 oz-in to 4500 lb-ft. Permanent magnet motors with horsepower ratings of over 15 hp are now available. The torque-to-weight ratios of PM motors equipped with rare-earth magnets can exceed those of conventional field wound motors by 40% to 90%. The power-to-weight ratios can exceed conventional motors by 50% to 200%. Permanent magnet motors of comparable horsepower ratings are smaller and lighter in weight than conventional field wound motors.

Operating Characteristics

Since the fields are permanent magnets, the field flux of PM motors remains constant at all times. This gives the motor operating characteristics

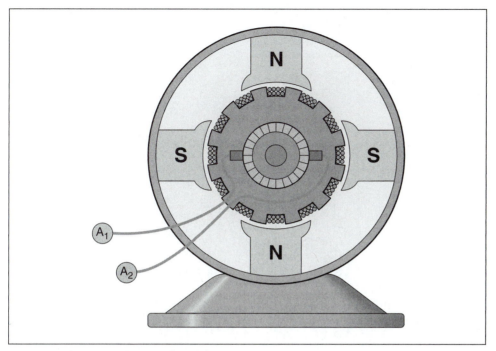

Figure 15-30 Permanent magnet motor.

very similar to those of conventional separately excited shunt motors. The speed can be controlled by the amount of voltage applied to the armature, and the direction of rotation is reversed by reversing the polarity of voltage applied to the armature leads *(Figure 15-31)*.

DC Servomotors

Small permanent magnet motors are used as **servomotors**. These motors have small, lightweight armatures that contain very little inertia. This permits servomotors to be operated at high speed and then stopped or reversed very quickly. Servomotors generally contain from two to six poles. They are used to operate tape drives on computers and to power the spindles on numerically controlled (NC) machines such as milling machines and lathes.

servomotors

DC ServoDisc® Motors

Another type of direct current servomotor that is totally different in design is the **ServoDisc® motor**. This motor uses permanent magnets to provide a constant magnetic field as do conventional servomotors, but the design of the armature is completely different. In a conventional servomotor, the

ServoDisc® motor

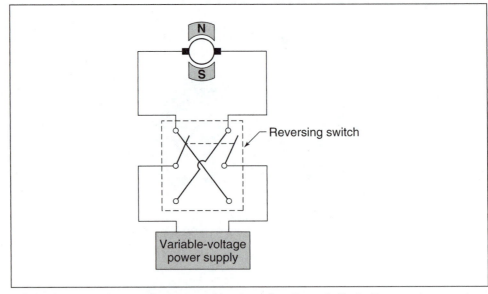

Figure 15-31 PM motors can be reversed by reversing the polarity to the armature, and the speed can be controlled by variable voltage.

armature is constructed in the same way as other DC motors by cutting slots in an iron core and winding wire through the slots. A commutator is then connected to one end to provide commutation, which switches the current path as the armature turns. This maintains a constant magnetic polarity for different sections of the armature. In the conventional servo-motor, the permanent magnets are mounted on the motor housing in such a manner that they create a radial magnetic field that is perpendicular to the windings of the armature *(Figure 15-32)*. The armature core is made of iron because it must conduct the magnetic lines of flux between the two pole pieces.

The armature of the ServoDisc® motor does not contain any iron. It is made of two to four layers of copper conductors formed into a thin disc. The conductors are "printed" on a fiberglass material in much the same way as a printed circuit. For this reason, the ServoDisc® motor is often called a **printed circuit motor**. Since the disc armature is very thin, it permits the permanent magnets to be mounted on either side of the disc and parallel to the shaft of the motor *(Figure 15-33)*. Since the disc is very thin, the air gap between the two magnets is small.

Torque is produced when current flowing though the copper conductors of the disc produces a magnetic field that reacts with the magnetic field of the permanent magnets. The permanent magnet pairs are arranged around the circumference of the motor housing in such a manner that they provide

printed circuit motor

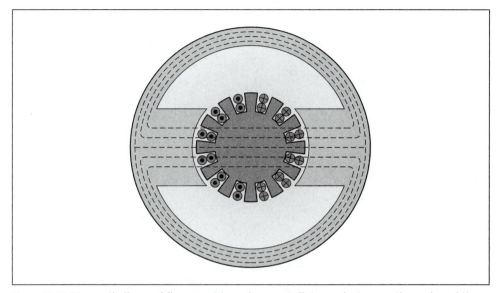

Figure 15-32 Magnetic lines of flux must travel a great distance between the poles of the permanent magnets.

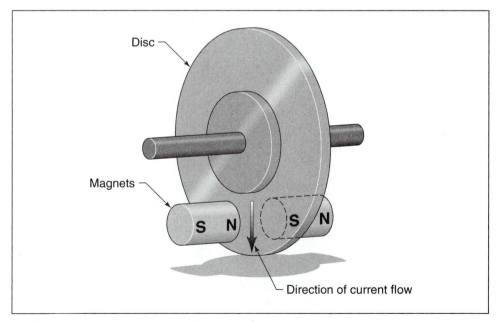

Figure 15-33 Basic construction of a ServoDisc® motor.

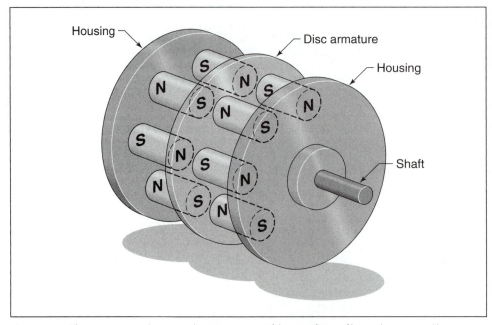

Figure 15-34 The permanent magnets are arranged to produce alternate magnetic polarities.

alternate magnetic fields *(Figure 15-34)*. The conductors on each side of the armature, upper and lower, are arranged in such a manner that when current flows through them, a force tangent to the pole piece is produced. This tangential force is the vector sum of the forces produced by the upper and lower conductors *(Figure 15-35)*. The ServoDisc® motor produces a relatively strong torque for its size and weight.

The conductors of the armature are so arranged on the fiberglass disc that they form a commutator on one side of the armature. Brushes riding against this commutator supply direct current to the armature conductors. Since the armature contains no iron, it has almost no inductance. This greatly reduces any arcing at the brushes, which results in extremely long brush life. A typical ServoDisc® motor is shown in *Figure 15-36*.

Characteristics of ServoDisc® Motors

The unique construction of the disc type servomotor gives it some operating characteristics that are different from those of other types of permanent magnetic DC motors. Permanent magnet direct current motors with iron armatures have a problem with cogging at low speeds. Cogging is caused by the reaction of the iron armature with the field of the permanent magnets. If

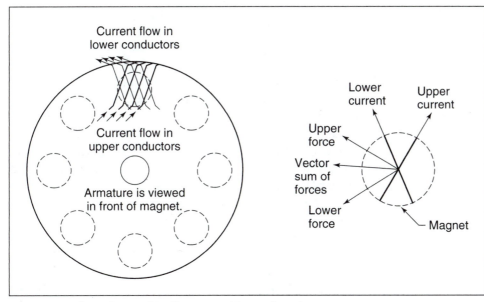

Figure 15-35 A force tangent to the magnet is produced.

Figure 15-36 DC ServoDisc® motors (courtesy of PMI Motion Technologies).

the armature is turned by hand, it will jump, or "cog," from one position to another. As the armature is turned, the magnetic flux lines of the pole pieces pass through the core of the armature *(Figure 15-32)*. Since the armature is slotted, certain areas will offer a path of lower reluctance than others. These areas are more strongly attracted to the pole pieces, which causes the armature to jump from one position to the next. Since the armature of the disc motor contains no iron, there is no cogging action. This permits very smooth operation at low speeds.

Another characteristic of disc motors is extremely fast acceleration. The thin, low-inertia disc armature permits an exceptional torque-inertia ratio. A typical ServoDisc® motor can accelerate from 0 to 3000 RPM in about 60° of rotation. In other words, the motor can accelerate from 0 to 3000 RPM in one sixth of a revolution. The low-inertia armature also permits rapid stops and reversals. Disc servomotors can operate at speeds over 4000 RPM.

The speed of the disc servomotor can be varied by changing the amount of voltage supplied to the armature. The voltage is generally varied using **pulse-width modulation**. Most amplifiers for the disc servomotor produce a pulsating DC voltage at a frequency of about 20 kHz. The average voltage supplied to the armature is determined by the length of time the voltage is turned on as compared with the time it is turned off (pulse-width). At a frequency of 20 kHz, the pulses have a width of 50 μs (microseconds) (pulse-width = 1/frequency). Assume that the pulses have a peak voltage of 24 V and are turned on for 40 μs and off for 10 μs during each pulse *(Figure 15-37)*. In this example, 24 V is supplied to the motor for 80% of the time, producing an average voltage of 19.2 V (24 × 0.80 = 19.2). Now assume that the pulse-width is changed so that the on time is 10 μs and the off time is 40 μs *(Figure 15-38)*. A voltage of 24 V is now supplied to the motor 20% of the time, resulting in an average value of 4.8 V (24 × 0.20 = 4.8).

Because of the basic design of the motor, the ServoDisc® motor can be constructed in a very small, thin case *(Figure 15-39)*. This makes it very useful in some applications where other types of servomotors could not be used.

pulse-width modulation

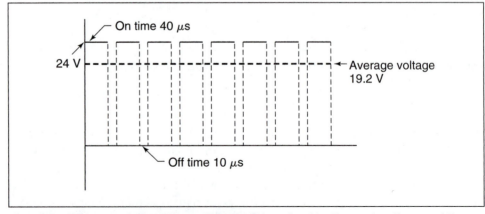

Figure 15-37 The amount of average voltage is determined by the peak voltage and the amount of time it is turned on or off.

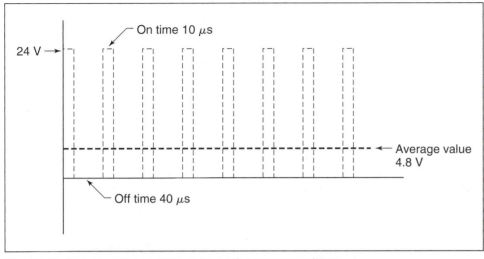

Figure 15-38 Reducing the on time reduces the average voltage.

Figure 15-39 The DC ServoDisc® motor can be constructed with small, thin cases (courtesy of PMI Motion Technologies).

RIGHT-HAND MOTOR RULE

In Unit 14 it was shown that it is possible to determine the direction of current flow through the armature of a generator by using the fingers of the left hand when the polarity of the field poles and the direction of rotation of the

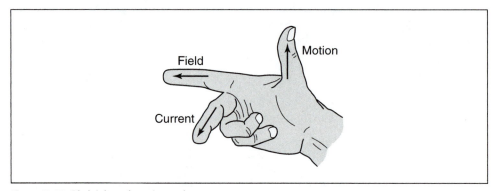

Figure 15-40 Right-hand motor rule.

armature are known. Similarly, the fingers of the right hand can be used to determine the direction of rotation of the armature when the magnetic field polarity of the pole pieces and the direction of current flow through the armature are known *(Figure 15-40)*. The thumb indicates the direction of thrust or movement of the armature. The forefinger indicates the direction of the field flux assuming that flux lines are in a direction of north to south, and the center finger indicates the direction of current flow through the armature. A simple method of remembering which finger represents which quantity is:

THumb = THrust (direction of armature rotation)

Forefinger = Field (direction of magnetic field)

Center finger = Current (direction of armature current)

Summary

1. A motor is a machine that converts electrical energy into mechanical energy.

2. Direct current motors operate on the principle of attraction and repulsion of magnetism.

3. Two factors that determine the torque produced by a motor are:
   A. Strength of the magnetic field of the pole pieces
   B. Strength of the magnetic field of the armature

4. Torque is proportional to armature current if field excitation current remains constant.

5. Current flow through the armature is limited by counter-EMF and armature resistance.

6. Three factors that determine the amount of counter-EMF produced by a motor are:
   A. Strength of the magnetic field of the pole pieces
   B. Number of turns of wire in the armature
   C. Speed of the armature

7. Three basic types of DC motors are series, shunt, and compound.

8. Shunt motors are sometimes known as constant speed motors.

9. Series motors must never be operated under a no-load condition.

10. Series motors can develop extremely high starting torque.

11. Compound motors contain both series and shunt field windings.

12. When full voltage is applied to both the armature and shunt field, the motor will operate at base speed.

13. When full voltage is applied to the field and reduced voltage is applied to the armature, the motor will operate below base speed.

14. When full voltage is applied to the armature and reduced voltage is applied to the shunt field, the motor will operate above base speed.

15. The direction of rotation of a direct current motor can be changed by reversing the connection of either the armature or the field leads.

16. It is common practice to reverse the connection of the armature leads to prevent changing a compound motor from a cumulative- to a differential-compound connection.

17. The shunt field relay is used to disconnect power to the armature if shunt field current drops below a certain level.

18. Brushless DC motors do not contain a wound armature, commutator, or brushes.

19. The stator windings for brushless motors are generally three-phase, but can be four-phase or two-phase.

20. Brushless DC motors require a converter to change the direct current into multiphase alternating current to operate the motor.

21. Brushless motors generally require less maintenance and have less down-time than conventional DC motors.

22. Permanent magnet motors use permanent magnets as the pole pieces and do not require series or shunt field windings.

23. Permanent magnet motors are more efficient than conventional DC motors.

24. The operating characteristics of a PM motor are similar to those of a shunt motor with external excitation.

25. Servomotors have lightweight armatures with low inertia.

26. Disc servomotors contain an armature that is made of several layers of copper conductors.

27. Since there is no iron in the armature of a disc servomotor, it does not have a problem with cogging.

28. Disc servomotors can be accelerated very rapidly.

Review Questions

1. What is the principle of operation of a direct current motor?

2. What is a motor?

3. What is the function of the commutator in a direct current motor?

4. What two factors determine the amount of torque developed by a DC motor?

5. What type of motor is known as a constant speed motor?

6. What is counter-EMF?

7. What three factors determine the amount of counter-EMF produced in the armature?

8. What limits the amount of current flow through the armature when power is first applied to the motor?

9. What factor determines the speed regulation of a DC motor?

10. What type of motor should never be operated at no load?

11. In general, what type of compound motor connection should be avoided?

12. What is the most common method of changing the direction of rotation of a compound motor?

13. How can a DC motor be made to operate at its base speed?

14. How can a DC motor be made to operate above its base speed?

15. What device is used to disconnect power to the armature if the shunt field current drops below a certain level?

16. Why do many industries leave power connected to the shunt field at all times even when the motor is not operating?

17. Who was the first person to establish a measurement for horsepower?

18. One horsepower is equal to how many watts?

19. A motor is operating a load that requires a torque of 750 lb–in and is turning at a speed of 1575 RPM. How much horsepower is the motor producing?

20. The motor in question 19 is connected to a 250-V DC line and has a current draw of 80 A. What is the efficiency of this motor?

UNIT **16**

Alternators

Objectives

After studying this unit, you should be able to:

- Discuss the operation of a three-phase alternator.
- Explain the effect of speed of rotation on frequency.
- Explain the effect of field excitation on output voltage.
- Connect a three-phase alternator and make measurements using test instruments.

Most of the electrical power in the world today is produced by alternating current generators or alternators. Electrical power companies use alternators rated in gigawatts (1 gigawatt = 1,000,000,000 W) to produce the power used throughout the United States and Canada. The entire North American continent is powered by alternating current generators connected together in parallel. These alternators are powered by steam turbines. The turbines, called prime movers, are powered by oil, coal, natural gas, or nuclear energy.

THREE-PHASE ALTERNATORS

alternators   **Alternators** operate on the same principle of electromagnetic induction as direct current generators, but they have no commutator to change the alternating current produced in the armature into direct current. There are two basic types of alternators: revolving armature type and revolving field type. Although there are some single-phase alternators that are used as

322

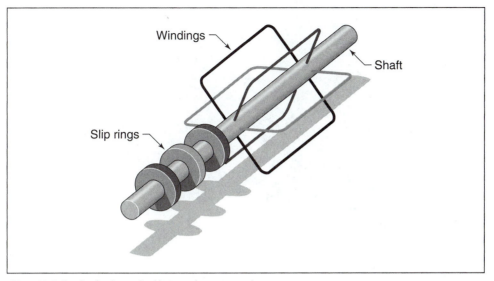

Figure 16-1 Basic design of a three-phase armature.

portable power units for emergency home use or to operate power tools in a remote location, most alternators are three-phase.

Revolving Armature Type Alternators

The **revolving armature** type alternator is the least used of the two basic types. This alternator uses an armature similar to that of a direct current machine with the exception that the loops of wire are connected to **slip rings** instead of to a commutator *(Figure 16-1)*. There are three separate windings, which are connected in either delta or wye. The armature windings are rotated inside a magnetic field *(Figure 16-2)*. Power is carried to the outside circuit via brushes riding against the slip rings. This alternator is the least used because it is very limited in the amount of output voltage and kilovolt-amp (kVA) capacity it can develop.

revolving armature

slip rings

Revolving Field Type Alternators

The **revolving field** type alternator uses a stationary armature called the **stator** and a rotating magnetic field. This design permits higher voltage and kVA ratings because the outside circuit is connected directly to the stator and is not routed through slip rings and brushes. This type of alternator is constructed by placing three sets of windings 120° apart *(Figure 16-3)*. In *Figure 16-3*, the winding of phase 1 winds around the top center pole piece. It then proceeds 180° around the stator and winds around the opposite pole piece in the opposite direction. The second phase winding winds around the

revolving field

stator

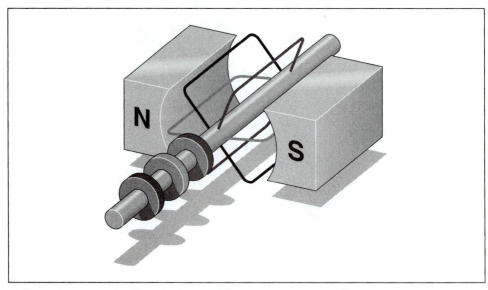

Figure 16-2 The armature conductors rotate inside a magnetic field.

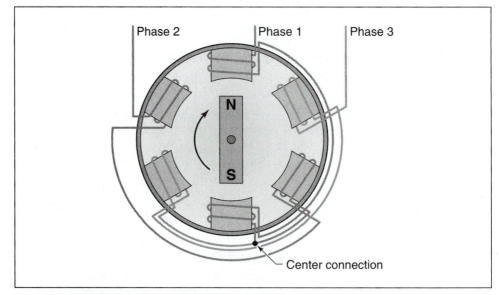

Figure 16-3 Basic design of a three-phase alternator.

top pole piece directly to the left of the top center pole piece. The second phase winding is wound in an opposite direction to the first. It then proceeds 180° around the stator housing and winds around the opposite pole piece in the opposite direction. The finish end of phase 2 connects to the finish end of phase 1. The start end of phase 3 winds around the top pole piece

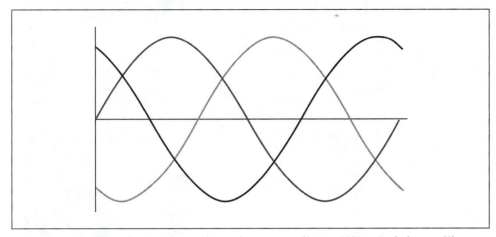

Figure 16-4 The alternator produces three sine wave voltages 120° out of phase with each other.

to the right of the top center pole piece. This winding is wound in an opposite direction than phase 1 also. The winding then proceeds 180° around the stator frame to its opposite pole piece and winds around it in an opposite direction. The finish end of phase 3 is then connected to the finish ends of phases 1 and 2. This forms a wye connection for the stator winding. When the magnet is rotated, voltage will be induced in the three windings. Since these windings are spaced 120° apart, the induced voltages will be 120° out of phase with each other *(Figure 16-4)*.

The stator shown in *Figure 16-3* is drawn in a manner to aid in understanding how the three phase windings are arranged and connected. In actual practice, the stator windings are placed in a smooth cylindrical core without projecting pole pieces *(Figure 16-5)*. This design provides a better path for magnetic lines of flux and increases the efficiency of the alternator.

THE ROTOR

The **rotor** is the rotating member of the machine. It provides the magnetism needed to induce voltage into the stator windings. The magnets of the rotor are electromagnets and require some source of external direct current to excite the alternator. This direct current is known as excitation current. The alternator cannot produce an output voltage until the rotor has been excited. Some alternators use slip rings and brushes to provide the excitation current to the rotor *(Figure 16-6)*. A good example of this type of rotor can be found in the alternator of most automobiles. The DC excitation current can be varied in order to change the strength of the magnetic field. A rotor with salient (projecting) poles is shown in *Figure 16-7.*

rotor

Figure 16-5 Wound stator (courtesy of Magnatek).

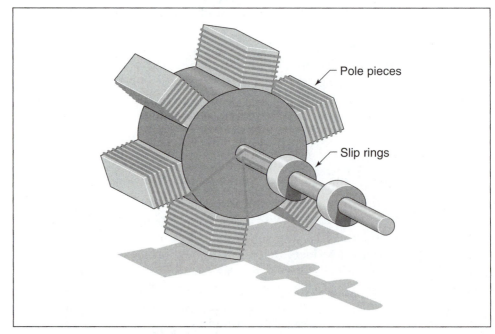

Figure 16-6 The rotor contains pole pieces that become electromagnets.

Figure 16-7 Rotor of the salient pole type.

THE BRUSHLESS EXCITER

Most large alternators use an exciter that contains no brushes. This is accomplished by adding a separate small alternator of the armature type on the same shaft of the rotor of the larger alternator. The armature rotates between wound electromagnets. The DC excitation current is connected to the wound stationary magnets *(Figure 16-8)*. The amount of voltage

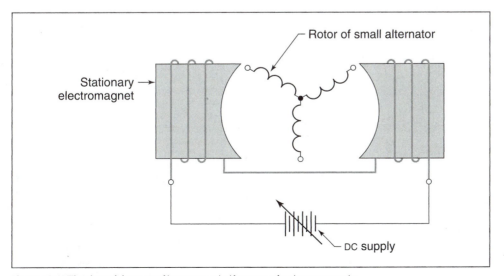

Figure 16-8 The brushless exciter uses stationary electromagnets.

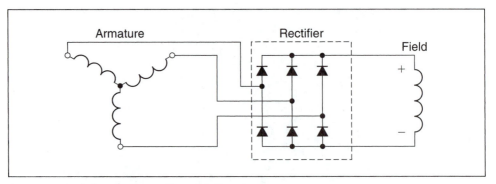

Figure 16-9 Basic brushless exciter circuit.

induced in the rotor can be varied by changing the amount of excitation current supplied to the electromagnets. The output voltage of the armature is connected to a three-phase bridge rectifier mounted on the rotor shaft *(Figure 16-9)*. The bridge rectifier converts the three-phase AC voltage produced in the armature into DC voltage before it is applied to the main rotor windings. Since the armature, rectifier, and rotor winding are connected to the main rotor shaft, they all rotate together and no brushes or slip rings are needed to provide excitation current for the large alternator. A photograph of the **brushless exciter** assembly is shown in *Figure 16-10*. The field winding is placed in slots cut in the core material of the rotor *(Figure 16-11)*.

brushless exciter

ALTERNATOR COOLING

cooling

There are two main methods of **cooling** alternators. Alternators of small kVA rating are generally air-cooled. Open spaces are left in the stator windings, and slots are often provided in the core material for the passage of air. Air-cooled alternators have a fan attached to one end of the shaft that circulates air through the entire assembly.

hydrogen

Large-capacity alternators are often enclosed and operate in a **hydrogen** atmosphere. There are several advantages in using hydrogen. Hydrogen is less dense than air at the same pressure. The lower density reduces the windage loss of the spinning rotor. A second advantage in operating an alternator in a hydrogen atmosphere is that hydrogen has the ability to absorb and remove heat much faster than air. At a pressure of one atmosphere, hydrogen has a specific heat of approximately 3.42. The specific heat of air at a pressure of one atmosphere is approximately 0.238. This means that hydrogen has the ability to absorb approximately 13.8 times more heat than

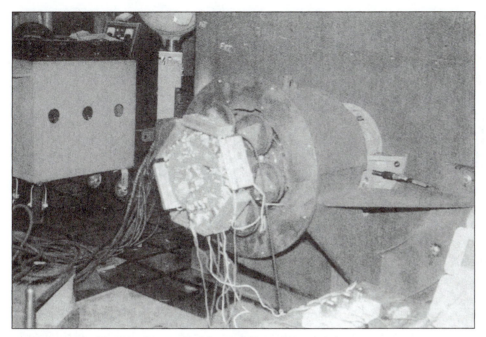

Figure 16-10 Brushless exciter assembly (courtesy of Magnatek).

Figure 16-11 Two-pole rotor slotting (courtesy of Reliant Energy).

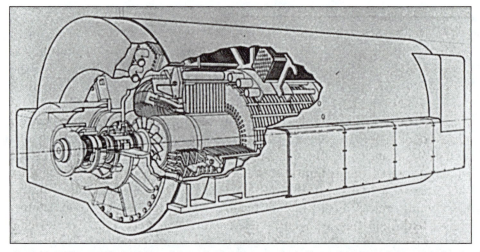

Figure 16-12 Two-pole turbine-driven hydrogen-cooled alternator (courtesy of Reliant Energy).

air. A cutaway drawing of an alternator intended to operate in a hydrogen atmosphere is shown in *Figure 16-12*.

FREQUENCY

frequency

The output **frequency** of an alternator is determined by two factors:
- Number of stator poles
- Speed of rotation of the rotor

Since the number of stator poles is constant for a particular machine, the output frequency is controlled by adjusting the speed of the rotor. The following chart shows the speed of rotation needed to produce 60 Hz for alternators with different numbers of poles.

RPM	Stator Poles
3600	2
1800	4
1200	6
900	8

The following formula can also be used to determine the frequency when the poles and RPM (revolutions per minute) are known:

$$f = \frac{PS}{120}$$

where

f = frequency in hertz

P = number of pairs of poles (one north and one south)

S = speed in RPM

120 = a constant

Example 1

What is the output frequency of an alternator that contains six poles per phase and is turning at a speed of 1000 RPM?

Solution

$$f = \frac{6 \times 1000}{120}$$

$$f = 50 \text{ Hz}$$

OUTPUT VOLTAGE

Three factors determine the amount of output voltage of an alternator:
1. Length of the armature or stator conductors (number of turns)
2. Strength of the magnetic field of the rotor
3. Speed of rotation of the rotor

The following formula can be used to compute the amount of voltage induced in the stator winding:

$$E = \frac{BLv}{10^8}$$

where

10^8 = flux lines equal to 1 weber

E = induced voltage

B = flux density in gauss

L = length of the conductor

v = velocity

One of the factors that determines the amount of induced voltage is the length of the conductor. This factor is often stated as number of turns of wire in the stator because the voltage induced in each turn adds. Increasing the number of turns of wire has the same effect as increasing the length of one conductor.

Controlling Output Voltage

The number of turns of wire in the stator cannot be changed in a particular machine without rewinding the stator, and the speed of rotation is generally maintained at a certain level to provide a constant output frequency. Therefore, the output voltage is controlled by increasing or decreasing the strength of the magnetic field of the rotor. The magnetic field strength can be controlled by controlling the DC excitation current to the rotor.

PARALLELING ALTERNATORS

When one alternator cannot produce all the power that is required, it often becomes necessary to use more than one machine. When more than one alternator is to be used, they are connected in parallel with each other. Several conditions must be met before **parallel alternators** can be used:

1. The phases must be connected in such a manner that the phase rotation of all the machines is the same.
2. Phases A, B, and C of one machine must be in sequence with phases A, B, and C of the other machine. For example, phase A of alternator 1 must reach its positive peak value of voltage at the same time phase A of alternator 2 does *(Figure 16-13)*.
3. The output voltage of the two alternators should be the same.

Determining Phase Rotation

The most common method of detecting when the **phase rotation** (the direction of magnetic field rotation) of one alternator is matched to the phase rotation of the other is with the use of three lights *(Figure 16-14)*. In *Figure 16-14*, the two alternators that are to be paralleled are connected together through a synchronizing switch. A set of lamps acts as a resistive load between the two machines when the switch contacts are in the open position. The voltage developed across the lamps is proportional to the difference in voltage between the two alternators. The lamps are used to indicate two conditions:

- The lamps indicate when the phase rotation of one machine is matched to the phase rotation of the other. When both alternators are operating, both are producing a voltage. The lamps will blink on and off when the phase rotation of one machine is not synchronized to the phase rotation of the other machine. If all three lamps blink on and off at the same

parallel alternators

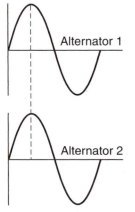

Alternator 1

Alternator 2

Figure 16-13 The voltages of both alternators must be in phase with each other.

phase rotation

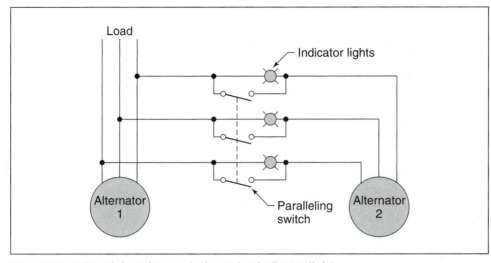

Figure 16-14 Determining phase rotation using indicator lights.

time, or in unison, the phase rotation of alternator 1 is correctly matched to the phase rotation of alternator 2. If the lamps blink on and off, but not in unison, the phase rotation between the two machines is not correctly matched, and two lines of alternator 2 should be switched.

- The lamps also indicate when the phase of one machine is synchronized with the phase of the other machine. If the positive peak of alternator 1 does not occur at the same time as the positive peak of alternator 2, there will be a potential difference between the two machines. This will permit the lamps to glow. The brightness of the lamps indicates how far out of synchronism the two machines are. When the peak voltages of the two alternators occur at the same time, there is no potential difference between them. The lamps should be off at this time. The synchronizing switch should never be closed when the lamps are glowing.

The Synchroscope

Another instrument often used for paralleling two alternators is the **synchroscope** *(Figure 16-15)*. The synchroscope measures the difference in voltage and frequency of the two alternators. The pointer of the synchroscope is free to rotate in a 360° arc. The alternator already connected to the load is considered to be the base machine. The synchroscope will indicate if the frequency of the alternator to be parallel to the base machine is fast or slow. When the voltages of the two alternators are in phase, the pointer will cover the shaded area on the face of the meter. When the two alternators are synchronized, the paralleling switch is closed.

Figure 16-15 Synchroscope.

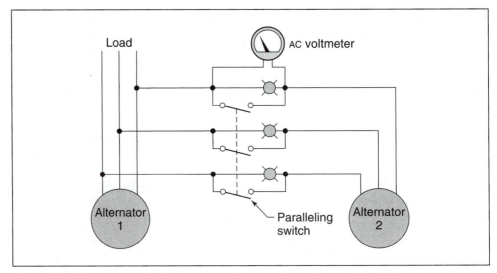

Figure 16-16 AC voltmeter indicates when the two alternators are in phase.

If a synchroscope is not available, the two alternators can be paralleled using three lamps as described earlier. If the three-lamp method is used, an AC voltmeter connected across the same phase of each machine will indicate when the potential difference between the two machines is zero *(Figure 16-16)*. That is the point at which the paralleling switch should be closed.

SHARING THE LOAD

After the alternators have been paralleled, the power input to alternator 2 must be increased to permit it to share part of the load. For example, if the alternator is being driven by a steam turbine, the power of the turbine would have to be increased. When this is done, the power to the load will remain constant. The power output of the base alternator will decrease, and the power output of the second alternator will increase.

FIELD DISCHARGE PROTECTION

field discharge resistor

When the DC excitation current is disconnected, the collapsing magnetic field can induce a high voltage in the rotor winding. This voltage can be high enough to arc contacts and damage the rotor winding or other circuit components. One method of preventing the induced voltage from becoming excessive is with the use of a **field discharge resistor**. A special double-pole, single-throw switch with a separate blade is used to connect the resistor to the field before the switch contacts open. When the switch is closed and direct current is connected to the field, the circuit connecting the resistor to the field

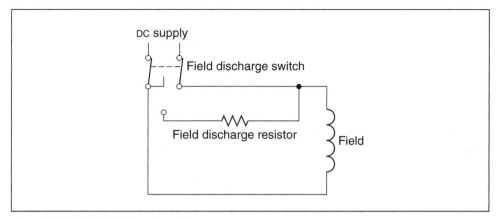

Figure 16-17 Switch in closed position.

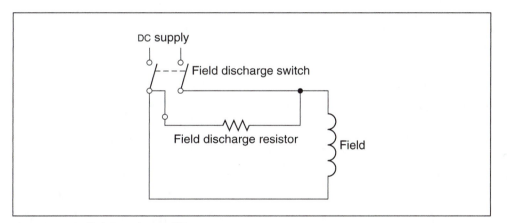

Figure 16-18 Switch in open position.

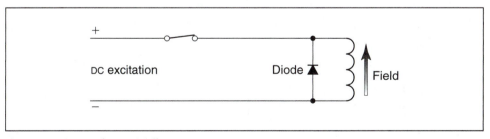

Figure 16-19 Normal current flow.

is open *(Figure 16-17)*. When the switch is opened, the special blade connects the resistor to the field before the main contacts open *(Figure 16-18)*.

Another method of preventing the high voltage discharge is to connect a diode in parallel with the field *(Figure 16-19)*. The diode is connected in

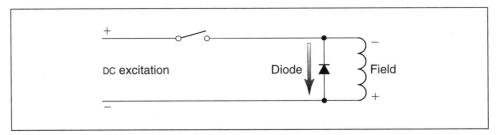

Figure 16-20 Induced current flow.

such a manner that when excitation current is flowing, the diode is reverse-biased and no current flows through the diode.

When the switch opens and the magnetic field collapses, the induced voltage is opposite in polarity to the applied voltage *(Figure 16-20)*. The diode in now forward-biased, permitting current to flow through the diode. The energy contained in the magnetic field is dissipated in the form of heat by the diode and field winding.

Summary

1. There are two basic types of three-phase alternators: rotating armature type and rotating field type.

2. The rotating armature type is the least used because of its limited voltage and power rating.

3. The rotor of the rotating field type alternator contains electromagnets.

4. Direct current must be supplied to the field before the alternator can produce an output voltage.

5. The direct current supplied to the field is called excitation current.

6. The output frequency of an alternator is determined by the number of stator poles and the speed of rotation.

7. Three factors that determine the output voltage of an alternator are:
   A. Length of the conductor of the armature or stator winding
   B. Strength of the magnetic field of the rotor
   C. Speed of the rotor

8. The output voltage is controlled by the amount of DC excitation current.

9. Before two alternators can be connected in parallel, the output voltage of the two machines should be the same, the phase rotation of the machines must be the same, and the output voltages of the two machines must be in phase.

10. Three lamps connected between the two alternators can be used to test for phase rotation.

11. A synchroscope can be used to determine phase rotation and difference of frequency between two alternators.

12. Two devices used to prevent a high voltage being induced in the rotor when the DC excitation current is stopped are a field discharge resistor and a diode.

13. Many large alternators use a brushless exciter to supply direct current to the rotor winding.

Review Questions

1. What conditions must be met before two alternators can be paralleled together?

2. How can the phase rotation of one alternator be changed in relationship to the other alternator?

3. What is the function of the synchronizing lamps?

4. What is a synchroscope?

5. Assume that alternator A is supplying power to a load and that alternator B is to be paralleled to A. After the paralleling has been completed, what must be done to permit alternator B to share the load with alternator A?

6. What two factors determine the output frequency of an alternator?

7. At what speed must a six-pole alternator turn to produce 60 Hz?

8. What three factors determine the output voltage of an alternator?

9. What are slip rings used for on a rotating field type alternator?

10. Is the rotor excitation current AC or DC?

11. When a brushless exciter is used, what converts the alternating current produced in the armature winding into direct current before it is supplied to the field winding?

12. What two devices are used to eliminate the induced voltage produced in the rotor when the field excitation current is stopped?

UNIT 17

Three-Phase Motors

Objectives

After studying this unit, you should be able to:

- Discuss the basic operating principles of three-phase motors.
- List factors that produce a rotating magnetic field.
- List different types of three-phase motors.
- Discuss the operating principles of squirrel-cage motors.
- Connect dual-voltage motors for proper operation on the desired voltage.
- Discuss the operation of consequent pole motors.
- Discuss the operation of wound rotor motors.
- Discuss the operation of synchronous motors.

Three-phase motors are used as the prime movers for industry throughout the United States and Canada. These motors convert the three-phase alternating current into mechanical energy to operate all types of machinery. Three-phase motors are smaller in size, lighter in weight, and have higher efficiencies per horsepower than single-phase motors. They are extremely rugged and require very little maintenance. Many of these motors operate 24 hours a day, seven days a week for many years without a problem.

THREE-PHASE MOTORS

There are three basic types of three-phase motors:
1. Squirrel-cage induction motor
2. Wound rotor induction motor
3. Synchronous motor

All three motors operate on the same principle, and they all use the same basic design for the stator windings. The difference between these motors is the type of rotor used. Two of the three motors are induction motors and operate on the principle of electromagnetic induction in a manner similar to transformers. In fact, AC induction motors were patented as **rotating transformers** by Nikola Tesla. The stator winding of a motor is often referred to as the motor primary, and the rotor is referred to as the motor secondary.

rotating transformers

THE ROTATING MAGNETIC FIELD

The principle of operation for all three-phase motors is the **rotating magnetic field**. There are three factors that cause the magnetic field to rotate:
1. The voltages in a three-phase system are 120° out of phase with each other.
2. The three voltages change polarity at regular intervals.
3. The arrangement of the stator windings around the inside of the motor.

rotating magnetic field

SYNCHRONOUS SPEED

The speed at which the magnetic field rotates is called the **synchronous speed**. Two factors that determine the speed of the rotating magnetic field are:
1. Number of stator poles (per phase)
2. Frequency of the applied voltage

The following chart shows the synchronous speed at 60 Hz for different numbers of stator poles.

synchronous speed

RPM	Stator Poles
3600	2
1800	4
1200	6
900	8

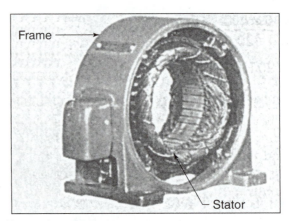

Figure 17-1 Stator of a three-phase motor.

The stator winding of a three-phase motor is shown in *Figure 17-1*. The synchronous speed can be calculated using the formula

$$S = \frac{120\ f}{P}$$

where

S = speed in RPM

f = frequency in Hz

P = number of stator poles (per phase)

Example 1

What is the synchronous speed of a four-pole motor connected to 50 Hz?

Solution

$$S = \frac{120 \times 50}{4}$$

$$S = 1500\ \text{RPM}$$

Example 2

What frequency should be applied to a six-pole motor to produce a synchronous speed of 400 RPM?

Solution

First, change the base formula to find frequency. Once that is done, known values can be substituted in the formula.

$$f = \frac{PS}{120}$$

$$f = \frac{6 \times 400}{120}$$

$$f = 20 \text{ Hz}$$

Direction of Rotation for Three-Phase Motors

On many types of machinery, the direction of rotation of the motor is critical. *The direction of rotation of any three-phase motor can be changed by reversing two of its stator leads.* This causes the direction of the rotating magnetic field to reverse. When a motor is connected to a machine that will not be damaged when its direction of rotation is reversed, power can be momentarily applied to the motor to observe its direction of rotation. If the rotation is incorrect, any two line leads can be interchanged to reverse the motor's rotation.

When a motor is to be connected to a machine that can be damaged by incorrect rotation, however, the direction of rotation must be determined before the motor is connected to its load. This can be accomplished in two basic ways. One way is to make electrical connection to the motor before it is mechanically connected to the load. The **direction of rotation** can then be tested by momentarily applying power to the motor before it is coupled to the load.

direction of rotation

There may be occasions when this is not practical or is very inconvenient. It is possible to determine the direction of rotation of a motor before power is connected to it with the use of a **phase rotation meter**. The phase rotation meter is used to compare the phase rotation of two different three-phase connections. The meter contains six terminal leads. Three of the leads are connected to one side of the meter and labeled MOTOR. These three motor leads are labeled A, B, or C. The LINE leads are located on the other side of the meter, and are labeled A, B, or C.

phase rotation meter

To determine the direction of rotation of the motor, first zero the meter by following the instructions provided by the manufacturer. Then set the meter selector switch to MOTOR and connect the three MOTOR leads of the meter to the T leads of the motor as shown in *Figure 17-2*. The phase rotation meter contains a zero center voltmeter. One side of the voltmeter is labeled INCORRECT, and the other side is labeled CORRECT. While observing the

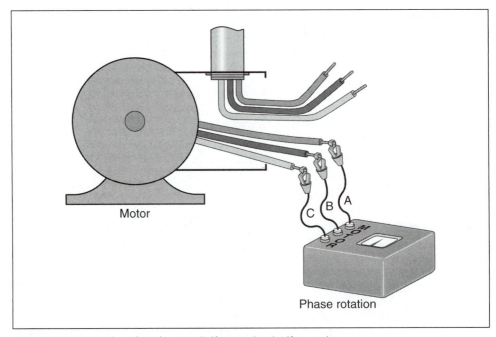

Figure 17-2 Connecting the phase rotation meter to the motor.

zero center voltmeter, manually turn the motor shaft in the direction of desired rotation. The zero center voltmeter will immediately swing in the CORRECT or INCORRECT direction. When the motor shaft stops turning, the needle may swing in the opposite direction. Use the first indication of the voltmeter.

If the voltmeter needle indicated CORRECT, label the motor T leads A, B, or C to correspond with the MOTOR leads from the phase rotation meter. If the voltmeter needle indicated INCORRECT, change any two of the MOTOR leads from the phase rotation meter and again turn the motor shaft. The voltmeter needle should now indicate CORRECT. The motor T leads can now be labeled to correspond with the MOTOR leads from the phase rotation meter.

After the motor T leads have been labeled A, B, or C to correspond with the leads of the phase rotation meter, the rotation of the line supplying power to the motor must be determined. Set the selector switch on the phase rotation meter to the LINE position. After making certain that the power has been turned off, connect the three LINE leads of the phase rotation meter to the incoming power line *(Figure 17-3)*. Turn on the power and observe the zero center voltmeter. If the meter is pointing in the CORRECT direction, turn off the power and label the line leads A, B, or C to correspond with the LINE leads of the phase rotation meter.

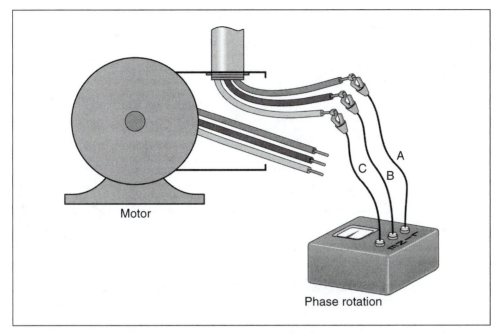

Figure 17-3 Connecting the phase rotation meter to the line.

If the voltmeter is pointing in the INCORRECT direction, turn off the power and change any two of the leads from the phase rotation meter. When the power is turned on, the voltmeter should point in the CORRECT direction. Turn off the power and label the line leads A, B, or C to correspond with the leads from the phase rotation meter.

Now that the motor T leads and the incoming power leads have been labeled, connect the line lead labeled A to the T lead labeled A, the line lead labeled B to the T lead labeled B, and the line lead labeled C to the T lead labeled C. When power is connected to the motor, it will operate in the proper direction.

CONNECTING DUAL-VOLTAGE THREE-PHASE MOTORS

Many of the three-phase motors used in industry are designed to be operated on two voltages, such as 240 or 480 V. Motors of this type contain two sets of windings per phase. Most dual-voltage motors bring out nine T leads at the terminal box. There is a standard method used to number these leads, as shown in *Figure 17-4*. Starting with terminal 1, the leads are numbered in a decreasing spiral as shown. Another method of determining the proper lead numbers is to add three to each terminal. For example, starting with

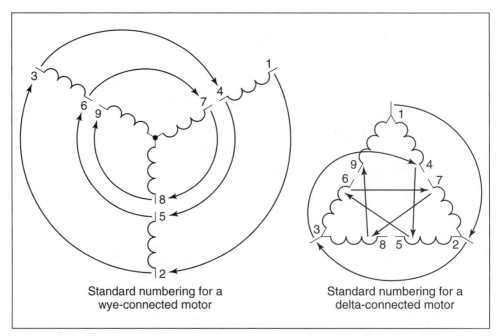

Standard numbering for a
wye-connected motor

Standard numbering for a
delta-connected motor

Figure 17-4 Standard numbering for three-phase motors.

lead 1, add three to one. Three plus one equals four. The phase winding that begins with 1 ends with 4. Now add three to four. Three plus four equals seven. The beginning of the second winding for phase one is seven. This method will work for the windings of all phases. If in doubt, draw a diagram of the phase windings and number them in a spiral.

High-Voltage Connections

Three-phase motors can be constructed to operate in either wye or delta. If a motor is to be connected to high voltage, the phase windings will be connected in series. In *Figure 17-5*, a schematic diagram and terminal connection chart for high voltage are shown for a wye-connected motor. In *Figure 17-6*, a schematic diagram and terminal connection chart for high voltage are shown for a delta-connected motor. Notice that in both cases the windings are connected in series.

Low-Voltage Connections

When a motor is to be connected for low-voltage operation, the phase windings must connect in parallel. *Figure 17-7* shows the basic schematic diagram for a wye-connected motor with parallel phase windings. In actual practice, however, it is not possible to make this exact connection with a

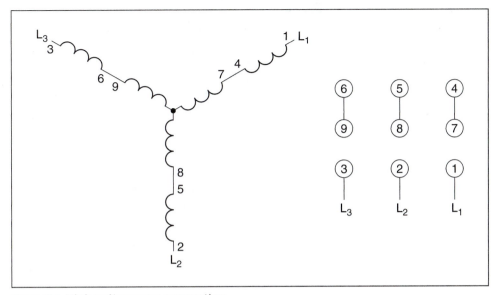

Figure 17-5 High-voltage wye connection.

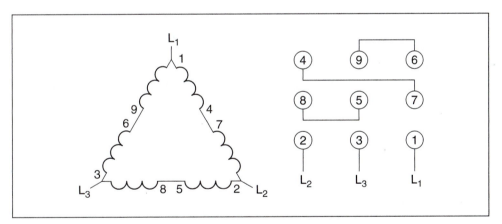

Figure 17-6 High-voltage delta connection.

nine-lead motor. The schematic shows that terminal 4 connects to the other end of the phase winding that starts with terminal 7. Terminal 5 connects to the other end of winding 8, and terminal 6 connects to the other end of winding 9. In actual motor construction, the opposite ends of windings 7, 8, and 9 are connected together inside the motor and are not brought outside the motor case. The problem is solved, however, by forming a second wye connection by connecting terminals 4, 5, and 6 as shown in *Figure 17-8.*

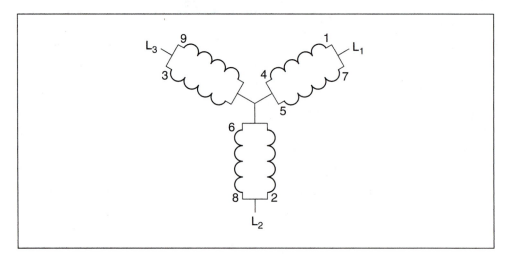

Figure 17-7 Stator windings connected in parallel.

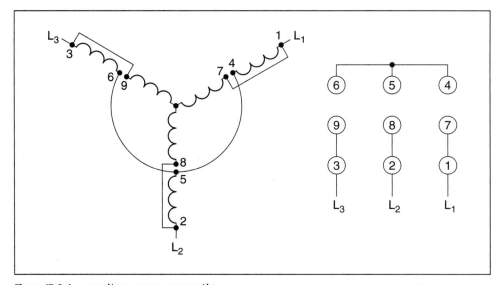

Figure 17-8 Low-voltage wye connection.

The phase windings of a delta-connected motor must also be connected in parallel for use on low voltage. A schematic for this connection is shown in *Figure 17-9*. A connection diagram and terminal connection chart for this hookup is shown in *Figure 17-10*.

Some dual-voltage motors will contain twelve T leads instead of nine. In this instance, the opposite ends of terminals 7, 8, and 9 are brought out for connection. *Figure 17-11* shows the standard numbering for both delta and

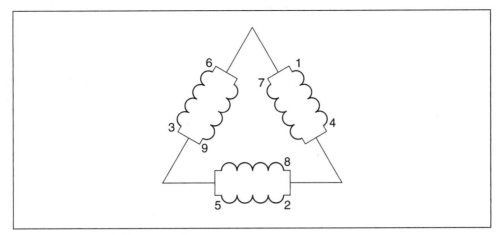

Figure 17-9 Parallel delta connection.

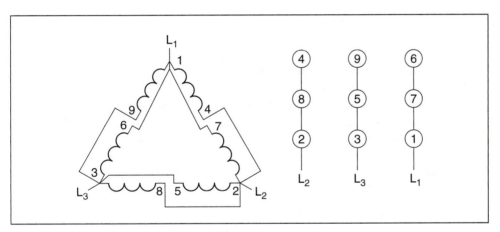

Figure 17-10 Low-voltage delta connection.

wye-connected motors. Twelve leads are brought out if the motor is intended to be used for wye-delta starting. When this is the case, the motor must be designed for normal operation with its windings connected in delta. If the windings are connected in wye during starting, the starting current of the motor is reduced to one-third of what it will be if the motor is started as a delta.

Voltage and Current Relationships for Dual-Voltage Motors

When a motor is connected to the higher voltage, the current flow will be half as much as when it is connected for low-voltage operation. The reason is that when the windings are connected in series for high-voltage operation,

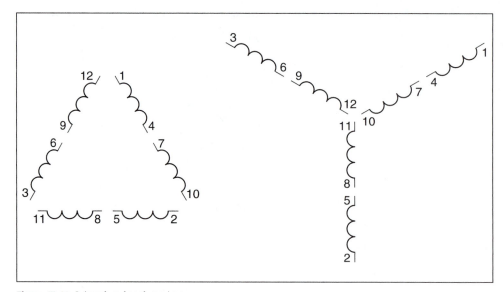

Figure 17-11 A twelve-lead motor.

the impedance will be four times greater than when the windings are connected for low-voltage operation. For example, assume that a dual-voltage motor is intended to operate on 480 or 240 V. Also assume that during full load, the motor's windings exhibit an impedance of 10 Ω each. When the winding is connected in series, *Figure 17-12*, the impedance per phase will be 20 Ω (10 + 10 = 20).

If a voltage of 480 V is connected to the motor, the phase voltage will be:

$$E_{Phase} = \frac{E_{Line}}{1.732}$$

$$E_{Phase} = \frac{480}{1.732}$$

$$E_{Phase} = 277\ V$$

The amount of current flow through the phase can be computed using Ohm's law.

$$I = \frac{E}{Z}$$

$$I = \frac{277}{20}$$

$$I = 13.85\ A$$

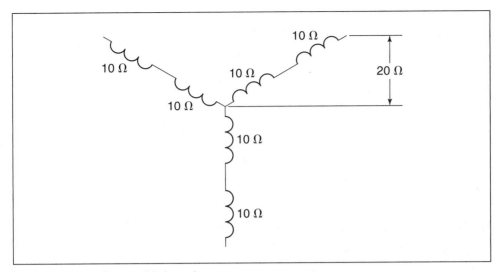

Figure 17-12 Impedance adds in series.

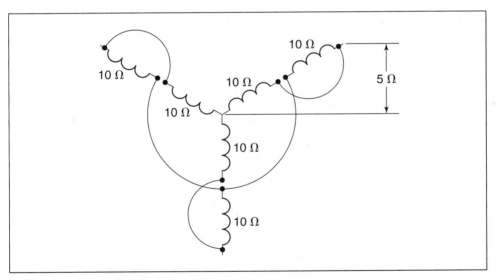

Figure 17-13 Impedance is less in parallel.

If the stator windings are connected in parallel, the total impedance will be found by adding the reciprocals of the impedances of the windings (*Figure 17-13*).

$$Z_T = \frac{1}{\dfrac{1}{Z_1} + \dfrac{1}{Z_2}}$$

$$Z_T = 5\ \Omega$$

If a voltage of 240 V is connected to the motor, the voltage applied across each phase will be 138.6 V (240/1.732 = 138.6). The amount of phase current can now be computed using Ohm's law.

$$I = \frac{E}{Z}$$

$$I = \frac{138.6}{5}$$

$$I = 27.12 \text{ A}$$

SQUIRREL-CAGE INDUCTION MOTORS

squirrel-cage motor

The **squirrel-cage motor** receives its name from the type of rotor used in the motor. A squirrel-cage rotor is made by connecting bars to two end rings. If the metal laminations were to be removed from the rotor, the result would look very similar to a squirrel cage, *(Figure 17-14)*. A squirrel cage is a cylindrical device constructed of heavy wire. A shaft is placed through the center of the cage. This permits the cage to spin around the shaft. A squirrel cage is placed inside the cage of small pets, such as squirrels and hamsters, to permit them to exercise by running inside of the squirrel cage. A cutaway view of a squirrel-cage motor is shown in *Figure 17-15*. A cutaway view of a squirrel-cage rotor is shown in *Figure 17-16*. The major parts of a squirrel-cage motor are shown in *Figure 17-17*.

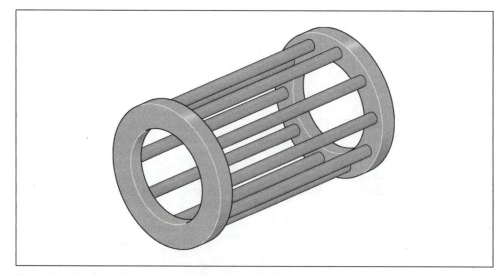

Figure 17-14 Basic squirrel-cage rotor without laminations.

Figure 17-15 Cutaway view of a squirrel-cage motor.

Figure 17-17 Squirrel-cage induction motor frame, stator winding, and rotor.

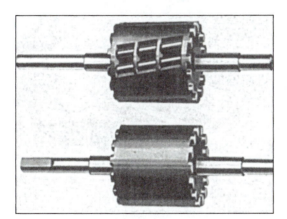

Figure 17-16 Cutaway view of a squirrel-cage rotor.

Principles of Operation

The squirrel-cage motor is an induction motor. This means that the current flow in the rotor is produced by induced voltage from the rotating magnetic field of the stator. In *Figure 17-18*, a squirrel-cage rotor is shown inside the stator of a three-phase motor. It will be assumed that the motor shown in *Figure 17-18* contains four poles per phase, which produces a rotating magnetic field with a synchronous speed of 1800 RPM when the stator is connected to a 60-Hz line. When power is first connected to the stator, the rotor is not turning. The magnetic field of the stator cuts the rotor bars at a rate of 1800 RPM. This cutting action induces a voltage into the rotor bars. This induced voltage will be the same frequency as the voltage applied to the

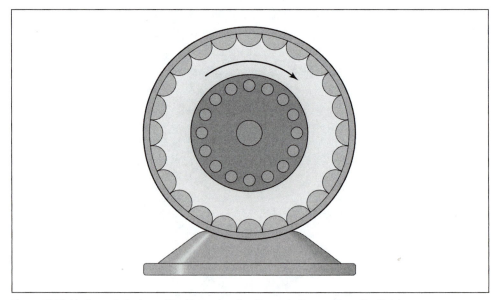

Figure 17-18 Voltage is induced in the rotor by the rotating magnetic field.

stator. The amount of induced voltage is determined by three factors:

1. Strength of the magnetic field of the stator
2. Number of turns of wire cut by the magnetic field (In the case of a squirrel-cage rotor this will be the number of bars in the rotor.)
3. Speed of the cutting action

Since the rotor is stationary at this time, maximum voltage is induced into the rotor. The induced voltage causes current to flow through the rotor bars. As current flows through the rotor, a magnetic field is produced around each bar *(Figure 17-19)*.

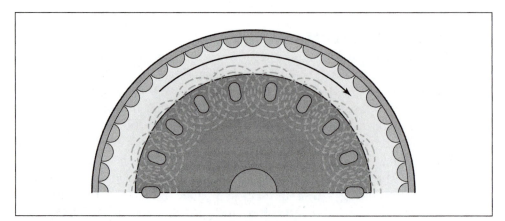

Figure 17-19 A magnetic field is produced around each rotor bar.

The magnetic field of the rotor is attracted to the magnetic field of the stator, and the rotor begins to turn in the same direction as the rotating magnetic field.

As the speed of the rotor increases, the rotating magnetic field cuts the rotor bars at a slower rate. For example, assume the rotor has accelerated to a speed of 600 RPM. The synchronous speed of the rotating magnetic field is 1800 RPM. Therefore, the rotor bars are now being cut at a rate of 1200 RPM instead of 1800 RPM (1800 RPM - 600 RPM = 1200 RPM). Since the rotor bars are being cut at a slower rate, less voltage is induced in the rotor, reducing rotor current. When the rotor current decreases, the stator current decreases also.

As the rotor continues to accelerate, the rotating magnetic field cuts the rotor bars at a decreasing rate. This reduces the amount of induced voltage and, therefore, the amount of rotor current. If the motor is operating without a load, the rotor will continue to accelerate until it reaches a speed close to that of the rotating magnetic field.

Torque

The amount of torque produced by an AC induction motor is determined by three factors:

1. Strength of the magnetic field of the stator
2. Strength of the magnetic field of the rotor
3. Phase angle difference between rotor and stator fields

$$T = K_T \times \varphi_S \times I_R \times COS\ \theta_R$$

where

T = torque in lb-ft

K_T = torque constant

φ_S = stator flux (constant at all speeds)

I_R = rotor current

$COS\ \theta_R$ = rotor power factor

Notice that one of the factors that determines the amount of torque produced by an induction motor is the strength of the magnetic field of the rotor. *An induction motor can never reach synchronous speed.* If the rotor were to turn at the same speed as the rotating magnetic field, there would be no induced voltage in the rotor and, consequently, no rotor current. Without rotor current there could be no magnetic field developed by the rotor and, therefore, no torque or turning force. A motor operating at no load will accelerate until the torque developed is proportional to the windage and bearing friction losses.

If a load is connected to the motor, it must furnish more torque to operate the load. This causes the motor to slow down. When the motor speed decreases, the rotating magnetic field cuts the rotor bars at a faster rate. This causes more voltage to be induced in the rotor and, therefore, more current. The increased current flow produces a stronger magnetic field in the rotor, which causes more torque to be produced. The increased current flow in the rotor causes increased current flow in the stator. This is why motor current will increase as load is added.

Another factor that determines the amount of torque developed by an induction motor is the phase angle difference between stator and rotor field flux. *Maximum torque is developed when the stator and rotor flux are in phase with each other.* Note in the preceding formula that one of the factors that determines the torque developed by an induction motor is the cosine of the rotor power factor. The cosine function reaches its maximum value of 1 when the phase angle is 0 (COS 0° = 1).

Starting Characteristics

When a squirrel-cage motor is first started, it will have a current draw several times greater than its normal running current. The actual amount of starting current is determined by the type of rotor bars, the horsepower rating of the motor, and the applied voltage. The type of rotor bars is indicated by the code letter found on the nameplate of a squirrel-cage motor. Table 430.7(B) of the *National Electrical Code®* can be used to compute the locked rotor current (starting current) of a squirrel-cage motor when the applied voltage, horsepower, and code letter are known. The table shown in *Figure 17-20* lists values for locked currents.

Example 3

An 800-hp three-phase squirrel-cage motor is connected to 2300 V. The motor has a code letter of J. What is the starting current of this motor?

Solution

The table in *Figure 17-20* lists a value of 7.1 to 7.99 kilovolt-amperes per horsepower as the locked rotor current of a motor with a code letter J. An average value of 7.5 will be used for this calculation. The apparent power can be computed by multiplying the 7.5 times the horsepower rating of the motor.

$$kVA = 7.5 \times 800$$

$$kVA = 6000$$

Code Letter	Kilovolt-Amperes per Horsepower with Locked Rotor
A	0 - 3.14
B	3.15 - 3.54
C	3.55 - 3.99
D	4.0 - 4.49
E	4.5 - 4.99
F	5.0 - 5.59
G	5.6 - 6.29
H	6.3 - 7.09
J	7.1 - 7.99
K	8.0 - 8.99
L	9.0 - 9.99
M	10.0 - 11.9
N	11.2 - 12.49
P	12.5 - 13.49
R	14.0 - 15.99
S	16.0 - 17.99
T	18.0 - 19.99
U	20.0 - 22.39
V	22.4 - and up

Figure 17-20 Table for determining starting current for squirrel-cage motors.

The line current supplying the motor can now be computed using the formula

$$I_{(Line)} = \frac{VA}{E_{(Line)} \times 1.732}$$

$$I_{(Line)} = \frac{6,000,000}{2300 \times 1.732}$$

$$I_{(Line)} = 1506.175 \text{ A}$$

This large starting current is caused by the fact that the rotor is not turning when power is first applied to the stator. Since the rotor is not turning, the squirrel-cage bars are cut by the rotating magnetic field at a fast rate. Remember that one of the factors that determines the amount of induced voltage is speed of the cutting action. This high induced voltage causes a large amount of current to flow in the rotor. The large current flow in the rotor causes a large

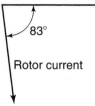

Figure 17-21 Rotor current is almost 90° out of phase with the induced voltage at the moment of starting.

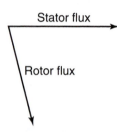

Figure 17-22 Rotor flux lags the stator flux by a large amount during starting.

percent slip

amount of current flow in the stator. Since a large amount of current flows in both the stator and rotor, a strong magnetic field is established in both.

The starting torque of a squirrel-cage motor is high since the magnetic field of both the stator and rotor are strong at this point. The starting torque of a typical squirrel-cage motor can range from 200% to 300% of its full-load rated torque. Although the starting torque is high, the squirrel-cage motor does not develop as much starting torque per ampere of starting current as other types of three-phase motors. Recall that the third factor for determining the torque developed by an induction motor is the difference in phase angle between stator flux and rotor flux. Since the rotor is being cut at a high rate of speed by the rotating stator field, the bars in the squirrel-cage rotor appear to be very inductive at this point because of the high frequency of the induced voltage. This causes the phase angle difference between the induced voltage in the rotor and rotor current to be almost 90° out of phase with each other, producing a lagging power factor for the rotor *(Figure 17-21)*. This causes the rotor flux to lag the stator flux by a large amount, and consequently a relatively weak starting torque per ampere of starting current, compared to other types of three-phase motors, is developed *(Figure 17-22)*. A typical torque curve for a squirrel-cage motor is shown in *Figure 17-23*.

Percent Slip

The speed performance of an induction motor is measured in **percent slip**. The percent slip can be determined by subtracting the synchronous

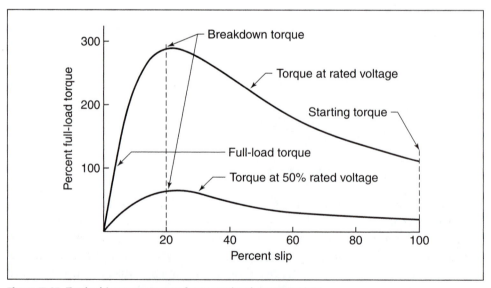

Figure 17-23 Typical torque curves for a squirrel-cage motor.

speed from the speed of the rotor. For example, assume that an induction motor has a synchronous speed of 1800 RPM and at full load the rotor turns at a speed of 1725 RPM. The difference between the two speeds is 75 RPM (1800 − 1725 = 75). The percent slip can be determined by using the formula

$$\text{percent slip} = \frac{\text{synchronous speed} - \text{rotor speed}}{\text{synchronous speed}} \times 100$$

$$\text{percent slip} = \frac{75}{1800} \times 100$$

$$\text{percent slip} = 4.16\%$$

A rotor slip of 2% to 5% is common for most squirrel-cage induction motors. The amount of slip for a particular motor is greatly affected by the type of rotor bars used in the construction of the rotor. Squirrel-cage motors are considered to be constant speed motors because there is a small difference between no-load speed and full-load speed.

Rotor Frequency

In the previous example, the rotor slips behind the rotating magnetic field by 75 RPM. This means that at full load, the bars of the rotor are being cut by magnetic lines of flux at a rate of 75 RPM. Therefore, the voltage being induced in the rotor at this point in time is at a much lower frequency than when the motor was started. The **rotor frequency** can be determined using the formula

rotor frequency

$$f = \frac{P \times S_R}{120}$$

where

f = frequency in Hz

P = number of stator poles

S_R = rotor slip in RPM

$$f = \frac{4 \times 75}{120}$$

$$f = 2.5 \text{ Hz}$$

Because the frequency of the current in the rotor decreases as the rotor approaches synchronous speed, the rotor bars become less inductive. The current flow through the rotor becomes limited more by the resistance of the

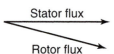

Figure 17-24 Rotor and stator flux become more in phase with each other as motor speed increases.

bars and less by inductive reactance. The current flow in the rotor becomes more in phase with the induced voltage, which causes less phase angle shift between stator and rotor flux *(Figure 17-24)*. This is the reason that squirrel-cage motors generally have a relatively poor starting torque per ampere of starting current as compared to other types of three-phase motors, but have a good running torque.

Reduced Voltage Starting

Due to the large amount of starting current for many squirrel-cage motors, it is sometimes necessary to reduce the voltage during the starting period. When the voltage is reduced, the starting torque is reduced also. If the applied voltage is reduced to 50% of its normal value, the magnetic fields of both the stator and rotor are reduced to 50% of normal. This causes the starting torque to be reduced to 25% of normal. A chart of typical torque curves for squirrel-cage motors is shown in *Figure 17-23*.

The torque formula given earlier can be used to show why this large reduction of torque occurs. Both the stator flux, φ_S, and the rotor current, I_R, are reduced to half of their normal value. The product of these two values, torque, is reduced to one-fourth. The torque varies as the square of the applied voltage for any given value of slip.

Code Letters

Squirrel-cage rotors are not all the same. Rotors are made with different types of bars. The type of rotor bars used in the construction of the rotor determines the operating characteristics of the motor. An AC squirrel-cage motor is given a code letter on its nameplate. The code letter indicates the type of bars used in the rotor. *Figure 17-25* shows a rotor with type A bars. A type A rotor has the highest resistance of any squirrel-cage rotor. This means that the starting torque per ampere of starting current will be high since the rotor current is closer to being in phase with the induced voltage than any other type of rotor. Also, the high resistance of the rotor bars limits the amount of current flow in the rotor when starting. This produces a low starting current for the motor. A rotor with type A bars has very poor running characteristics, however. Since the bars are resistive, a large amount of voltage will have to be induced into the rotor to produce an increase in rotor current and, therefore, an increase in the rotor magnetic field. This means that when load is added to the motor, the rotor must slow down a great amount to produce enough current in the rotor to increase the torque. Motors with type A rotors have the highest percentage slip of any squirrel-cage motor. Motors with type A rotors are generally used in applications where starting is a problem, such as a motor that must accelerate a large

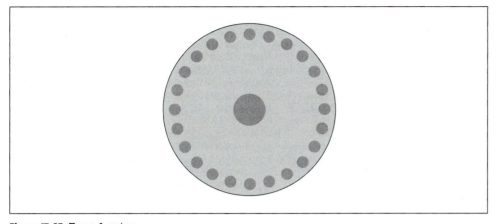

Figure 17-25 Type A rotor.

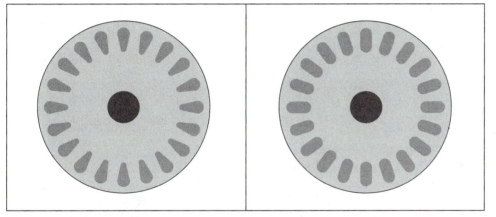

Figure 17-26 Type B – E rotor.   **Figure 17-27** Type F – V rotor.

flywheel from 0 RPM to its full speed. Flywheels can have a very large amount of inertia, which may require several minutes to accelerate them to their running speed when they are started.

Figure 17-26 shows a rotor with bars similar to those found in rotors with code letters B through E. These rotor bars have lower resistance than the type A rotor. Rotors of this type have fair starting torque, low starting current, and fair speed regulation.

Figure 17-27 shows a rotor with bars similar to those found in rotors with code letters F through V. This rotor has low starting torque, high starting current, and good running torque. Motors containing rotors of this type generally have very good speed regulation and low percent slip.

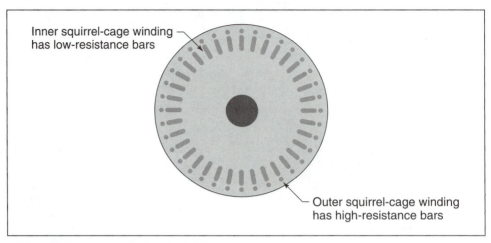

Inner squirrel-cage winding has low-resistance bars

Outer squirrel-cage winding has high-resistance bars

Figure 17-28 Double squirrel-cage rotor.

The Double Squirrel-Cage Rotor

Some motors use a rotor that contains two sets of squirrel-cage windings *(Figure 17-28)*. The outer winding consists of bars with a relatively high resistance located close to the top of the iron core. Since these bars are located close to the surface, they have a relatively low reactance. The inner winding consists of bars with a large cross-sectional area, which gives them a low resistance. The inner winding is placed deeper in the core material, which causes it to have a much higher reactance.

When the double squirrel-cage motor is started, the rotor frequency is high. Since the inner winding is inductive, its impedance will be high as compared to the resistance of the outer winding. During this period, most of the rotor current flows through the outer winding. The resistance of the outer winding limits the current flow through the rotor, which limits the starting current to a relatively low value. Since the current is close to being in phase with the induced voltage, the rotor flux and stator flux are close to being in phase with each other, and a strong starting torque is developed. The starting torque of a double squirrel-cage motor can be as high as 250% of rated full-load torque.

When the rotor reaches its full-load speed, rotor frequency decreases to 2 or 3 Hz. The inductive reactance of the inner winding has now decreased to a low value. Most of the rotor current now flows through the low-resistance inner winding. This type of motor has good running torque and excellent speed regulation.

Single-Phasing

Three lines supply power to a three-phase motor. If one of these lines should open, the motor will be connected to single-phase power *(Figure 17-29)*. This condition is known as **single-phasing**.

single-phasing

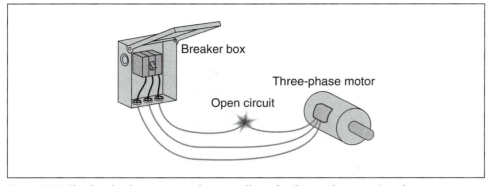

Figure 17-29 Single-phasing occurs when one line of a three-phase system is open.

If the motor is not running and single-phase power is applied to the motor, the induced voltage in the rotor sets up a magnetic field in the rotor. This magnetic field opposes the magnetic field of the stator (Lenz's law). As a result, practically no torque is developed in either the clockwise or counterclockwise direction, and the motor will not start. The current supplying the motor will be excessive, however, and damage to the stator windings can occur.

If the motor is operating under load at the time the single-phasing condition occurs, the rotor will continue to turn at a reduced speed. The moving bars of the rotor cut the stator field flux, which continues to induce voltage and current in the bars. Due to reduced speed, the rotor has high reactive and low resistive components, causing the rotor current to lag the induced voltage by almost 90°. This lagging current creates rotor fields midway between the stator poles, resulting in greatly reduced torque. The reduction in rotor speed causes high current flow and will most likely damage the stator winding if the motor is not disconnected from the power line.

The Nameplate

Electric motors have nameplates that give a great deal of information about the motor. *Figure 17-30* illustrates the nameplate of a three-phase squirrel-cage induction motor. The nameplate shows that the motor is 10 hp, it is a three-phase motor, and operates on 240 or 480 V. The full-load running current of the motor is 28 A when operated on 240 V and 14 A when operated on 480 V. The motor is designed to be operated on a 60-Hz AC voltage and has a full-load speed of 1745 RPM. This speed indicates that the motor has four poles per phase. Since the full-load speed is 1745 RPM, the synchronous speed would be 1800 RPM. The motor contains a type J squirrel-cage rotor and has a service factor of 1.25. The service factor is used to determine the amperage rating of the overload protection for the motor. Some motors indicate a marked temperature rise in Celsius degrees instead of a service factor. The frame number indicates the type of mounting the

Manufacturer	
HP 10	Phase 3
Volts 240/480	Amps 28/14
Hz 60	Fl speed 1745 RPM
Code J	SF 1.25
Frame XXXX	Model No. XXXX

Figure 17-30 Motor nameplate.

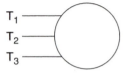

Figure 17-31 Schematic symbol of a three-phase squirrel-cage induction motor.

consequent pole

motor has. *Figure 17-31* shows the schematic symbol used to represent a three-phase squirrel-cage motor.

Consequent Pole Squirrel-Cage Motors

Consequent pole squirrel-cage motors permit the synchronous speed to be changed by changing the number of stator poles. If the number of poles is doubled, the synchronous speed will be reduced by one-half. A two-pole motor has a synchronous speed of 3600 RPM when operated at 60 Hz. If the number of poles is doubled to four, the synchronous speed becomes 1800 RPM. The number of stator poles can be changed by changing the direction of current flow through alternate pairs of poles.

Figure 17-32 illustrates this concept. In *Figure 17-32A*, two coils are connected in such a manner that current flows through them in the same direction. Both poles will produce the same magnetic polarity and are essentially one pole. In *Figure 17-32B*, the coils have been reconnected in such a manner that current flows through them in opposite directions. Each coil now produces the opposite magnetic polarity and are essentially two different poles.

Consequent pole motors with one stator winding bring out six leads labeled T_1 through T_6. Depending on the application, the windings will be connected as a series delta or a parallel wye. If it is intended that the motor maintain the same horsepower rating for both high and low speed, the high speed connection will be a series delta *(Figure 17-33)*. The low speed connection will be a parallel wye *(Figure 17-34)*.

If it is intended that the motor maintain constant torque for both low and high speeds, the series delta connection will provide low speed, and the parallel wye will provide high speed.

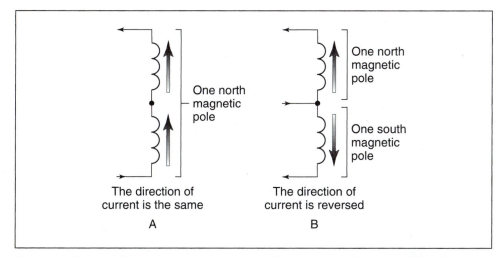

Figure 17-32 The number of poles can be changed by reversing the current flow through alternate poles.

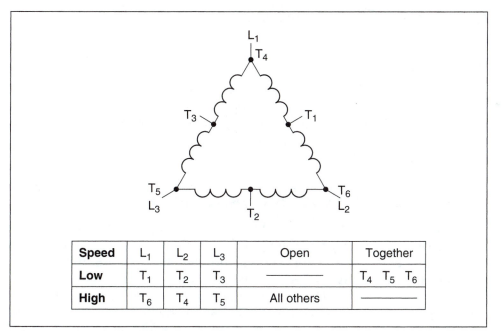

Speed	L_1	L_2	L_3	Open	Together
Low	T_1	T_2	T_3	———	T_4 T_5 T_6
High	T_6	T_4	T_5	All others	———

Figure 17-33 High-speed series delta connection.

Since the speed range of a consequent pole motor is limited to a 1:2 ratio, motors intended to operate at more than two speeds contain more than one stator winding. A consequent pole motor with three speeds, for example, will have one stator winding for one speed only and a second winding with

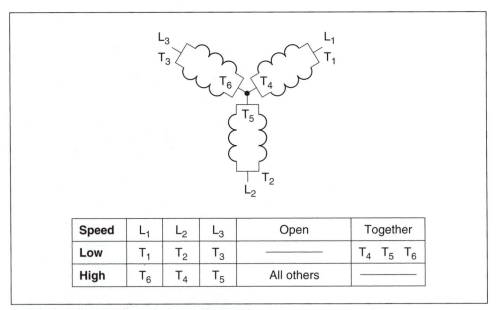

Speed	L_1	L_2	L_3	Open	Together
Low	T_1	T_2	T_3	——————	T_4 T_5 T_6
High	T_6	T_4	T_5	All others	——————

Figure 17-34 Low-speed parallel wye connection.

taps. The tapped winding may provide synchronous speeds of 1800 and 900 RPM, and the separate winding may provide a speed of 1200 RPM. Consequent pole motors with four speeds contain two separate stator windings with taps. If the second stator winding of the motor in this example were to be tapped, the motor would provide synchronous speeds of 1800, 1200, 900, and 600 RPM.

THREE-SPEED CONSEQUENT POLE MOTORS

Consequent pole motors that are intended to operate with three speeds, contain two separate stator windings. One winding is reconnectable like the winding in a two-speed motor. The second winding is wound for a certain number of poles and is not reconnectable. If one stator winding is wound with six poles and the second is reconnectable for two or four poles, the motor would develop synchronous speeds of 3600, 1800, or 1200 RPM when connected to a 60-Hz line. If the reconnectable winding is wound for four- or eight-pole connection, the motor would develop synchronous speeds of 1800, 1200, or 900 RPM. Three-speed consequent pole motors can be wound to produce constant horsepower, constant torque, or variable torque. Examples of different connection diagrams for three-speed two-winding consequent pole motors are shown in *Figures 14-35A* through *14-35I*.

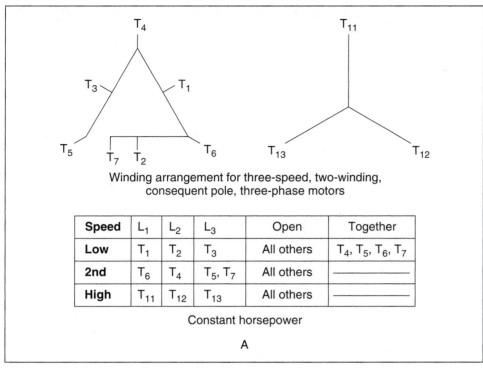

Winding arrangement for three-speed, two-winding,
consequent pole, three-phase motors

Speed	L_1	L_2	L_3	Open	Together
Low	T_1	T_2	T_3	All others	T_4, T_5, T_6, T_7
2nd	T_6	T_4	T_5, T_7	All others	————————
High	T_{11}	T_{12}	T_{13}	All others	————————

Constant horsepower

A

Figure 17-35A through Figure 17-35I Stator winding connections for three-speed consequent pole motors with one reconnectable and one nonreconnectable stator winding.

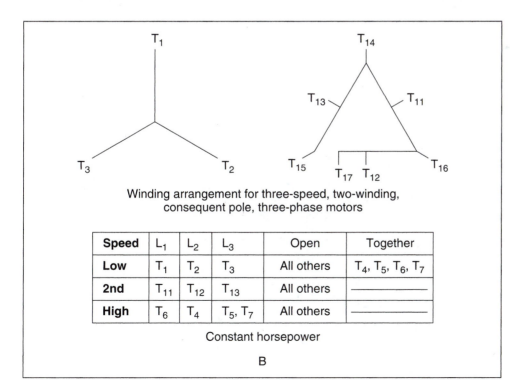

Winding arrangement for three-speed, two-winding,
consequent pole, three-phase motors

Speed	L_1	L_2	L_3	Open	Together
Low	T_1	T_2	T_3	All others	T_4, T_5, T_6, T_7
2nd	T_{11}	T_{12}	T_{13}	All others	————————
High	T_6	T_4	T_5, T_7	All others	————————

Constant horsepower

B

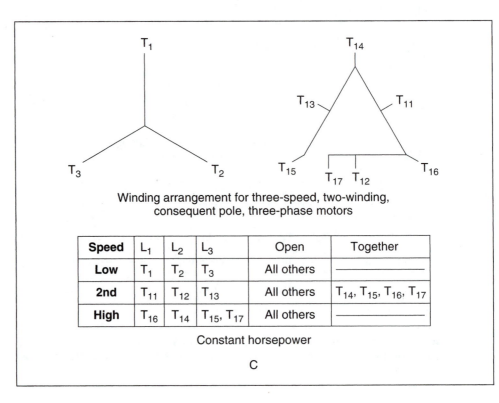

Winding arrangement for three-speed, two-winding,
consequent pole, three-phase motors

Speed	L₁	L₂	L₃	Open	Together
Low	T_1	T_2	T_3	All others	——————
2nd	T_{11}	T_{12}	T_{13}	All others	$T_{14}, T_{15}, T_{16}, T_{17}$
High	T_{16}	T_{14}	T_{15}, T_{17}	All others	——————

Constant horsepower

C

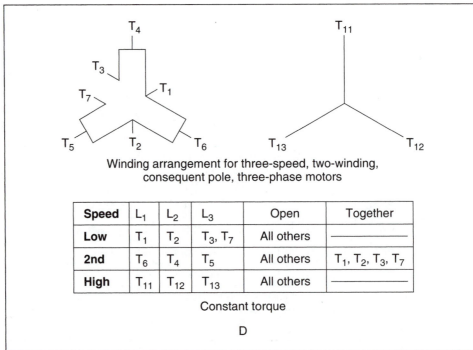

Winding arrangement for three-speed, two-winding,
consequent pole, three-phase motors

Speed	L₁	L₂	L₃	Open	Together
Low	T_1	T_2	T_3, T_7	All others	——————
2nd	T_6	T_4	T_5	All others	T_1, T_2, T_3, T_7
High	T_{11}	T_{12}	T_{13}	All others	——————

Constant torque

D

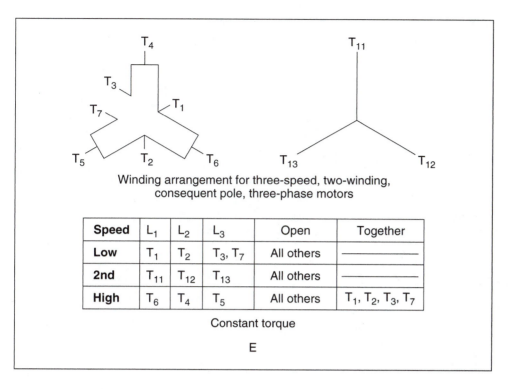

Winding arrangement for three-speed, two-winding,
consequent pole, three-phase motors

Speed	L_1	L_2	L_3	Open	Together
Low	T_1	T_2	T_3, T_7	All others	————
2nd	T_{11}	T_{12}	T_{13}	All others	————
High	T_6	T_4	T_5	All others	T_1, T_2, T_3, T_7

Constant torque

E

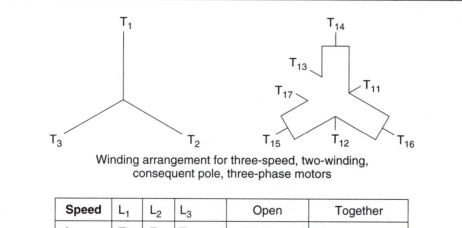

Winding arrangement for three-speed, two-winding,
consequent pole, three-phase motors

Speed	L_1	L_2	L_3	Open	Together
Low	T_1	T_2	T_3	All others	————
2nd	T_{11}	T_{12}	T_{13}, T_{17}	All others	————
High	T_{16}	T_{14}	T_{15}	All others	T_{11}, T_{12}, T_{13}, T_{17}

Constant torque

F

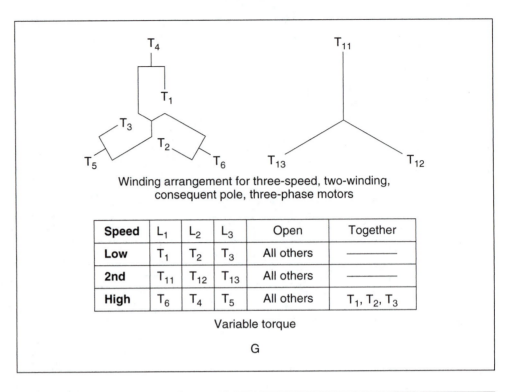

Winding arrangement for three-speed, two-winding,
consequent pole, three-phase motors

Speed	L_1	L_2	L_3	Open	Together
Low	T_1	T_2	T_3	All others	————
2nd	T_{11}	T_{12}	T_{13}	All others	————
High	T_6	T_4	T_5	All others	T_1, T_2, T_3

Variable torque

G

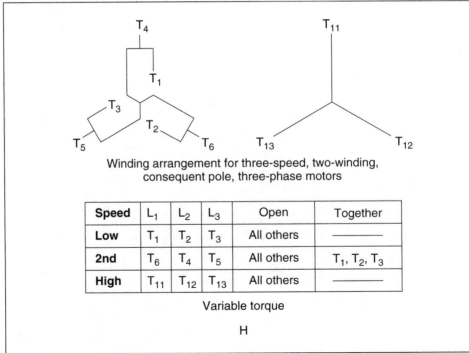

Winding arrangement for three-speed, two-winding,
consequent pole, three-phase motors

Speed	L_1	L_2	L_3	Open	Together
Low	T_1	T_2	T_3	All others	————
2nd	T_6	T_4	T_5	All others	T_1, T_2, T_3
High	T_{11}	T_{12}	T_{13}	All others	————

Variable torque

H

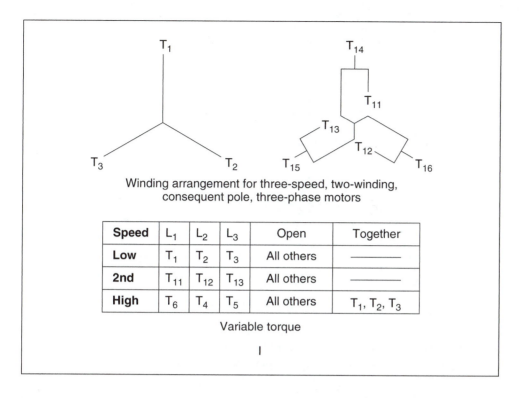

Winding arrangement for three-speed, two-winding,
consequent pole, three-phase motors

Speed	L_1	L_2	L_3	Open	Together
Low	T_1	T_2	T_3	All others	————
2nd	T_{11}	T_{12}	T_{13}	All others	————
High	T_6	T_4	T_5	All others	T_1, T_2, T_3

Variable torque

I

FOUR-SPEED CONSEQUENT POLE MOTORS

Consequent pole motors intended to operate with four speeds use two reconnectable windings. Like two-speed or three-speed motors, four-speed motors can be wound to operate at constant horsepower, constant torque, or variable torque. Some examples of winding connections for four-speed two-winding three-phase consequent pole motors are shown in *Figures 17-36A* through *17-36F.*

WOUND ROTOR INDUCTION MOTORS

The **wound-rotor** induction motor is very popular in industry because of its high starting torque and low starting current. The stator winding of the wound rotor motor is the same as the squirrel-cage motor. The difference between the two motors lies in the construction of the rotor. Recall that the squirrel-cage rotor is constructed of bars connected together at each end by a shorting ring as shown in *Figure 17-14.*

The rotor of a wound rotor motor is constructed by winding three separate coils on the rotor 120° apart. The rotor will contain as many poles per phase as the stator winding. These coils are then connected to three slip rings

wound-rotor

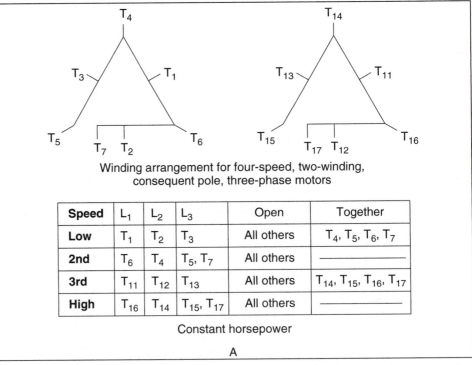

Winding arrangement for four-speed, two-winding,
consequent pole, three-phase motors

Speed	L_1	L_2	L_3	Open	Together
Low	T_1	T_2	T_3	All others	T_4, T_5, T_6, T_7
2nd	T_6	T_4	T_5, T_7	All others	————
3rd	T_{11}	T_{12}	T_{13}	All others	T_{14}, T_{15}, T_{16}, T_{17}
High	T_{16}	T_{14}	T_{15}, T_{17}	All others	————

Constant horsepower

A

Figure 17-36A through Figure 17-36F Stator winding connections for four-speed consequent pole motors with two reconnectable stator windings.

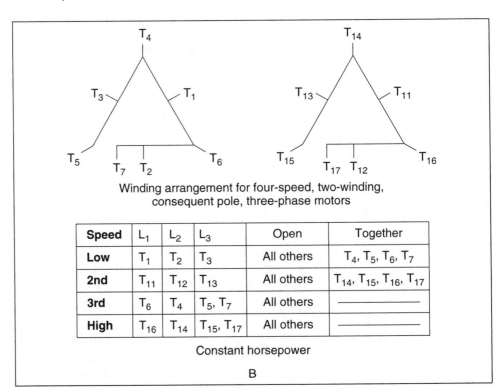

Winding arrangement for four-speed, two-winding,
consequent pole, three-phase motors

Speed	L_1	L_2	L_3	Open	Together
Low	T_1	T_2	T_3	All others	T_4, T_5, T_6, T_7
2nd	T_{11}	T_{12}	T_{13}	All others	T_{14}, T_{15}, T_{16}, T_{17}
3rd	T_6	T_4	T_5, T_7	All others	————
High	T_{16}	T_{14}	T_{15}, T_{17}	All others	————

Constant horsepower

B

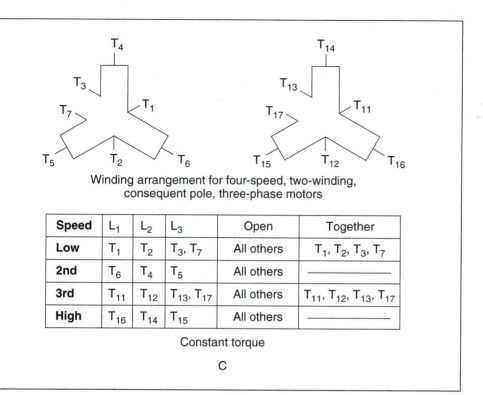

Winding arrangement for four-speed, two-winding,
consequent pole, three-phase motors

Speed	L_1	L_2	L_3	Open	Together
Low	T_1	T_2	T_3, T_7	All others	T_1, T_2, T_3, T_7
2nd	T_6	T_4	T_5	All others	————
3rd	T_{11}	T_{12}	T_{13}, T_{17}	All others	$T_{11}, T_{12}, T_{13}, T_{17}$
High	T_{16}	T_{14}	T_{15}	All others	————

Constant torque

C

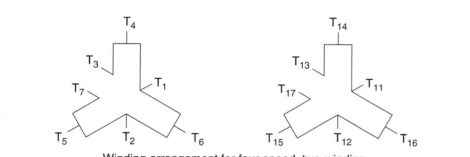

Winding arrangement for four-speed, two-winding,
consequent pole, three-phase motors

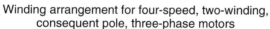

Speed	L_1	L_2	L_3	Open	Together
Low	T_1	T_2	T_3, T_7	All others	————
2nd	T_{11}	T_{12}	T_{13}, T_{17}	All others	————
3rd	T_6	T_4	T_5	All others	T_1, T_2, T_3, T_7
High	T_{16}	T_{14}	T_{15}	All others	$T_{11}, T_{12}, T_{13}, T_{17}$

Constant torque

D

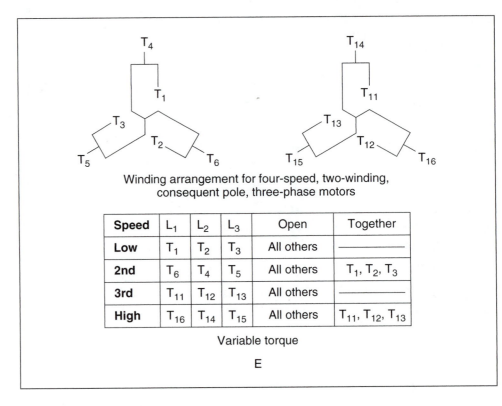

Winding arrangement for four-speed, two-winding,
consequent pole, three-phase motors

Speed	L_1	L_2	L_3	Open	Together
Low	T_1	T_2	T_3	All others	——————
2nd	T_6	T_4	T_5	All others	T_1, T_2, T_3
3rd	T_{11}	T_{12}	T_{13}	All others	——————
High	T_{16}	T_{14}	T_{15}	All others	T_{11}, T_{12}, T_{13}

Variable torque

E

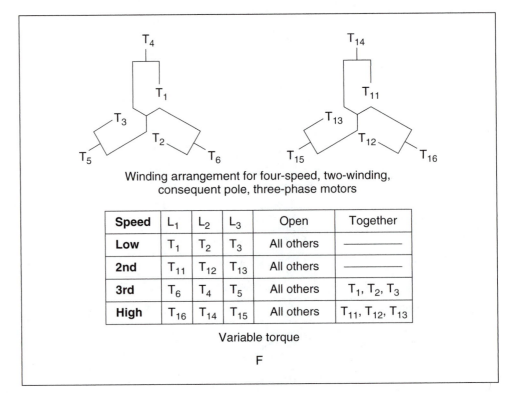

Winding arrangement for four-speed, two-winding,
consequent pole, three-phase motors

Speed	L_1	L_2	L_3	Open	Together
Low	T_1	T_2	T_3	All others	——————
2nd	T_{11}	T_{12}	T_{13}	All others	——————
3rd	T_6	T_4	T_5	All others	T_1, T_2, T_3
High	T_{16}	T_{14}	T_{15}	All others	T_{11}, T_{12}, T_{13}

Variable torque

F

Figure 17-37 Rotor of a wound-rotor induction motor.

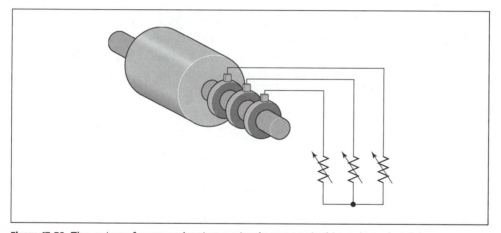

Figure 17-38 The rotor of a wound-rotor motor is connected to external resistors.

located on the rotor shaft as shown in *Figure 17-37*. Brushes that are connected to the slip rings provide external connection to the rotor. This permits the rotor circuit to be connected to a set of resistors as shown in *Figure 17-38*.

The stator terminal connections are generally labeled T_1, T_2, and T_3. The rotor connections are commonly labeled M_1, M_2, and M_3. The M_2 lead is generally connected to the middle slip ring, and the M_3 lead is connected close to the rotor windings. The direction of rotation for the wound rotor motor is reversed by changing any two stator leads. Changing the M leads will have no effect on the direction of rotation. The schematic symbol for a wound-rotor motor is shown in *Figure 17-39*.

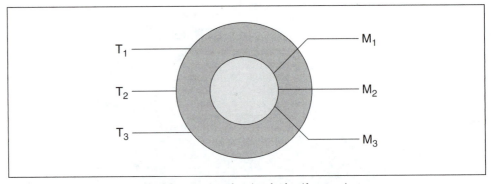

Figure 17-39 Schematic symbol for a wound-rotor induction motor.

Wound-Rotor Motor Operation

When power is applied to the stator winding, a rotating magnetic field is created in the motor. This magnetic field cuts through the windings of the rotor and induces a voltage into them. The amount of current flow in the rotor is determined by the amount of induced voltage and the total impedance of the rotor circuit (I = E/Z). The rotor impedance is a combination of inductive reactance created in the rotor windings and the external resistance. The impedance could be calculated by using the formula for resistance and inductive reactance connected in series.

$$Z = \sqrt{R^2 + X_L^2}$$

As the rotor speed increases, the frequency of the induced voltage will decrease just as it does in the squirrel-cage motor. The reduction in frequency causes the rotor circuit to become more resistive and less inductive, decreasing the phase angle between induced voltage and rotor current.

When current flows through the rotor, a magnetic field is produced. This magnetic field is attracted to the rotating magnetic field of the stator. As the rotor speed increases, the induced voltage decreases because of less cutting action between the rotor windings and rotating magnetic field. This produces less current flow in the rotor and, therefore, less torque. If rotor circuit resistance is reduced, more current can flow, which will increase motor torque, and the rotor will increase in speed. This action continues until all external resistance has been removed from the rotor circuit by shorting the M leads together and the motor is operating at maximum speed. At this point, the wound-rotor motor is operating in the same manner as a squirrel-cage motor.

Starting Characteristics of a Wound-Rotor Motor

The wound-rotor motor will have a higher starting torque and lower starting current per horsepower than a squirrel-cage motor. The starting current is less because resistance is connected in the rotor circuit during starting. This resistance limits the amount of current that can flow in the rotor circuit. Since the stator current is proportional to rotor current because of transformer action, the stator current is less also. The starting torque is high because of the resistance in the rotor circuit. Recall that one of the factors that determines motor torque is the phase angle difference between stator flux and rotor flux. Since resistance is connected in the rotor circuit, stator and rotor flux are close to being in phase with each other, producing a high starting torque for the wound-rotor induction motor. If an attempt is made to start the motor with no circuit connected to the rotor, the motor cannot start. If no resistance is connected to the rotor circuit, there can be no current flow, and consequently, no magnetic field developed in the rotor.

Speed Control

The speed of a wound-rotor motor can be controlled by permitting resistance to remain in the rotor circuit during operation. When this is done, the rotor and stator current is limited, which reduces the strength of both magnetic fields. The reduced magnetic field strength permits the rotor to slip behind the rotating magnetic field of the stator. The resistors of speed controllers must have higher power ratings than the resistors of starters because they operate for extended periods of time.

The operating characteristics of a wound-rotor motor with the slip rings shorted are almost identical to those of a squirrel-cage motor. The percent slip, power factor, and efficiency are very similar for motors of equal horsepower rating.

SYNCHRONOUS MOTORS

The three-phase **synchronous** motor has several characteristics that separate it from other types of three-phase motors. Some of these characteristics are:

    synchronous

1. The synchronous motor is not an induction motor. It does not depend on induced current in the rotor to produce torque.
2. It will operate at a constant speed from full load to no load.
3. The synchronous motor must have DC excitation to operate.
4. It will operate at the speed of the rotating magnetic field (synchronous speed).
5. It has the ability to correct its own power factor and the power factor of other devices connected to the same line.

Figure 17-40 Rotor of a synchronous motor.

Rotor Construction

The synchronous motor has the same type of stator windings as the other two three-phase motors. The rotor of a synchronous motor, however, contains both wound pole pieces and a squirrel-cage winding *(Figure 17-40)*. The wound pole pieces become electromagnets when direct current is applied to them. The excitation current can be applied to the rotor through two slip rings located on the rotor shaft or by a brushless exciter.

Starting a Synchronous Motor

The rotor of a synchronous motor contains a set of squirrel-cage bars similar to those found in a type A rotor. This set of squirrel-cage bars is used to start the motor and is known as the **amortisseur winding** *(Figure 17-40)*. When power is first connected to the stator, the rotating magnetic field cuts through the squirrel-cage bars. The cutting action of the field induces a current into the squirrel-cage winding. The current flow through the amortisseur winding produces a rotor magnetic field that is attracted to the rotating magnetic field of the stator. This causes the rotor to begin turning in the direction of rotation of the stator field. When the rotor has accelerated to a speed that is close to the synchronous speed of the field, direct current is connected to the rotor through the slip rings on the rotor shaft or by a brushless exciter *(Figure 17-41)*.

When DC current is applied to the rotor, the windings of the rotor become electromagnets. The electromagnetic field of the rotor locks in step with the rotating magnetic field of the stator. The rotor will now turn at the same speed as the rotating magnetic field. When the rotor turns at the synchronous speed of the field, there is no more cutting action between the stator

amortisseur winding

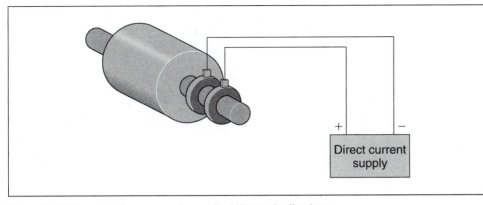

Figure 17-41 DC excitation current supplied through slip rings.

field and the amortisseur winding. This causes the current flow in the amortisseur winding to cease.

Notice that the synchronous motor starts as a squirrel-cage induction motor. Since the rotor uses bars that are similar to those used in a type A rotor, they have a relatively high resistance, which gives the motor good starting torque and low starting current. *A synchronous motor must never be started with DC current connected to the rotor.* If DC current is applied to the rotor, the field poles of the rotor become electromagnets. When the stator is energized, the rotating magnetic field begins turning at synchronous speed. The electromagnets are alternately attracted and repelled by the stator field. As a result, the rotor does not turn. The rotor and power supply can be damaged by high induced voltages, however.

The Field Discharge Resistor

When the stator winding is first energized, the rotating magnetic field cuts through the rotor winding at a fast rate of speed. This causes a large amount of voltage to be induced into the winding of the rotor. To prevent this from becoming excessive, a resistor is connected across the winding. This resistor is known as the field discharge resistor *(Figure 17-42)*. It also helps to reduce the voltage induced into the rotor by the collapsing magnetic field when the DC current is disconnected from the rotor. The field discharge resistor is connected in parallel with the rotor winding during starting. If the motor is manually started, a field discharge switch is used to connect the excitation current to the rotor. If the motor is automatically started, a special type of relay is used to connect excitation current to the rotor and disconnect the field discharge resistor.

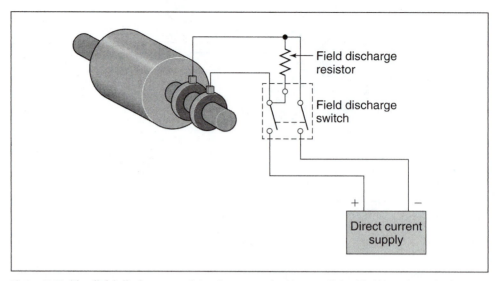

Figure 17-42 The field discharge resistor is connected in parallel with the rotor winding during starting.

Constant Speed Operation

Although the synchronous motor starts as an induction motor, it does not operate as one. After the amortisseur winding has been used to accelerate the rotor to about 95% of the speed of the rotating magnetic field, direct current is connected to the rotor, and the electromagnets lock in step with the rotating field. Notice that the synchronous motor does not depend on induced voltage from the stator field to produce a magnetic field in the rotor. The magnetic field of the rotor is produced by external DC current applied to the rotor. This is the reason that the synchronous motor has the ability to operate at the speed of the rotating magnetic field.

As load is added to the motor, the magnetic field of the rotor remains locked with the rotating magnetic field of the stator, and the rotor continues to turn at the same speed. The added load, however, causes the magnetic fields of the rotor and stator to become stressed *(Figure 17-43)*. The action is similar to connecting the north and south ends of two magnets and then trying to pull them apart. If the force being used to pull the magnets apart becomes greater than the strength of the magnetic attraction, the magnetic coupling will be broken, and the magnets can be separated. The same is true for the synchronous motor. If the load on the motor becomes too great, the rotor will be pulled out of sync with the rotating magnetic field. The amount of torque necessary to cause this condition is called the **pull-out torque**. The pull-out torque for most synchronous motors will range from 150% to 200% of rated full-load torque. If this should happen, the motor must be

pull-out torque

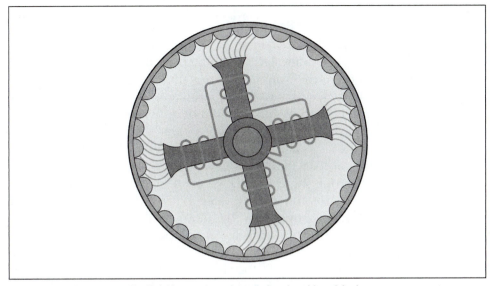

Figure 17-43 The magnetic field becomes stressed as load is added.

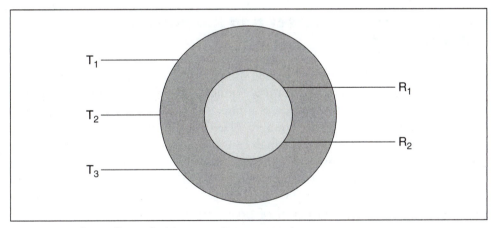

Figure 17-44 Schematic symbol for a synchronous motor.

stopped and restarted. The schematic symbol for a synchronous motor is shown in *Figure 17-44*.

The Power Supply

The DC power of a synchronous motor can be provided by several methods. The most common of these methods is either a small DC generator mounted to the shaft of the motor or an electronic power supply that converts the AC line voltage into DC voltage.

Power Factor Correction

The synchronous motor has the ability to correct its own power factor and the power factor of other devices connected to the same line. The amount of power factor correction is controlled by the amount of excitation current in the rotor. If the rotor of a synchronous motor is underexcited, the motor will have a lagging power factor as a common induction motor. As rotor excitation current is increased, the synchronous motor appears to be more capacitive. When the excitation current reaches a point that the power factor of the motor is at unity or 100%, it is at the normal excitation level. At this point, the current supplying the motor will drop to its lowest value.

If the excitation current is increased above the normal level, the motor will have a leading power factor and appear as a capacitive load. When the rotor is overexcited, the current supplying the motor will increase due to the change in power factor. The power factor at this point, however, is leading and not lagging. Since capacitance has now been added to the line, it will correct the lagging power factor of other inductive devices connected to the same line. Changes in the amount of excitation current will not affect the speed of the motor.

Interaction of the Direct and Alternating Current Fields

Figure 17-45 illustrates how the magnetic flux of the AC field aids or opposes the DC field. In this example, it is assumed that the DC field is held stationary and the rotating armature is connected to the AC source. Although most synchronous motors have a stationary AC field and a rotating DC field, the principle of operation is the same. When the DC excitation current is less than the amount required for normal excitation, the AC current must supply some portion of the magnetizing current to aid the weak DC current *(Figure 17-45A)*. This portion of magnetizing current lags the applied voltage by 90°. The current waveform shown in *Figure 17-45A* depicts only the portion of magnetizing current that is out of phase with the voltage. The remaining part of the AC current is used to produce the torque necessary to operate the load. The synchronous motor will have a lagging power factor at this time.

In *Figure 17-45B*, the DC excitation current has been increased to the normal excitation value. All of the AC current is now used to produce the torque necessary to operate the load. Since the AC current no longer supplies any of the magnetizing current, it is in phase with the voltage, and the motor power factor is at unity or 100%. The amount of AC current supplied to the motor will be at its lowest value during this period.

In *Figure 17-45C* the DC excitation current is greater than that needed for normal excitation. The AC current now supplies a demagnetizing component

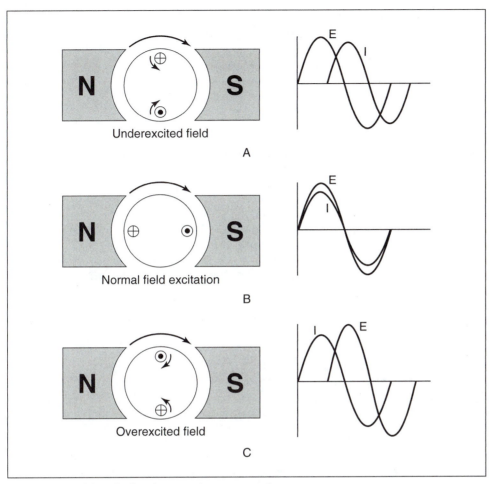

Figure 17-45 Field excitation in a synchronous motor.

of current. The portion of AC current used to demagnetize the overexcited DC field will lead the applied voltage by 90°. The current waveform shown in *Figure 17-45C* illustrates only the portion of AC current used to demagnetize the DC field and does not take into account the amount of AC current used to produce torque for the load. The synchronous motor now has a leading power factor.

Synchronous Motor Applications

Synchronous motors are very popular in industry, especially in the large horsepower ratings (motors up to 5000 hp are not uncommon). They have a low starting current per horsepower and a high starting torque. They operate

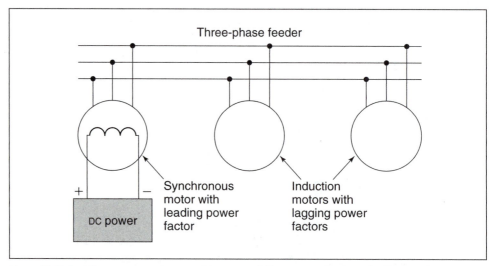

Figure 17-46 Synchronous motor used to correct the power factor of other motors.

at a constant speed from no load to full load and maintain maximum efficiency. Synchronous motors are used to operate DC generators, fans, blowers, pumps, and centrifuges. They correct their own power factor and can correct the power factor of other inductive loads connected to the same feeder *(Figure 17-46)*. Synchronous motors are sometimes operated at no load and are used for power factor correction only. When this is done, the motor is referred to as a **synchronous condenser**.

synchronous condenser

Advantages of the Synchronous Condenser

The advantage of using a synchronous condenser over a bank of capacitors for power factor correction is that the amount of correction is easily controlled. When a bank of capacitors is used for correcting power factor, capacitors must be added to or removed from the bank if a change in the amount of correction is needed. When a synchronous condenser is used, only the excitation current must be changed to cause an alteration of power factor.

SELSYN MOTORS

selsyn motors

The word *selsyn* is a contraction derived from *self-synchronous*. **Selsyn motors** are used to provide position control and angular feedback information in industrial applications. Although selsyn motors are actually operated on single-phase alternating current, they do contain three-phase windings *(Figure 17-47)*. The schematic symbol for a selsyn motor is shown in *Figure 17-48*. This symbol is very similar to the symbol used to represent a

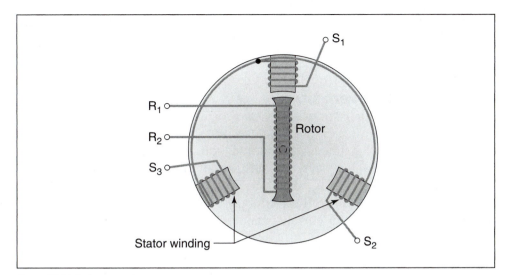

Figure 17-47 Selsyn motor.

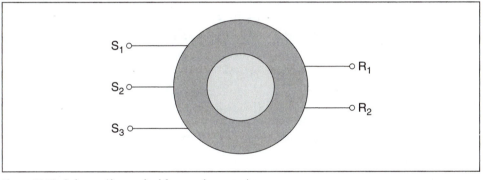

Figure 17-48 Schematic symbol for a selsyn motor.

three-phase synchronous motor. The stator windings are labeled S_1, S_2, and S_3. The rotor leads are labeled R_1 and R_2. The rotor leads are connected to the rotor winding by means of slip rings and brushes.

When selsyn motors are employed, at least two are used together. One motor is referred to as the transmitter and the other is called the receiver. It makes no difference which motor acts as the transmitter and which acts as the receiver. Connection is made by connecting S_1 of the transmitter to S_1 of the receiver, S_2 of the transmitter to S_2 of the receiver, and S_3 of the transmitter to S_3 of the receiver. The rotor leads of each motor are connected to a source of single-phase alternating current *(Figure 17-49)*. If the stator winding leads of the two selsyn motors are connected improperly, the receiver will rotate in a direction opposite that of the transmitter. If the rotor leads are connected improperly, the rotor of the transmitter and the rotor of the receiver will have an angle difference of 180°.

Selsyn Motor Operation

Selsyn motors actually operate as transformers. The rotor winding is the primary, and the stator winding is the secondary. In *Figure 17-49*, the rotor of the transmitter is in line with stator winding S_1. Since the rotor is connected to a source of alternating current, an alternating magnetic field exists in the rotor. This alternating magnetic field will induce a voltage into the windings of the stator. Since the rotors of both motors are connected to the same source of alternating current, magnetic fields of identical strength and polarity exist in both motors.

Since the rotor of the transmitter is in line with stator winding S_1, maximum voltage and current are being induced in stator S_1, and less than maximum voltage and current are being induced in stator windings S_2 and S_3. Since the stator windings of the receiver are connected to the stator windings of the transmitter, the same current will flow through the receiver, producing a magnetic field in the receiver. This magnetic field will attract or repel the magnetic field of the rotor depending on the relative polarity of the two fields. When the rotor of the receiver is in the same position as the rotor of the transmitter, an equal amount of voltage will be induced in the stator windings of the receiver, causing stator winding current to become zero.

If the rotor of the transmitter is turned to a different position, the magnetic field of the stator will change, resulting in a change of the magnetic field in the stator of the receiver. This will cause the rotor of the receiver to rotate to

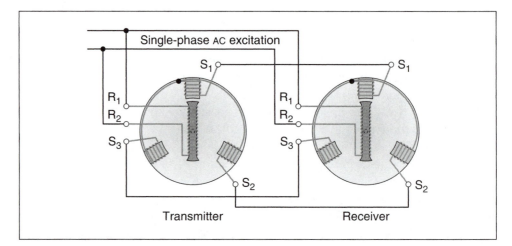

Figure 17-49 Connection of two selsyn motors.

a new position, where the two stator magnetic fields again cancel each other. Each time the rotor position of the transmitter is changed, the rotor of the receiver will change by the same amount.

The Differential Selsyn

The **differential selsyn** is used to produce the algebraic sum of the rotation of two other selsyn units. Differential selsyns are constructed in a manner different from other selsyn motors. The differential selsyn contains three rotor windings connected in wye, as well as three stator windings connected in wye. The rotor windings are brought out through three slip rings and brushes in a manner very similar to a wound-rotor induction motor. The differential selsyn is not connected to a source of power. Power must be provided by one of the other selsyn motors connected to it *(Figure 17-50)*.

differential selsyn

If any one of the selsyn units is held in place and a second unit is turned, the third will turn by the same amount. If any two of the selsyn units are turned at the same time, the third will turn an amount equal to the sum of the angle of rotation of the other two.

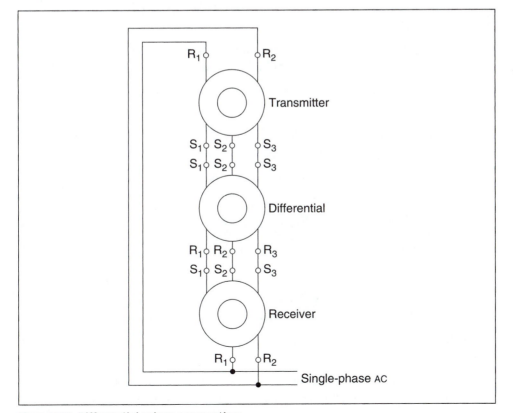

Figure 17-50 Differential selsyn connection.

Summary

1. Three basic types of three-phase motors are:
   A. Squirrel-cage induction motor
   B. Wound-rotor induction motor
   C. Synchronous motor

2. All three-phase motors operate on the principle of a rotating magnetic field.

3. Three factors that cause a magnetic field to rotate are:
   A. The fact that the voltages of a three-phase system are 120° out of phase with each other
   B. The fact that voltages change polarity at regular intervals
   C. The arrangement of the stator windings

4. The speed of the rotating magnetic field is called the synchronous speed.

5. Two factors that determine the synchronous speed are:
   A. Number of stator poles per phase
   B. Frequency of the applied voltage

6. The direction of rotation of any three-phase motor can be changed by reversing the connection of any two stator leads.

7. The direction of rotation of a three-phase motor can be determined with a phase rotation meter before power is applied to the motor.

8. Dual-voltage motors will have 9 or 12 leads brought out at the terminal connection box.

9. Dual-voltage motors intended for high-voltage connection have their phase windings connected in series.

10. Dual-voltage motors intended for low-voltage connection have their phase windings connected in parallel.

11. Motors that bring out 12 leads are generally intended for wye-delta starting.

12. Three factors that determine the torque produced by an induction motor are:
    A. Strength of the magnetic field of the stator
    B. Strength of the magnetic field of the rotor
    C. Phase angle difference between rotor and stator flux

13. Maximum torque is developed when stator and rotor flux are in phase with each other.

14. The code letter on the nameplate of a squirrel-cage motor indicates the type of rotor bars used in the construction of the rotor.

15. The type A rotor has the lowest starting current, highest starting torque, and poorest speed regulation of any squirrel-cage rotor.

16. The double squirrel-cage rotor contains two sets of squirrel-cage windings in the same rotor.

17. Consequent pole squirrel-cage motors change speed by changing the number of stator poles.

18. Wound-rotor induction motors have wound-rotors that contain three-phase windings.

19. wound-rotor motors have three slip rings on the rotor shaft to provide external connection to the rotor.

20. wound-rotor motors have higher starting torque and lower starting current than squirrel-cage motors of the same horsepower.

21. The speed of a wound-rotor motor can be controlled by permitting resistance to remain in the rotor circuit during operation.

22. Synchronous motors operate at synchronous speed.

23. Synchronous motors operate at a constant speed from no load to full load.

24. When load is connected to a synchronous motor, stress develops between the magnetic fields of the rotor and stator.

25. Synchronous motors must have DC excitation from an external source.

26. DC excitation is provided to some synchronous motors through two slip rings located on the rotor shaft, and other motors use a brushless exciter.

27. Synchronous motors have the ability to produce a leading power factor by overexcitation of the DC current supplied to the rotor.

28. Synchronous motors have a set of type A squirrel-cage bars used for starting. This squirrel-cage winding is called the amortisseur winding.

29. A field discharge resistor is connected across the rotor winding during starting to prevent high voltage in the rotor due to induction.

30. Changing the DC excitation current does not affect the speed of the motor.

31. Selsyn motors are used to provide position control and angular feedback information.

32. Although selsyn motors contain three-phase windings, they operate on single-phase AC.

33. A differential selsyn unit can be used to determine the algebraic sum of the rotation of two other selsyn units.

REVIEW QUESTIONS

1. What are the three basic types of three-phase motors?

2. What is the principle of operation of all three-phase motors?

3. What is synchronous speed?

4. What two factors determine synchronous speed?

5. Name three factors that cause the magnetic field to rotate.

6. Name three factors that determine the torque produced by an induction motor.

7. Is the synchronous motor an induction motor?

8. What is the amortisseur winding?

9. Why must a synchronous motor never be started when DC excitation current is applied to the rotor?

10. Name three characteristics that make the synchronous motor different from an induction motor.

11. What is the function of the field discharge resistor?

12. Why can an induction motor never operate at synchronous speed?

13. What is the difference between a squirrel-cage motor and a wound-rotor motor?

14. What is the advantage of the wound-rotor motor over the squirrel-cage motor?

15. Name three factors that determine the amount of voltage induced in the rotor of a wound-rotor motor.

16. Why will the rotor of a wound-rotor motor not turn if the rotor circuit is left open with no resistance connected to it?

17. Why is the starting torque of a wound-rotor motor higher per ampere of starting current than that of a squirrel-cage motor?

18. What determines when a synchronous motor is at normal excitation?

19. How can a synchronous motor be made to have a leading power factor?

20. Is the excitation current of a synchronous motor AC or DC?

21. How is the speed of a consequent pole squirrel-cage motor changed?

22. A three-phase squirrel-cage motor is connected to a 60-Hz line. The full-load speed is 870 RPM. How many poles per phase does the stator have?

UNIT 18

Single-Phase Motors

Objectives

After studying this unit, you should be able to:

- List the different types of split-phase motors.
- Discuss the operation of split-phase motors.
- Reverse the direction of rotation of a split-phase motor.
- Discuss the operation of multispeed split-phase motors.
- Discuss the operation of shaded-pole motors.
- Discuss the operation of universal motors.

Although most of the large motors used in industry are three-phase, there are times when single-phase motors must be used. Single-phase motors are used almost exclusively to operate home appliances such as air conditioners, refrigerators, well pumps, and fans. They are designed to operate on 120 V or 240 V. They range in size from fractional horsepower to several horsepower depending on the application.

SINGLE-PHASE MOTORS

In unit 17, it was stated that there are three basic types of three-phase motors and that all operate on the principle of a rotating magnetic field. While this is true for three-phase motors, it is not true for single-phase motors. There are not only many different types of single-phase motors, but they also have different operating principles.

SPLIT-PHASE MOTORS

Split-phase motors fall into three general classifications:
1. Resistance-start induction-run motor
2. Capacitor-start induction-run motor
3. Capacitor-start capacitor-run motor (This motor is also known as a permanent-split capacitor motor in the air conditioning and refrigeration industry.)

split-phase motors

Although all of these motors have different operating characteristics, they are similar in construction and use the same operating principle. **Split-phase motors** receive their name from the manner in which they operate. As do three-phase motors, split-phase motors operate on the principle of a rotating magnetic field. A rotating magnetic field, however, cannot be produced with only one phase. Split-phase motors, therefore, split the current flow through two separate windings to simulate a two-phase power system. A rotating magnetic field can be produced with a two-phase system.

The Two-Phase System

two-phase

In some parts of the world two-phase power is produced. A **two-phase** system is produced by having an alternator with two sets of coils wound 90° apart *(Figure 18-1)*. The voltages of a two-phase system are, therefore, 90°

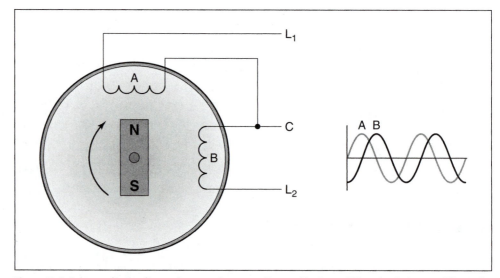

Figure 18-1 A two-phase alternator produces voltages that are 90° out of phase with each other.

out of phase with each other. These two out-of-phase voltages can be used to produce a rotating magnetic field in a manner similar to that of producing a rotating magnetic field with the voltages of a three-phase system. Since there have to be two voltages or currents out of phase with each other to produce a rotating magnetic field, split-phase motors use two separate windings to create a phase difference between the currents in each of these windings. These motors literally split one phase and produce a second phase, hence the name split-phase motor.

Stator Windings

The stator of a split-phase motor contains two separate windings, the **start winding** and the **run winding**. The start winding is made of small wire and is placed near the top of the stator core. The run winding is made of relatively large wire and is placed in the bottom of the stator core. *Figure 18-2* shows a photograph of two split-phase stators. The stator on the left is used for a resistance-start induction-run motor or a capacitor-start induction-run motor. The stator on the right is used for a capacitor-start capacitor-run motor. Both stators contain four poles, and the start winding is placed at a 90° angle from the run winding.

Notice the difference in size and position of the two windings of the stator shown on the left. The start winding is made from small wire and is placed near the top of the stator core. This causes it to have a higher resistance than the run winding. The start winding is located between the poles of the run winding. The run winding is made with larger wire and placed near the bottom of the core. This gives it higher inductive reactance and less resistance than the start winding. These two windings are connected in parallel with each other *(Figure 18-3)*.

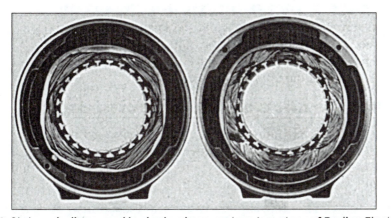

Figure 18-2 Stator windings used in single-phase motors (courtesy of Bodine Electric Co.).

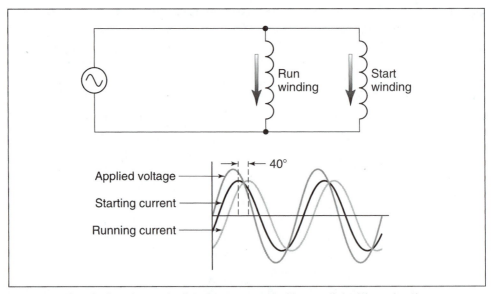

Figure 18-3 The start and run windings are connected in parallel with each other.

When power is applied to the stator, current will flow through both windings. Since the start winding is more resistive, the current flow will be more in phase with the applied voltage. The current flow through the run winding will lag the applied voltage due to inductive reactance. These two out-of-phase currents are used to create a rotating magnetic field in the stator. The speed of this rotating magnetic field is called synchronous speed and is determined by the same two factors that determined the synchronous speed for a three-phase motor:

- Number of stator poles per phase
- Frequency of the applied voltage

RESISTANCE-START INDUCTION-RUN MOTORS

The resistance-start induction-run motor receives its name from the fact that the out-of-phase condition between start- and run-winding current is caused by the start winding being more resistive than the run winding. The amount of starting torque produced by a split-phase motor is determined by three factors:

1. Strength of the magnetic field of the stator
2. Strength of the magnetic field of the rotor
3. Phase angle difference between current in the start winding and current in the run winding (Maximum torque is produced when these two currents are 90° out of phase with each other.)

Although these two currents are out of phase with each other, they are not 90° out of phase. The run winding is more inductive than the start winding, but it does have some resistance, which prevents the current from being 90°

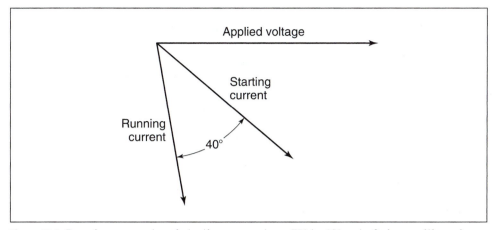

Figure 18-4 Running current and starting current are 35° to 40° out of phase with each other.

out of phase with the voltage. The start winding is more resistive than the run winding, but it does have some inductive reactance, preventing the current from being in phase with the applied voltage. Therefore, a phase angle difference of 35° to 40° is produced between these two currents, resulting in a rather poor starting torque *(Figure 18-4)*.

Disconnecting the Start Winding

A stator rotating magnetic field is necessary only to start the rotor turning in both the resistance-start induction-run and capacitor-start induction-run motors. Once the rotor has accelerated to approximately 75% of rated speed, the start winding can be disconnected from the circuit, and the motor will continue to operate with only the run winding energized. Motors that are not hermetically sealed (most refrigeration and air conditioning compressors are hermetically sealed), use a **centrifugal switch** to disconnect the start windings from the circuit. The contacts of the centrifugal switch are connected in series with the start winding *(Figure 18-5)*. The centrifugal switch contains a set of spring-loaded weights that control the operation of a fiber bushing and a switch assembly *(Figure 18-6)*. When the shaft is not turning, the springs hold a fiber washer in contact with the movable contact of the switch *(Figure 18-7)*. The fiber washer causes the movable contact to complete a circuit with a stationary contact.

When the rotor accelerates to about 75% of rated speed, centrifugal force causes the weights to overcome the force of the springs. The fiber washer retracts and permits the contacts to open and disconnect the start winding from the circuit. The start winding of this type of motor is intended to be energized only during the period of time that the motor is actually starting. If the start winding is not disconnected, it will be damaged by excessive current flow.

centrifugal switch

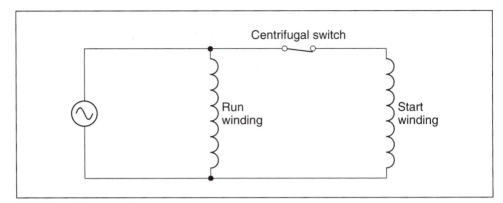

Figure 18-5 A centrifugal switch is used to disconnect the start winding from the circuit.

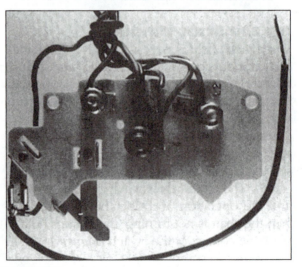

Figure 18-6 Centrifugal switch.

Figure 18-7 The centrifugal switch is closed when the rotor is not turning.

Starting Hermetically Sealed Motors

Resistance-start induction-run and capacitor-start induction-run motors are sometimes hermetically sealed as is the case with air conditioning and refrigeration compressors. When these motors are hermetically sealed, a centrifugal switch cannot be used to disconnect the start winding. A device that can be mounted externally must be used to disconnect the start windings from the circuit. Starting relays are used to perform this function. There are two basic types of starting relays used with the resistance-start and capacitor-start motors:

- Hot wire relay
- Current relay

Although the *hot wire relay* is seldom used, it is still found on some older units that are still in service. The hot wire relay functions as both a starting relay and an overload relay. In the circuit shown in *Figure 18-8*, it is assumed that a thermostat controls the operation of the motor. When the thermostat closes, current flows through a resistive wire and two normally closed contacts connected to the start and run windings of the motor. The starting current of the motor is high and rapidly heats the resistive wire, causing it to expand. The expansion of the wire causes the spring-loaded start winding contact to open and disconnect the start winding from the circuit, reducing motor current. If the motor is not overloaded, the resistive wire never becomes hot enough to cause the overload contact to open and the motor continues to run. If the motor should become overloaded, however, the resistive wire will expand enough to open the overload contact and

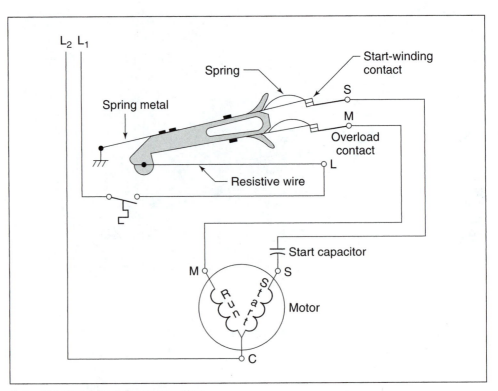

Figure 18-8 Hot wire relay connection.

disconnect the motor from the line. A photograph of a hot wire starting relay is shown in *Figure 18-9*.

current relay   The **current relay** also operates by sensing the amount of current flow in the circuit. This type of relay operates on the principle of a magnetic field instead of expanding metal. The current relay contains a coil with a few turns of large wire and a set of normally open contacts *(Figure 18-10)*. The coil of the relay is connected in series with the run winding of the motor, and the contacts are connected in series with the start winding as shown in *Figure 18-11*. When the thermostat contact closes, power is applied to the run winding of the motor. Since the start winding is open, the motor cannot start. This causes a high current to flow in the run winding circuit. This high current flow produces a strong magnetic field in the coil of the relay, causing the normally open contacts to close and connect the start winding to the circuit. When the motor starts, the run winding current is greatly reduced, permitting the start contacts to reopen and disconnect the start winding from the circuit.

Relationship of Stator and Rotor Fields

The split-phase motor contains a squirrel-cage rotor very similar to those used with three-phase squirrel-cage motors. When power is connected to

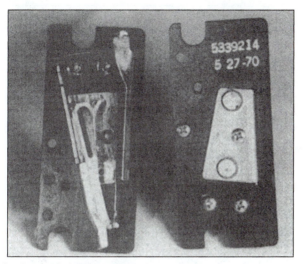

Figure 18-9 Hot wire type of starting relay.

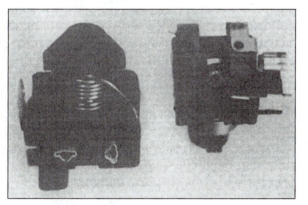

Figure 18-10 Current-type starting relay.

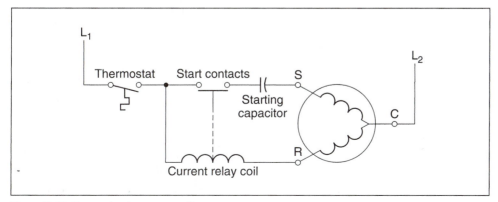

Figure 18-11 Current relay connection.

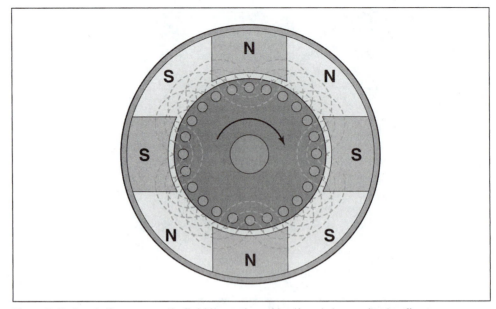

Figure 18-12 A rotating magnetic field is produced by the stator and rotor flux.

the stator windings, the rotating magnetic field induces a voltage into the bars of the squirrel-cage rotor. The induced voltage causes current to flow in the rotor, and a magnetic field is produced around the rotor bars. The magnetic field of the rotor is attracted to the stator field, and the rotor begins to turn in the direction of the rotating magnetic field. After the centrifugal switch opens, only the run winding induces voltage into the rotor. This induced voltage is in phase with the stator current. The inductive reactance of the rotor is high, causing the rotor current to be almost 90° out of phase with the induced voltage. This causes the pulsating magnetic field of the rotor to lag the pulsating magnetic field of the stator by 90°. In the rotor, magnetic poles are created midway between the stator poles *(Figure 18-12)*. These two pulsating magnetic fields produce a rotating magnetic field of their own, and the rotor continues to rotate.

Direction of Rotation

The direction of rotation for the motor is determined by the direction of rotation of the magnetic field created by the run and start windings when the motor is first started. The direction of motor rotation can be changed by reversing the connection of either the start winding or the run winding, but not both. If the start winding is disconnected, the motor can be operated in either direction by manually turning the rotor shaft in the desired direction of rotation.

CAPACITOR-START INDUCTION-RUN MOTORS

The capacitor-start induction-run motor is very similar in construction and operation to the resistance-start induction-run motor. The capacitor-start induction-run motor, however, has an AC electrolytic capacitor connected in series with the centrifugal switch and start winding *(Figure 18-13)*. Although the running characteristics of both the capacitor-start induction-run motor and the resistance-start induction-run motor are identical, the starting characteristics are not. The capacitor-start induction-run motor produces a starting torque that is substantially higher than the resistance-start induction-run motor. Recall that one of the factors that determines the starting torque for a split-phase motor is the phase angle difference between start winding current and run winding current. The starting torque of a resistance-start induction-run motor is low because the phase angle difference between these two currents is only about 40° *(Figure 18-4)*.

When a capacitor of the proper size is connected in series with the start winding, it causes the start winding current to lead the applied voltage. This leading current produces a 90° phase shift between run winding current and start winding current *(Figure 18-14)*. Maximum starting torque is developed at this point.

Although the capacitor-start induction-run motor has a high starting torque, the motor should not be started more than about eight times per hour. Frequent starting can damage the start capacitor by causing it to overheat. If the capacitor must be replaced, care should be taken to use a capacitor of the correct microfarad rating. If a capacitor with too little capacitance is used, the starting current will be less than 90° out of phase with the running current, and the starting torque will be reduced. If the capacitance value is too great, the starting current will be more than 90° out of phase with the running

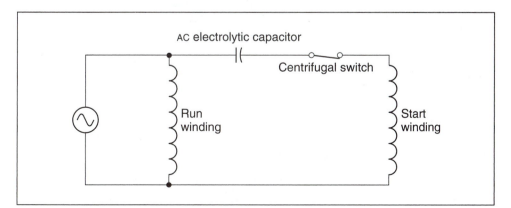

Figure 18-13 An AC electrolytic capacitor is connected in series with the start winding.

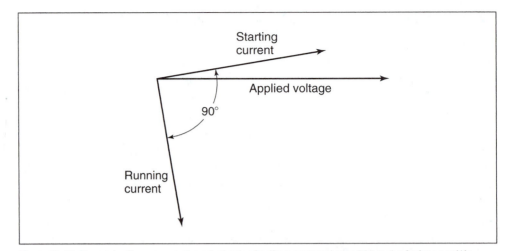

Figure 18-14 Run winding current and start winding current are 90° out of phase with each other.

Figure 18-15 Capacitor-start induction-run motor.

current, and the starting torque will again be reduced. A capacitor-start induction-run motor is shown in *Figure 18-15*.

DUAL-VOLTAGE SPLIT-PHASE MOTORS

Many split-phase motors are designed for operation on 120 or 240 V. *Figure 18-16* shows the schematic diagram of a split-phase motor designed for dual voltage operation. This particular motor contains two run windings and

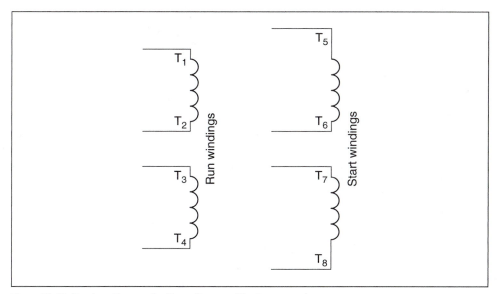

Figure 18-16 Dual-voltage windings for a split-phase motor.

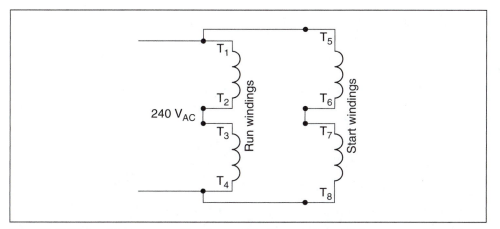

Figure 18-17 High-voltage connection for a split-phase motor with two run and two start windings.

two start windings. The lead numbers for single-phase motors are numbered in a standard manner also. One of the run windings has lead numbers of T_1 and T_2. The other run winding has its leads numbered T_3 and T_4. This particular motor uses two different sets of start winding leads. One set is labeled T_5 and T_6, and the other set is labeled T_7 and T_8.

If the motor is to be connected for high-voltage operation, the run windings and start windings will be connected in series as shown in *Figure 18-17*.

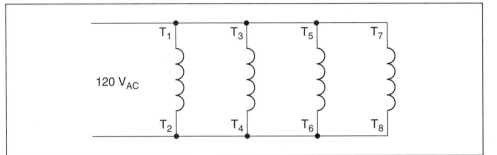

Figure 18-18 Low-voltage connection for a split-phase motor with two run and two start windings.

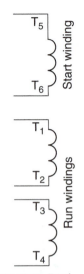

Figure 18-19 Dual-voltage motor with one start winding labeled T_5 and T_6.

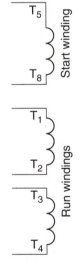

Figure 18-20 Dual-voltage motor with one start winding labeled T_5 and T_8.

The start windings are then connected in parallel with the run windings. It should be noted that if the opposite direction of rotation is desired, T_5 and T_8 will be changed.

For low-voltage operation, the windings must be connected in parallel as shown in *Figure 18-18*. This connection is made by first connecting the run windings in parallel by hooking T_1 and T_3 together and T_2 and T_4 together. The start windings are paralleled by connecting T_5 and T_7 together and T_6 and T_8 together. The start windings are then connected in parallel with the run windings. If the opposite direction of rotation is desired, T_5 and T_6 should be reversed.

Not all dual-voltage single-phase motors contain two sets of start windings. *Figure 18-19* shows the schematic diagram of a motor that contains two sets of run windings and only one start winding. In this illustration, the start winding is labeled T_5 and T_6. It should be noted, however, that some motors identify the start winding by labeling it T_5 and T_8 as shown in *Figure 18-20*.

Regardless of which method is used to label the terminal leads of the start winding, the connection will be the same. If the motor is to be connected for high-voltage operation, the run windings will be connected in series, and the start winding will be connected in parallel with one of the run windings, as shown in *Figure 18-21*. In this type of motor, each winding is rated at 120 V. If the run windings are connected in series across 240 V, each winding will have a voltage drop of 120 V. By connecting the start winding in parallel across only one run winding, it will receive only 120 V when power is applied to the motor. If the opposite direction of rotation is desired, T_5 and T_8 should be swapped.

If the motor is to be operated on low voltage, the windings are connected in parallel as shown in *Figure 18-22*. Since all windings are connected in parallel, each will receive 120 V when power is applied to the motor.

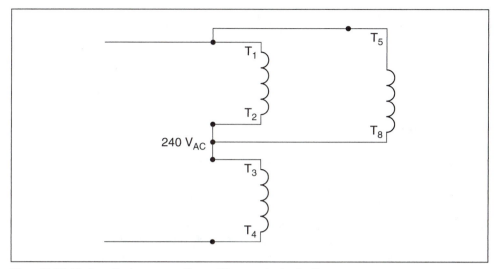

Figure 18-21 High-voltage connection with one start winding.

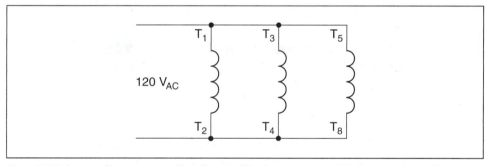

Figure 18-22 Low-voltage connection for a split-phase motor with one start winding.

DETERMINING DIRECTION OF ROTATION FOR SPLIT-PHASE MOTORS

The direction of rotation of a single-phase motor can generally be determined when the motor is connected. The direction of rotation is determined by facing the back or rear of the motor. *Figure 18-23* shows a connection diagram for rotation. If clockwise rotation is desired, T_5 should be connected to T_1. If counterclockwise rotation is desired, T_8 or T_6 should be connected to T_1. It should be noted that this connection diagram assumes that the motor contains two sets of run and two sets of start windings. The type of motor used will determine the actual connection. For example, *Figure 18-21* shows the connection of a motor with two run windings and only one start winding. If this motor were to be connected for clockwise rotation, terminal T_5

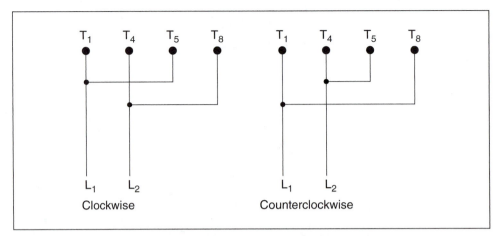

Figure 18-23 Determining direction of rotation for a split-phase motor.

would be connected to T_1, and terminal T_8 would be connected to T_2 and T_3. If counterclockwise rotation is desired, terminal T_8 would be connected to T_1, and terminal T_5 would be connected to T_2 and T_3.

CAPACITOR-START CAPACITOR-RUN MOTORS (PERMANENT-SPLIT CAPACITOR MOTORS)

Although the capacitor-start capacitor-run motor is a split-phase motor, it operates on a different principle than the resistance-start induction-run motor or the capacitor-start induction-run motor. The capacitor-start capacitor-run motor is designed so that its start winding remains energized at all times. A capacitor is connected in series with the winding to provide a continuous leading current in the start winding *(Figure 18-24)*. Since the start

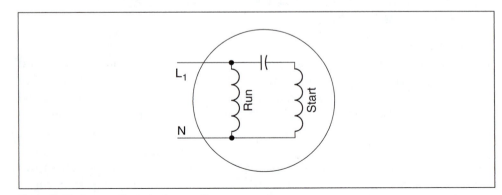

Figure 18-24 A capacitor-start capacitor-run motor.

winding remains energized at all times, no centrifugal switch is needed to disconnect the start winding as the motor approaches full speed. The capacitor used in this type of motor will generally be of the oil-filled type since it is intended for continuous use. An exception to this general rule is a small fractional horsepower motor used in reversible ceiling fans. These fans have a low current draw and use an AC electrolytic capacitor to help save space.

The capacitor-start capacitor-run motor actually operates on the principle of a rotating magnetic field in the stator. Since both run and start windings remain energized at all times, the stator magnetic field continues to rotate, and the motor operates as a two-phase motor. This motor has excellent starting and running torque. It is quiet in operation and has a high efficiency. Since the capacitor remains connected in the circuit at all times, the motor power factor is close to unity.

Although the capacitor-start capacitor-run motor does not require a centrifugal switch to disconnect the capacitor from the start winding, there are some motors that use a second capacitor during the starting period to help improve starting torque *(Figure 18-25)*. Capacitor-start capacitor-run motors that are not hermetically sealed contain a centrifugal switch to disconnect the start capacitor when the motor reaches about 75% of its full-load speed.

Hermetically sealed motors, such as the compressor of a central air conditioning unit designed for operation on single-phase power, use an external starting relay to disconnect the starting capacitor. This type of motor generally employs a potential starting relay *(Figure 18-26)*. The potential starting relay operates by sensing an increase in the voltage developed in the start winding

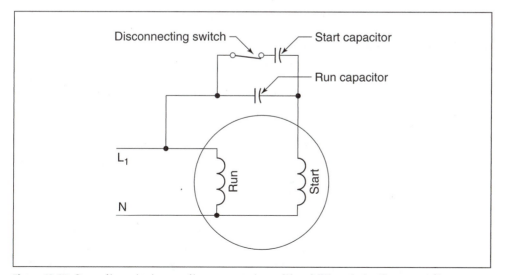

Figure 18-25 Capacitor-start capacitor-run motor with additional starting capacitor.

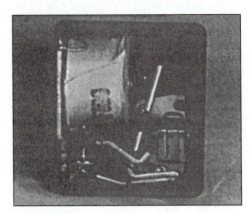

Figure 18-26 Potential starting relay.

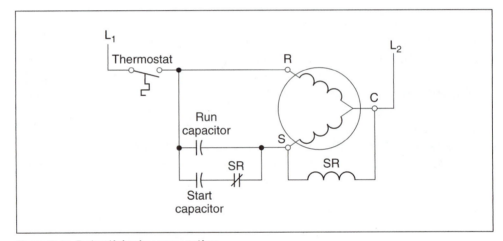

Figure 18-27 Potential relay connection.

when the motor is operating. A schematic diagram of a potential starting relay circuit is shown in *Figure 18-27.* In this circuit, the potential relay is used to disconnect the starting capacitor from the circuit when the motor reaches about 75% of its full speed. The starting relay coil, SR, is connected in parallel with the start winding of the motor. A normally closed SR contact is connected in series with the starting capacitor. When the thermostat contact closes, power is applied to both the run and start windings. At this point in time, both the start and run capacitors are connected in the circuit.

As the rotor begins to turn, its magnetic field induces a voltage into the start winding, producing a higher voltage across the start winding than the applied voltage. When the motor has accelerated to about 75% of its full speed, the voltage across the start winding is high enough to energize the coil of the potential relay. This causes the normally closed SR contact to open and disconnect the start capacitor from the circuit.

SHADED-POLE INDUCTION MOTORS

The **shaded-pole induction motor** is popular because of its simplicity and long life. This motor contains no start windings or centrifugal switch. It contains a squirrel-cage rotor and operates on the principle of a rotating magnetic field. The rotating magnetic field is created by a **shading coil** wound on one side of each pole piece. Shaded-pole motors are generally fractional horsepower motors that are used for low-torque applications such as operating fans and blowers.

shaded-pole induction motor

shading coil

The Shading Coil

The shading coil is wound around one end of the pole piece *(Figure 18-28)*. The shading coil is actually a large loop of copper wire or a copper band. Both ends are connected to form a complete circuit. The shading coil acts in the same manner as a transformer with a shorted secondary winding. When the current of the AC waveform increases from zero toward its positive peak, a magnetic field is created in the pole piece. As magnetic lines of flux cut through the shading coil, a voltage is induced in the coil. Since the coil is a low-resistance short circuit, a large amount of current flows in the loop. This current causes an opposition to the change of magnetic flux *(Figure 18-29)*. As long as voltage is induced into the shading coil, there will be an opposition to the change of magnetic flux.

When the AC current reaches its peak value, it is no longer changing, and no voltage is being induced into the shaded coil. Since there is no current flow in the shading coil, there is no opposition to the magnetic flux. The magnetic flux of the pole piece is now uniform across the pole face *(Figure 18-30)*.

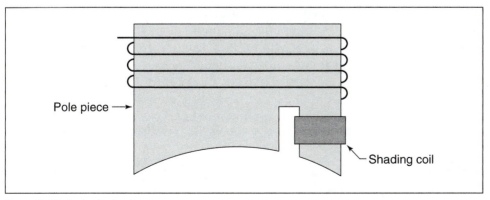

Figure 18-28 A shaded pole.

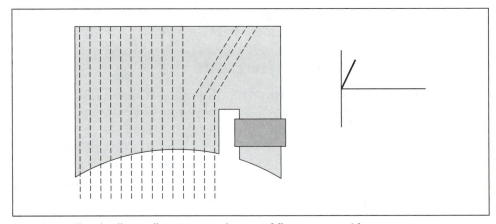

Figure 18-29 The shading coil opposes a change of flux as current increases.

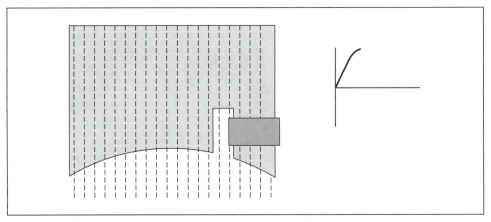

Figure 18-30 There is no opposition to magnetic flux when the current is not changing.

When the AC current begins to decrease from its peak value back toward zero, the magnetic field of the pole piece begins to collapse. A voltage is again induced into the shading coil. This induced voltage creates a current that opposes the change of magnetic flux *(Figure 18-31)*. This causes the magnetic flux to be concentrated in the shaded section of the pole piece.

When the AC current passes through zero and begins to increase in the negative direction, the same set of events happens, except that the polarity of the magnetic field is reversed. If these events were to be viewed in rapid order, the magnetic field would be seen to rotate across the face of the pole piece.

Speed

The speed of the shaded-pole induction motor is determined by the same factors that determine the synchronous speed of other induction motors—frequency and number of stator poles. Shaded-pole motors are

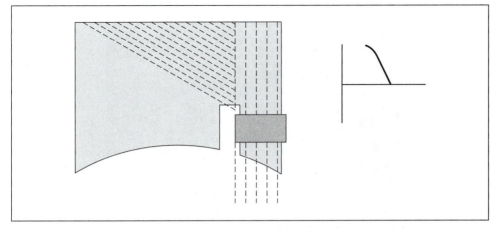

Figure 18-31 The shading coil opposes a change of flux when the current decreases.

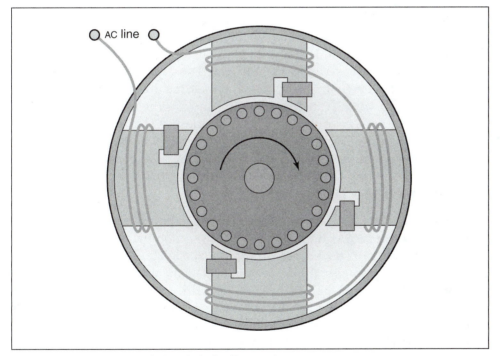

Figure 18-32 Four-pole shaded-pole induction motor.

commonly wound as four- or six-pole motors. *Figure 18-32* shows a drawing of a four-pole shaded-pole induction motor.

General Operating Characteristics

The shaded-pole motor contains a standard squirrel-cage rotor. The amount of torque produced is determined by the strength of the magnetic

Figure 18-33 Stator winding and rotor of a shaded-pole induction motor.

field of the stator, the strength of the magnetic field of the rotor, and the phase angle difference between rotor and stator flux. The shaded-pole induction motor has low starting and running torque.

The direction of rotation is determined by the direction in which the rotating magnetic field moves across the pole face. The rotor will turn in the direction shown by the arrow in *Figure 18-32*. If the direction must be changed, it can be done by removing the stator winding and turning it around. This is not a common practice, however. As a general rule the shaded-pole induction motor is considered to be nonreversible. *Figure 18-33* shows a photograph of the stator winding and rotor of a shaded-pole induction motor.

MULTISPEED MOTORS

**multispeed
motors
consequent pole
motor**

There are two basic types of **multispeed** single-phase **motors**. One is the **consequent pole motor** and the other is a specially wound capacitor-start capacitor-run motor or shaded-pole induction motor. The consequent pole single-phase motor operates in the same basic way as the three-phase consequent pole motor discussed in Unit 17. The speed is changed by reversing the current flow through alternate poles and increasing or decreasing the total number of stator poles. The consequent pole motor is used where high running torque must be maintained at different speeds. A good example of where this type of motor is used is in two-speed compressors for central air conditioning units.

Multispeed Fan Motors

Multispeed fan motors have been used for many years. These motors are generally wound for two to five steps of speed and operate fans and squirrel-cage blowers. A schematic drawing of a three-speed motor is shown in *Figure 18-34*. Notice that the run winding has been tapped to produce low,

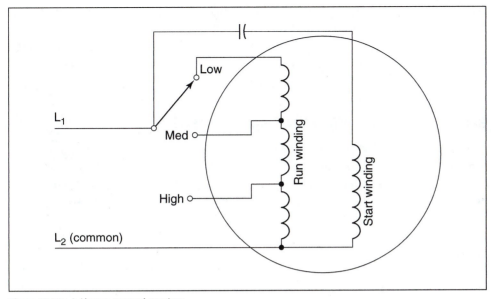

Figure 18-34 A three-speed motor.

medium, and high speed. The start winding is connected in parallel with the run winding section. The other end of the start winding lead is connected to an external oil-filled capacitor. This motor obtains a change of speed by inserting inductance in series with the run winding. The actual run winding for this motor is between the terminals marked *high* and *common*. The winding shown between *high* and *medium* is connected in series with the main run winding. When the rotary switch is connected to the medium speed position, the inductive reactance of this coil limits the amount of current flow through the run winding. When the current of the run winding is reduced, the strength of the magnetic field of the run winding is reduced, and the motor produces less torque. This causes a greater amount of slip and a decrease in motor speed.

If the rotary switch is changed to the low position, more inductance is inserted in series with the run winding. This causes less current to flow through the run winding and another reduction in torque. When the torque is reduced, the motor speed decreases again.

Common speeds for a four-pole motor of this type are 1625, 1500, and 1350 RPM. Notice that this motor does not have wide ranges between speeds as would be the case with a consequent pole motor. Most induction motors would overheat and damage the motor winding if the speed were to be reduced to this extent. This type of motor, however, has much higher impedance windings than most motors. The run windings of most split-phase motors have a wire resistance of 1–4 Ω. This motor will generally have a resistance of

10–15 Ω in its run winding. It is the high impedance of the windings that permits the motor to be operated in this manner without damage.

Since this motor is designed to slow down when load is added, it is not used to operate high-torque loads. This type of motor is generally used to operate only low-torque loads such as fans and blowers.

SINGLE-PHASE SYNCHRONOUS MOTORS

synchronous motors

Single phase **synchronous motors** are small and develop only fractional horsepower. They operate on the principle of a rotating magnetic field developed by a shaded-pole stator. Although they will operate at synchronous speed, they do not require DC excitation current. They are used in applications where constant speed is required such as clock motors, timers, and recording instruments. They also are used as the driving force for small fans because they are small and inexpensive to manufacture. There are two basic types of synchronous motors: Warren or General Electric motor and Holtz motor. These motors are also referred to as hysteresis motors.

Warren Motors

Warren motor

The **Warren motor** is constructed with a laminated stator core and a single coil. The coil is generally wound for 120 V AC operation. The core contains two poles, which are divided into two sections each. One-half of each pole piece contains a shading coil to produce a rotating magnetic field *(Figure 18-35)*. Since the stator is divided into two poles, the synchronous field speed will be 3600 RPM when connected to 60 Hz.

The difference between the Warren and Holtz motors is the type of rotor used. The rotor of the Warren motor is constructed by stacking hardened steel laminations onto the rotor shaft. These disks have high hysteresis loss. The laminations form two crossbars for the rotor. When power is connected to the motor, the rotating magnetic field induces a voltage into the rotor, and a strong starting torque is developed, causing the rotor to accelerate to near synchronous speed. Once the motor has accelerated to near synchronous speed, the flux of the rotating magnetic field will follow the path of minimum reluctance (magnetic resistance) through the two crossbars. This causes the rotor to lock in step with the rotating magnetic field and the motor to operate at 3600 RPM. These motors are often used with small gear trains to reduce the speed to the desired level.

Holtz Motors

Holtz motor

The **Holtz motor** uses a different type of rotor *(Figure 18-36)*. This rotor is cut in such a manner that six slots are formed. These slots form six salient (projecting or jutting) poles for the rotor. A squirrel-cage winding is constructed by

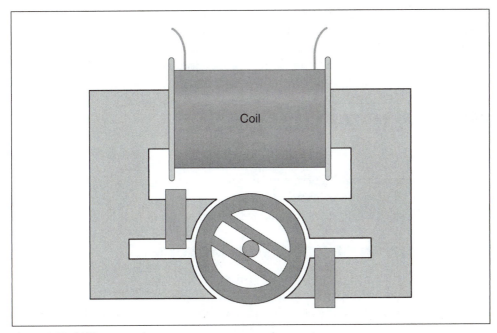

Figure 18-35 A Warren motor.

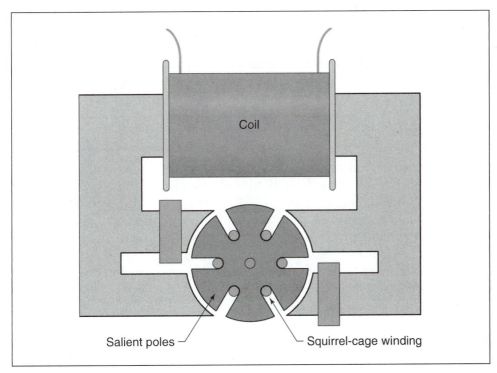

Figure 18-36 A Holtz motor.

inserting a metal bar at the bottom of each slot. When power is connected to the motor, the squirrel-cage winding provides the torque necessary to start the rotor turning. When the rotor approaches synchronous speed, the salient poles will lock in step with the field poles each half cycle. This produces a rotor speed of 1200 RPM (one-third of synchronous speed) for the motor.

REPULSION MOTORS

There are three basic repulsion motors:
1. Repulsion motor
2. Repulsion-start induction-run motor
3. Repulsion-induction motor

Each of these three motors has different operating characteristics.

CONSTRUCTION OF REPULSION MOTORS

repulsion motor

A repulsion motor operates on the principle that like magnetic poles repel each other, not on the principle of a rotating magnetic field. The stator of a **repulsion motor** contains only a run winding very similar to that used in the split-phase motor. Start windings are not necessary. The rotor is actually called an armature because it contains a slotted metal core with windings placed in the slots. The windings are connected to a commutator. A set of brushes makes contact with the surface of the commutator bars. The entire assembly looks very much like a DC armature and brush assembly. One difference, however, is that the brushes of the repulsion motor are shorted together. Their function is to provide a current path through certain parts of the armature, not to provide power to the armature from an external source.

Operation

Although the repulsion motor does not operate on the principle of a rotating magnetic field, it is an induction motor. When AC power is connected to the stator winding, a magnetic field with alternating polarities is produced in the poles. This alternating field induces a voltage into the windings of the armature. When the brushes are placed in the proper position, current flows through the armature windings, producing a magnetic field of the same polarity in the armature. The armature magnetic field is repelled by the stator magnetic field, causing the armature to rotate. Repulsion motors will contain the same number of brushes as there are stator poles. Repulsion motors are commonly wound for four, six, or eight poles.

Brush Position

The position of the brushes is very important. Maximum torque is developed when the brushes are placed 15° on either side of the pole pieces. *Figure 18-37*

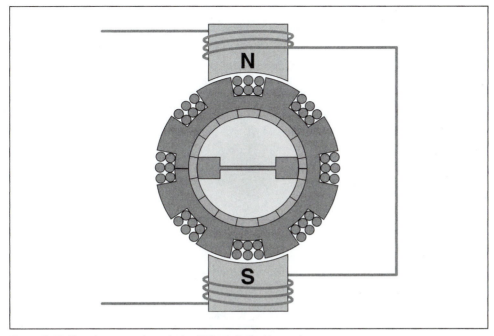

Figure 18-37 Brushes are placed at a 90° angle to the poles.

shows the effect of having the brushes placed at a 90° angle to the pole pieces. When the brushes are in this position, a circuit is completed between the coils located at a right angle to the poles. In this position, there is no induced voltage in the armature windings, and no torque is produced by the motor.

In *Figure 18-38*, the brushes have been moved to a position so that they are in line with the pole pieces. In this position, a large amount of current flows through the coils directly under the pole pieces. This current produces a magnetic field of the same polarity as the pole piece. Since the magnetic field produced in the armature is at a 0° angle to the magnetic field of the pole piece, no twisting or turning force is developed, and the armature does not turn.

In *Figure 18-39*, the brushes have been shifted in a clockwise direction so that they are located 15° from the pole piece. The induced voltage in the armature winding produces a magnetic field of the same polarity as the pole piece. The magnetic field of the armature is repelled by the magnetic field of the pole piece, and the armature turns in the clockwise direction.

In *Figure 18-40*, the brushes have been shifted counterclockwise to a position 15° from the center of the pole piece. The magnetic field developed in the armature again repels the magnetic field of the pole piece, and the armature turns in the counterclockwise direction.

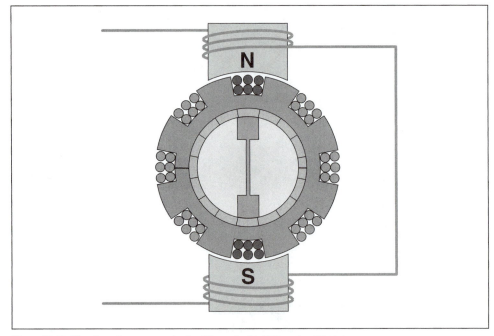

Figure 18-38 The brushes are set at a 0° angle to the pole pieces.

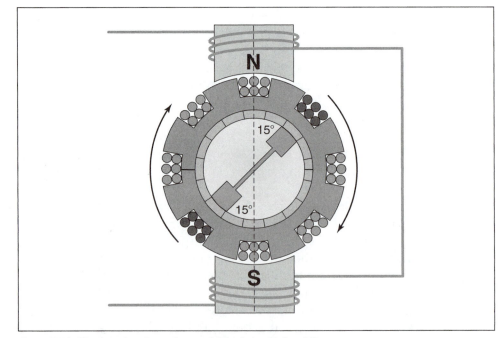

Figure 18-39 The brushes have been shifted clockwise 15°.

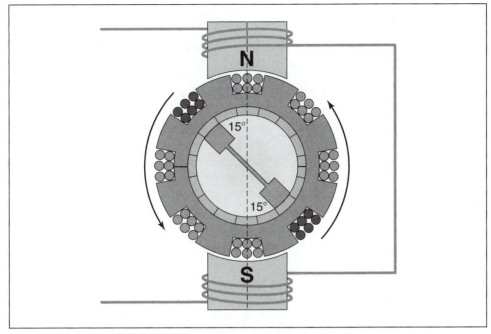

Figure 18-40 The brushes have been shifted counterclockwise 15°.

The direction of armature rotation is determined by the setting of the brushes. The direction of rotation for any type of repulsion motor is changed by setting the brushes 15° on either side of the pole pieces. Repulsion motors have the highest starting torque of any single-phase motor. The speed of a repulsion motor, not to be confused with the repulsion-start induction-run motor or the repulsion-induction motor, can be varied by changing the AC voltage supplying power for the motor. The repulsion motor has excellent starting and running torque but can exhibit unstable speed characteristics. The repulsion motor can race to very high speed if operated with no mechanical load connected to the shaft.

REPULSION-START INDUCTION-RUN MOTORS

The repulsion-start induction-run motor starts as a repulsion motor, but runs as a squirrel-cage motor. There are two types of repulsion-start induction-run motors:
- Brush-riding
- Brush-lifting

The brush-riding motor uses an axial commutator *(Figure 18-41)*. The brushes ride against the commutator segments at all times when the motor is in operation. After the motor has accelerated to approximately 75% of its

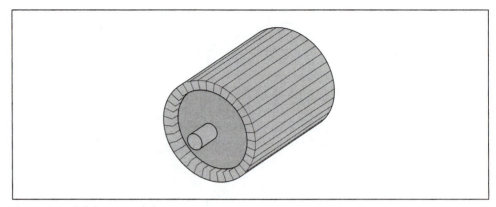

Figure 18-41 Axial commutator.

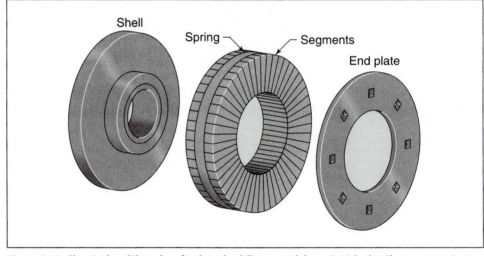

Figure 18-42 Short-circuiting ring for brush-riding repulsion-start induction-run motor.

full-load speed, centrifugal force causes copper segments of a short-circuiting ring to overcome the force of a spring *(Figure 18-42)*. The segments sling out and make contact with the segments of the commutator. This effectively short-circuits all the commutator segments together, and the motor operates in the same manner as a squirrel-cage motor.

The brush-lifting motor uses a radial commutator *(Figure 18-43)*. Weights are mounted at the front of the armature. When the motor reaches about 75% of full speed, these weights swing outward due to centrifugal force and cause two push rods to act against a spring barrel and short-circuiting necklace. The weights overcome the force of the spring and cause the entire spring barrel and brush holder assembly to move toward the back of the motor *(Figure 18-44)*. The motor is designed so that the short-circuiting necklace will short-circuit the commutator bars before the brushes lift off

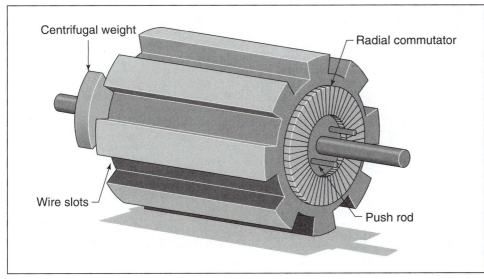

Figure 18-43 A radial commutator is used with the brush-lifting motor.

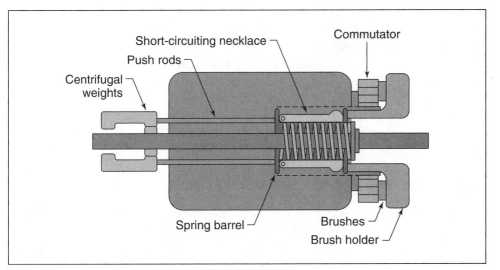

Figure 18-44 Brush-lifting repulsion-start induction-run motor.

the surface of the radial commutator. The motor will now operate as a squirrel-cage induction motor. The brush-lifting motor has several advantages over the brush-riding motor. Since the brushes lift away from the commutator surface during operation, wear on both the commutator and brushes is greatly reduced. Also, the motor does not have to overcome the friction of the brushes riding against the commutator surface during operation. As a result the brush-lifting motor is quieter in operation.

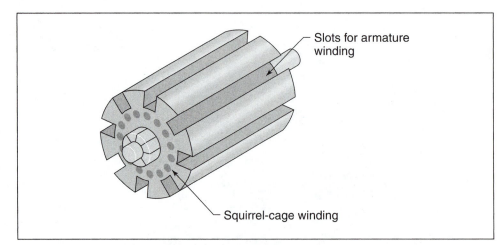

Slots for armature winding

Squirrel-cage winding

Figure 18-45 Repulsion-induction motors contain both armature and squirrel-cage windings.

T₁

T₂

Figure 18-46 Schematic symbol for a repulsion motor.

REPULSION-INDUCTION MOTORS

The repulsion-induction motor is basically the same as the repulsion motor except that a set of squirrel-cage windings is added to the armature *(Figure 18-45)*. This type of motor contains no centrifugal mechanism or short-circuiting device. The brushes ride against the commutator at all times. The repulsion-induction motor has very high starting torque because it starts as a repulsion motor. The squirrel-cage winding, however, gives it much better speed characteristics than a standard repulsion motor. This motor has very good speed regulation between no load and full load. Its running characteristics are similar to a DC compound motor. The schematic symbol for a repulsion motor is shown in *Figure 18-46*.

STEPPING MOTORS

stepping motors

Stepping motors are devices that convert electrical impulses into mechanical movement. Stepping motors differ from other types of DC or AC motors in that their output shaft moves through a specific angular rotation each time the motor receives a pulse. The stepping motor allows a load to be controlled as to speed, distance, or position. These motors are very accurate in their control performance. There is generally less than 5% error per angle of rotation, and this error is not cumulative regardless of the number of rotations. Stepping motors are operated on DC power, but can be used as a two-phase synchronous motor when connected to AC power.

Theory of Operation

Stepping motors operate on the theory that like magnetic poles repel and unlike magnetic poles attract. Consider the circuit shown in *Figure 18-47*. In this illustration, the rotor is a permanent magnet and the stator windings consist of two electromagnets. If current flows through the winding of stator pole A in such a direction that it creates a north magnetic pole and through B in such a direction that it creates a south magnetic pole, it would be impossible to determine the direction of rotation. In this condition, the rotor could turn in either direction.

Now consider the circuit shown in *Figure 18-48*. In this circuit, the motor contains four stator poles instead of two. The direction of current flow through stator pole A is still in such a direction as to produce a north magnetic field, and the current flow through pole B produces a south magnetic field. The current flow through stator pole C, however, produces a south magnetic field, and the current flow through pole D produces a north magnetic field. In this illustration, there is no doubt as to the direction or angle of rotation. In this example, the rotor shaft will turn 90° in a counterclockwise direction.

Figure 18-49 shows yet another condition. In this example, the current flow through poles A and C is in such a direction as to form a north magnetic pole, and the direction of current flow through poles B and D forms south magnetic poles. In this illustration, the permanent magnetic rotor has rotated to a position between the actual pole pieces.

To allow for better stepping resolution, most stepping motors have eight stator poles, and the pole pieces and rotor have teeth machined into them as shown in *Figure 18-50*. In actual practice the number of teeth machined in the stator and rotor determines the angular rotation achieved each time the motor is stepped. The stator-rotor tooth configuration shown in *Figure 18-50* produces an angular rotation of 1.8° per step.

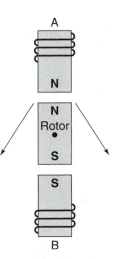

Figure 18-47 The rotor could turn in either direction.

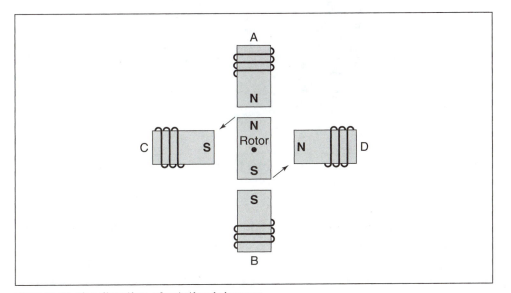

Figure 18-48 The direction of rotation is known.

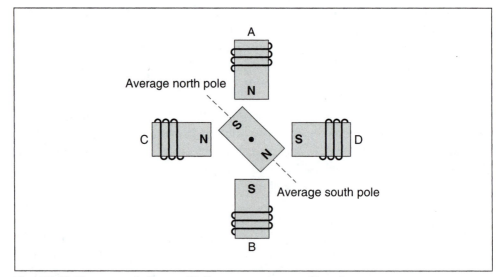

Figure 18-49 The magnet aligns with the average magnetic pole.

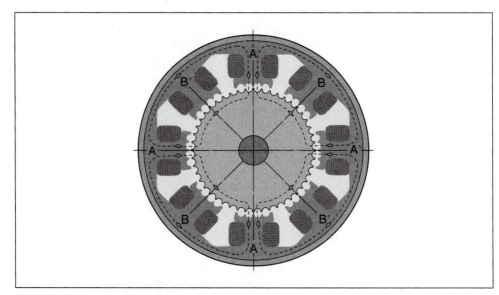

Figure 18-50 Construction of a stepping motor (courtesy of The Superior Electric Co.).

Windings

There are different methods of winding stepping motors. A standard three-lead motor is shown in *Figure 18-51*. The common terminal of the two windings is connected to ground of an aboveground and belowground power supply. Terminal 1 is connected to the common of a single-pole

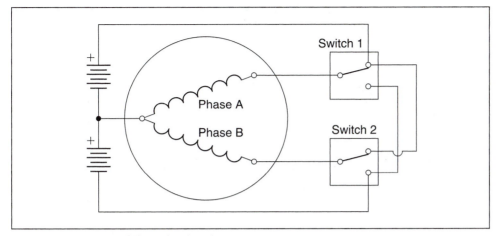

Figure 18-51 A standard three-lead motor.

double-throw switch (switch 1) and terminal 3 is connected to the common of another single-pole double-throw switch (switch 2). One of the stationary contacts of each switch is connected to the positive, or aboveground, voltage, and the other stationary contact is connected to the negative, or belowground, voltage. The polarity of each winding is determined by the position setting of its control switch.

Stepping motors can also be wound bifilar as shown in *Figure 18-52.* The term **bifilar** means that two windings are wound together. This is similar to a transformer winding with a center-tap lead. Bifilar stepping motors have twice as many windings as the three-lead type, which makes it necessary to use smaller wire in the windings. This results in higher wire resistance in the winding, producing a better inductive-resistive (LR) time constant for the bifilar wound motor. The increased LR time constant results in better motor performance. The use of a bifilar stepping motor also simplifies the drive circuitry requirements. Notice that the bifilar motor does not require an aboveground and belowground power supply. As a general rule, the power supply voltage should be about five times greater than the motor voltage. A current-limiting resistance is used in the common lead of the motor. This current-limiting resistor also helps to improve the LR time constant.

bifilar

Four-Step Switching (Full-Stepping)

The switching arrangement shown in *Figure 18-52* can be used for a four-step switching sequence (full-stepping). Each time one of the switches changes position, the rotor will advance one-fourth of a tooth. After four steps, the rotor has turned the angular rotation of one "full" tooth. If the rotor and stator have 50 teeth, it will require 200 steps for the motor to rotate one

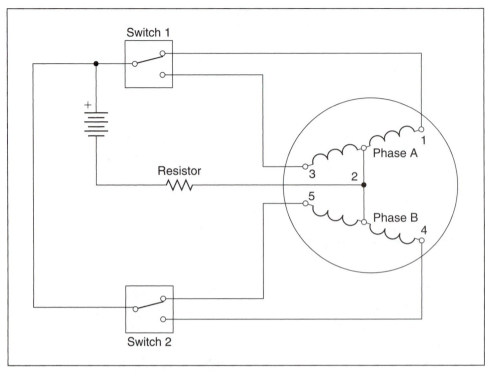

Figure 18-52 Bifilar-wound stepping motor.

Step	Switch 1	Switch 2
1	1	5
2	1	4
3	3	4
4	3	5
1	1	5

Figure 18-53 Four-step switching sequence.

full revolution. This corresponds to an angular rotation of 1.8° per step (360°/200 steps = 1.8° per step). The chart shown in *Figure 18-53* illustrates the switch positions for each step.

Eight-Step Switching (Half-Stepping)

Figure 18-54 illustrates the connections for an eight-step switching sequence (half-stepping). In this arrangement, the center-tap leads for phases A

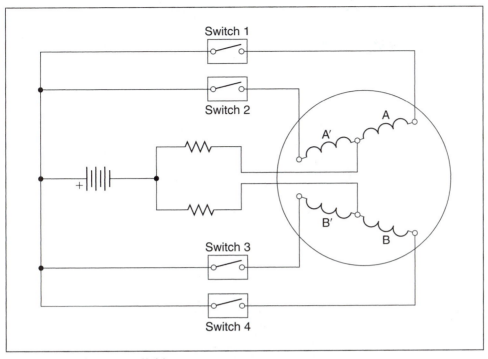

Figure 18-54 Eight-step switching.

and B are connected through their own separate current-limiting resistors back to the negative terminal of the power supply. This circuit contains four separate single-pole switches instead of two switches. The advantage of this arrangement is that each step causes the motor to rotate one-eighth of a tooth instead of one-fourth of a tooth. The motor now requires 400 steps to produce one revolution, which produces an angular rotation of 0.9° per step. This results in better stepping resolution and greater speed capability. The chart in *Figure 18-55* illustrates the switch position for each step. A stepping motor is shown in *Figure 18-56*.

AC Operation

Stepping motors can be operated on AC voltage. In this mode of operation, they become two-phase, AC, synchronous, constant-speed motors and are classified as a *permanent magnet induction motor*. Refer to the exploded diagram of a stepping motor shown in *Figure 18-57*. Notice that this motor has no brushes, slip rings, commutator, gears, or belts. Bearings maintain a constant air gap between the permanent magnet rotor and the stator windings. A typical eight stator pole stepping motor will have a synchronous speed of 72 RPM when connected to a 60-Hz, two-phase AC power line.

Step	Switch 1	Switch 2	Switch 3	Switch 4
1	On	Off	On	Off
2	On	Off	Off	Off
3	On	Off	Off	On
4	Off	Off	Off	On
5	Off	On	Off	On
6	Off	On	Off	Off
7	Off	On	On	Off
8	Off	Off	On	Off
1	On	Off	On	Off

Figure 18-55 Eight-step switching sequence.

Figure 18-56 Stepping motor (courtesy of The Superior Electric Co.).

A resistive-capacitive network can be used to provide the 90° phase shift needed to change single-phase AC into two-phase AC. A simple forward-off-reverse switch can be added to provide directional control. A sample circuit of this type is shown in *Figure 18-58*. The correct values of resistance and capacitance are necessary for proper operation. Incorrect values can result in random direction of rotation when the motor is started, change of direction when the load is varied, erratic and unstable operation, and failure of the motor to start. The correct values of resistance and capacitance will be different with different stepping motors. The manufacturer's recommendations should be followed for the particular type of stepping motor used.

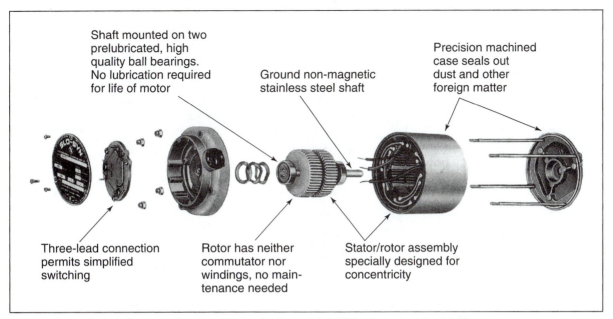

Shaft mounted on two prelubricated, high quality ball bearings. No lubrication required for life of motor

Ground non-magnetic stainless steel shaft

Precision machined case seals out dust and other foreign matter

Three-lead connection permits simplified switching

Rotor has neither commutator nor windings, no maintenance needed

Stator/rotor assembly specially designed for concentricity

Figure 18-57 Exploded diagram of a stepping motor (courtesy of The Superior Electric Co.).

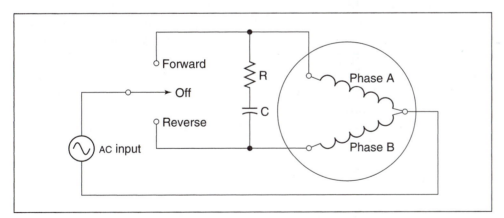

Figure 18-58 Phase-shift circuit converts single-phase into two-phase.

Stepping Motor Characteristics

When stepping motors are used as two-phase synchronous motors they can start, stop, or reverse direction of rotation virtually instantly. The motor will start within about 1-1/2 cycles of the applied voltage and will stop within 5-25 ms. The motor can maintain a stalled condition without harm to the motor. Since the rotor is a permanent magnet, there is no

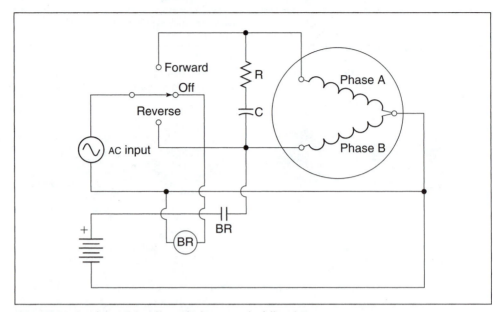

Figure 18-59 Applying DC voltage to increase holding torque.

induced current in the rotor. There is no high inrush of current when the motor is started. The starting and running currents are the same. This simplifies the power requirements of the circuit used to supply the motor. Due to the permanent magnetic structure of the rotor, the motor does provide holding torque when turned off. If more holding torque is needed, DC voltage can be applied to one or both windings when the motor is turned off. An example circuit of this type is shown in *Figure 18-59*. If DC is applied to one winding, the holding torque will be approximately 20% greater than the rated torque of the motor. If DC is applied to both windings, the holding torque will be about 1-1/2 times greater than the rated torque.

UNIVERSAL MOTORS

universal motor

The **universal motor** is often referred to as an AC series motor. This motor is very similar to a DC series motor in its construction in that it contains a wound armature and brushes *(Figure 18-60)*. The universal motor, however, has the addition of a compensating winding. If a DC series motor was connected to alternating current, the motor would operate poorly for several reasons. The armature windings would have a large amount of inductive reactance when connected to alternating current. Another reason for poor operation is that the field poles of most DC machines contain solid

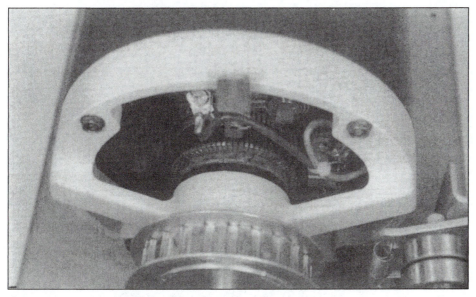

Figure 18-60 Armature and brushes of a universal motor.

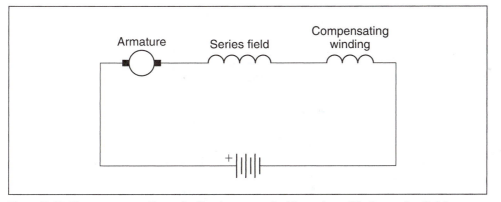

Figure 18-61 The compensating winding is connected in series with the series field winding.

metal pole pieces. If the field was connected to AC, a large amount of power would be lost to eddy current induction in the pole pieces. Universal motors contain a laminated core to help prevent this problem. The compensating winding is wound around the stator and functions to counteract the inductive reactance in the armature winding.

The universal motor is so named because it can be operated on AC or DC voltage. When the motor is operated on direct current, the compensating winding is connected in series with the series field winding *(Figure 18-61)*.

conductive compensation

inductive compensation

Connecting the Compensating Winding for AC Current

When the universal motor is operated with AC power, the compensating winding can be connected in two ways. If it is connected in series with the armature as shown in *Figure 18-62,* it is known as **conductive compensation**.

The compensating winding can also be connected by shorting its leads together as shown in *Figure 18-63*. When connected in this manner, the winding acts as a shorted secondary winding of a transformer. Induced current permits the winding to operate when connected in this manner. This connection is known as **inductive compensation**. Inductive compensation cannot be used when the motor is connected to direct current.

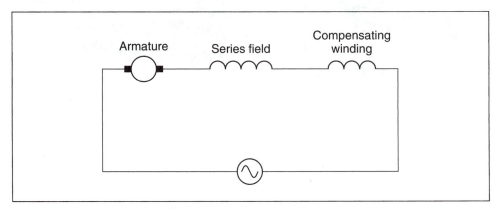

Figure 18-62 Conductive compensation.

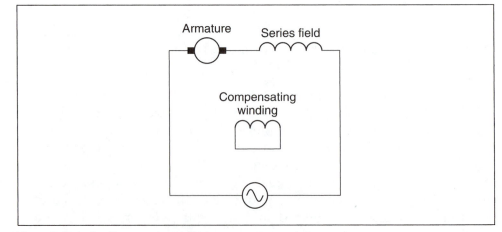

Figure 18-63 Inductive compensation.

The Neutral Plane

Since the universal motor contains a wound armature, commutator, and brushes, the brushes should be set at the neutral plane position. This can be done in the universal motor in a manner similar to that of setting the neutral plane of a DC machine. When setting the brushes to the neutral plane position in a universal motor, either the series or compensating winding can be used. To set the brushes to the neutral plane position by using the series winding *(Figure 18-64)*, alternating current is connected to the armature leads. A voltmeter is connected to the series winding. Voltage is then applied to the armature. The brush position is then moved until the voltmeter connected to the series field reaches a null position. (The null position is reached when the voltmeter reaches its lowest point.)

If the compensating winding is used to set the neutral plane, alternating current is again connected to the armature and a voltmeter is connected to the compensating winding *(Figure 18-65)*. Alternating current is then applied to the armature. The brushes are then moved until the voltmeter indicates its highest or peak voltage.

Speed Regulation

The speed regulation of the universal motor is very poor. Since this motor is a series motor, it has the same poor speed regulation as a DC series motor. If the universal motor is connected to a light load or no load, its speed is almost unlimited. It is not unusual for this motor to be operated at several thousand revolutions per minute. Universal motors are used in a number of portable appliances where high horsepower and light weight are needed, such as drill motors, skill saws, and vacuum cleaners. The universal motor is able to produce a high horsepower for its size and weight because of its high operating speed.

Changing the Direction of Rotation

The direction of rotation of the universal motor can be changed in the same manner as changing the direction of rotation of a DC series motor. To change the direction of rotation, change the armature leads with respect to the field leads.

Summary

1. Not all single-phase motors operate on the principle of a rotating magnetic field.

2. Split-phase motors start as two-phase motors by producing an out-of-phase condition for the current in the run winding and the current in the start winding.

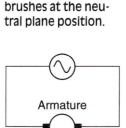

Figure 18-64 Using the series field to set the brushes at the neutral plane position.

Figure 18-65 Using the compensating winding to set the brushes to the neutral plane position.

3. The resistance of the wire in the start winding of a resistance-start induction-run motor is used to produce a phase angle difference between the current in the start winding and the current in the run winding.

4. The capacitor-start induction-run motor uses an AC electrolytic capacitor to increase the phase angle difference between starting and running current. This causes an increase in starting torque.

5. Maximum starting torque for a split-phase motor is developed when the start winding current and run winding current are 90° out of phase with each other.

6. Most resistance-start induction run motors and capacitor-start induction-run motors use a centrifugal switch to disconnect the start windings when the motor reaches approximately 75% of full-load speed.

7. The capacitor-start capacitor-run motor operates as a two-phase motor because both the start and run windings remain energized during motor operation.

8. Most capacitor-start capacitor-run motors use an AC oil-filled capacitor connected in series with the start winding.

9. The capacitor of the capacitor-start capacitor-run motor does help to correct the power factor.

10. Shaded-pole induction motors operate on the principle of a rotating magnetic field.

11. The rotating magnetic field of a shaded-pole induction motor is produced by placing shading loops or coils on one side of the pole piece.

12. The synchronous field speed of a single-phase motor is determined by the number of stator poles and the frequency of the applied voltage.

13. Consequent pole motors are used when a change of motor speed is desired and high torque must be maintained.

14. Multispeed fan motors are constructed by connecting windings in series with the main run winding.

15. Multispeed fan motors have high-impedance stator windings to prevent them from overheating when their speed is reduced.

16. There are three basic repulsion motors: repulsion motor, repulsion-start induction-run motor, and repulsion-induction motor.

17. Repulsion motors have the highest starting torque of any single-phase motor.

18. The direction of rotation of repulsion motors is changed by setting the brushes 15° on either side of the pole pieces.

19. The direction of rotation for split-phase motors is changed by reversing the start winding in relation to the run winding.

20. Shaded-pole motors are generally considered to be nonreversible.

21. There are two types of repulsion-start induction-run motors: brush-riding and brush-lifting.

22. The brush-riding motor uses an axial commutator and a short-circuiting device, which short-circuits the commutator segments when the motor reaches approximately 75% of full-load speed.

23. The brush-lifting repulsion-start induction-run motor uses a radial commutator. A centrifugal device causes the brushes to move away from the commutator and a short-circuiting necklace to short-circuit the commutator when the motor reaches about 75% of full-load speed.

24. The repulsion-induction motor contains both a wound armature and squirrel-cage windings.

25. There are two types of single-phase synchronous motors: Warren and Holtz.

26. Single-phase synchronous motors are sometimes called hysteresis motors.

27. The Warren motor operates at a speed of 3600 RPM.

28. The Holtz motor operates at a speed of 1200 RPM.

29. Stepping motors generally operate on direct current and are used to produce angular movements in steps.

30. Stepping motors are generally used for position control.

31. Stepping motors can be used as synchronous motors when connected to two-phase alternating current.

32. Stepping motors operate at a speed of 72 RPM when connected to 60–Hz power.

33. Stepping motors can produce a holding torque when direct current is connected to their windings.

34. Universal motors operate on direct or alternating current.

35. Universal motors contain a wound armature and brushes.

36. Universal motors are also called AC series motors.

37. Universal motors have a compensating winding that helps overcome inductive reactance.

38. The direction of rotation for a universal motor can be changed by reversing the armature leads with respect to the field leads.

Review Questions

1. What are the three basic types of split-phase motors?

2. The voltages of a two-phase system are how many degrees out of phase with each other?

3. How are the start and run windings of a split-phase motor connected in relation to each other?

4. In order to produce maximum starting torque in a split-phase motor, how many degrees out of phase should the start and run winding currents be with each other?

5. What is the advantage of the capacitor-start induction-run motor over the resistance-start induction-run motor?

6. On average, how many degrees out of phase with each other are the start and run winding currents in a resistance-start induction-run motor?

7. What device is used to disconnect the start windings for the circuit in most nonhermetically sealed capacitor-start induction-run motors?

8. Why does a split-phase motor continue to operate after the start windings have been disconnected from the circuit?

9. How can the direction of rotation of a split-phase motor be reversed?

10. If a dual-voltage split-phase motor is to be operated on high voltage, how are the run windings connected in relation to each other?

11. When determining the direction of rotation for a split-phase motor, should you face the motor from the front or from the rear?

12. What type of split-phase motor does not generally contain a centrifugal switch?

13. What type of single-phase motor develops the highest starting torque?

14. What is the principle of operation of a repulsion motor?

15. What type of commutator is used with a brush-lifting repulsion-start induction-run motor?

16. When a repulsion-start induction-run motor reaches about 75% of rated full-load speed, it stops operating as a repulsion motor and starts operating

as a squirrel-cage motor. What must be done to cause the motor to begin operating as a squirrel-cage motor?

17. What is the principle of operation of a capacitor-start capacitor-run motor?

18. What causes the magnetic field to rotate in a shaded-pole induction motor?

19. How can the direction of rotation of a shaded-pole induction motor be changed?

20. How is the speed of a consequent pole motor changed?

21. Why can a multispeed fan motor be operated at lower speed than most induction motors without harm to the motor windings?

22. What is the speed of operation of the Warren motor?

23. What is the speed of operation of the Holtz motor?

24. Explain the difference in operation between a stepping motor and a common DC motor.

25. What is the principle of operation of a stepping motor?

26. What does the term *bifilar* mean?

27. Why do stepping motors have teeth machined in the stator poles and rotor?

28. When a stepping motor is connected to AC power, how many phases must be applied to the motor?

29. What is the synchronous speed of an eight-pole stepping motor when connected to a two-phase 60-Hz AC line?

30. How can the holding torque of a stepping motor be increased?

31. Why is the AC series motor often referred to as a universal motor?

32. What is the function of the compensating winding?

33. How is the direction of rotation of the universal motor reversed?

34. When the motor is connected to DC voltage, how must the compensating winding be connected?

35. Explain how to set the neutral plane position of the brushes by using the series field.

36. Explain how to set the neutral plane position by using the compensating winding.

UNIT 19

Motor Maintenance and Troubleshooting

Objectives

After studying this unit, you should be able to:

- Describe maintenance procedures for motors.
- Discuss the necessity of keeping accurate maintenance records.
- Test motor windings to determine if they are open or grounded.
- Discuss differences in the maintenance required for DC machines as compared with AC machines.

As a general rule, electric motors require a limited amount of maintenance. Most companies keep maintenance records that list the nature of equipment failure, date, repair time, and replacement parts for a particular machine. Many times these records reveal certain problems that occur at regular intervals, such as bearing failure, brush failure, and so on. Many companies elect to replace bearings and brushes at a time when the plant is not in operation instead of waiting for the machine to fail. If periodic preventive maintenance is not desired or feasible, vibration sensors on the motor shaft can warn of impending bearing failure before it occurs. This permits the bearing to be changed at a time the machine is not in operation instead of having it fail while in production.

MOTOR BEARINGS

Some motors contain sleeve bearings, which require oiling at periodic intervals. Other motors contain open ball or roller bearings, which are lubricated with grease. These motors are often equipped with a grease fitting to permit lubrication with a grease gun. Other motors contain sealed bearings, which do not require lubricating. A lubrication schedule should be maintained for motors that require lubrication. A lubrication schedule is the best method of ensuring that motors are not overlubricated in some cases and never lubricated in other cases.

If vibration sensors are not used, testing motor bearings is generally a matter of feel. Rotating the rotor shaft by hand will generally reveal a bad bearing. If the shaft binds, is hard to turn, or feels rough, it is a good indication of a bad bearing. Lubricating the bearing may solve the problem, but if it continues to feel rough or is hard to turn after lubrication, the bearing is probably bad.

DIRECT CURRENT MACHINES

The maintenance for direct current machines is greater than that required for alternating current machines because DC machines contain a wound armature with a commutator and brushes. The segments between commutator bars can become shorted due to a buildup of carbon caused by the wearing of the brushes *(Figure 19-1)*. Periodic maintenance for direct current machines includes cleaning between the commutator segments with a commutator pick.

Periodic maintenance on DC machines also includes checking the brushes for wear and checking the spring tension on the brushes with a gage. Less than normal spring tension can cause a bad connection between the brush and the commutator segments, resulting in arcing, overheating, and pitting of the commutator bars. Excessive spring tension can result in abnormal wear of both the brushes and the commutator bars.

Testing Field Winding Insulation

Periodic maintenance also includes testing the insulation of the field windings and the armature with a megohmmeter. The series and shunt fields can be tested by connecting one of the megohmmeter leads to either the series field lead or the shunt field lead and connecting the other megohmmeter lead to the case of the motor *(Figure 19-2)*. The winding is tested at its rated voltage or slightly greater. Both the series and shunt field windings should be tested. There is no set standard for determining if a motor winding is good or bad. Some industries use a standard of 1 megohm per 100 volts.

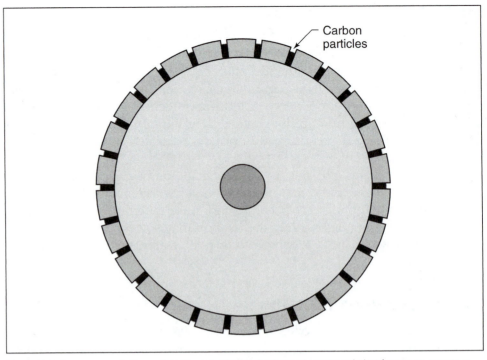

Figure 19-1 An accumulation of carbon particles between commutator bars can cause a shorted armature.

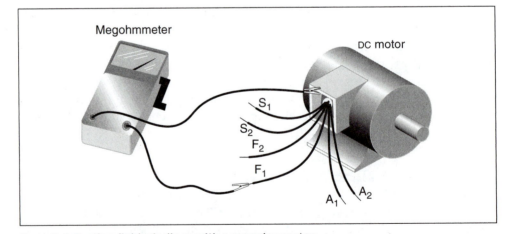

Figure 19-2 Testing field windings with a megohmmeter.

Others use 2 megohms per 100 volts. Other industries use a flat standard such as 1.5 to 15 megohms = Poor, 15 to 50 megohms = Fair, and 50 to 200 megohms = Good. Whatever the resistance of the winding being tested, record it on a maintenance card for future reference.

Testing an Armature

Testing an armature is a bit more difficult. An armature can be tested for grounds and opens with an ohmmeter or a megohmmeter. To test for a grounded winding, connect one lead of an ohmmeter to the shaft of the motor. Each of the commutator bars is touched with the other ohmmeter lead to see if there is a ground path. If the ohmmeter does not reveal a problem, a megohmmeter can then be used in the same way to check the condition of the insulation.

To test the armature for an open winding, connect the leads of an ohmmeter to two adjacent commutator bars *(Figure 19-3)*. The ohmmeter should indicate approximately the same resistance between each set of bars. An infinite amount of resistance indicates an open winding, and a very low resistance indicates a possible shorted winding.

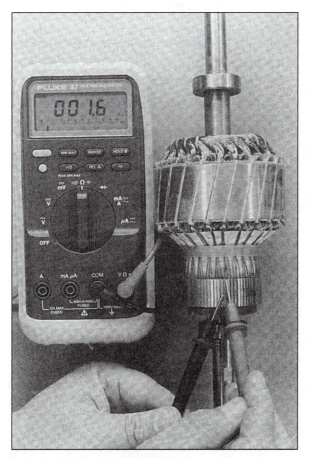

Figure 19-3 Testing an armature for an open winding.

Another test very similar to the ohmmeter test can be performed with a source of low-voltage direct current and a milliammeter. Instead of an ohmmeter, a source of low-voltage DC is used to produce a current flow through the armature winding. A milliammeter is connected in series to indicate the amount of current flow. The test is made by touching adjacent bars and measuring the current flow. The current flow should be approximately the same for each pair of bars tested. A high ampere reading indicates shorted windings, and a zero ampere reading indicates an open winding.

growler

Another device that is often used to test an armature is called a **growler**. A growler is a device constructed by wrapping a coil of wire around a set of V-shaped laminated cores *(Figure 19-4)*. The coil of wire is connected to a

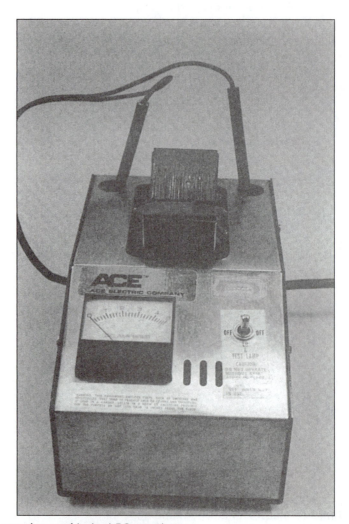

Figure 19-4 A growler used to test DC armatures.

source of alternating current. To use the growler, place an armature on the V-shaped laminated core and place a piece of thin metal, such as a hacksaw blade, on top of the armature *(Figure 19-5)*. Turn on the AC power and observe the action of the hacksaw blade. The changing magnetic field of the AC coil induces a voltage into the armature windings by transformer action. If the armature coil is not shorted, the hacksaw blade remains stationary. A shorted winding, however, causes the hacksaw blade to vibrate rapidly and

Figure 19-5 Testing an armature with a growler.

make a growling sound. Each slot of the armature is tested by rotating the armature and moving the hacksaw blade across it. A shorted coil on a lap- or wave-wound armature causes the hacksaw blade to vibrate over two armature slots.

TESTING ALTERNATING CURRENT MOTORS

Testing the stator windings of an AC motor is very similar to testing the field windings in a direct current machine. Most of the AC motors in industry are three-phase. The stator windings of these motors are generally tested with a megohmmeter. The arrangement of the stator windings dictates the manner in which testing is done. Single voltage motors have their stators connected in either a wye or delta configuration *(Figure 19-6)*. Either of these motors is tested for grounds by connecting one lead of the megohmmeter to the case of the motor and the other lead to one of the T leads. Each T lead is tested for a ground.

Testing for an Open Winding

An open winding in a wye-connected motor is located with an ohmmeter. Assume that winding C of the wye-connected motor in *Figure 19-6* is open. An ohmmeter connected between T1 and T3 will indicate continuity. An

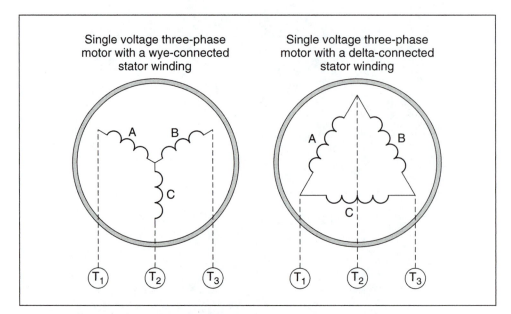

Figure 19-6 Single voltage three-phase motors.

ohmmeter connected between T2 and either of the other two T leads, however, will indicate no continuity.

Determining that one winding in a delta-connected motor is open is not as simple. Assume that winding C of the delta-connected motor is open. An ohmmeter connected between T1 and T3 indicates continuity because a current path exists through windings A and B. If the ohmmeter is connected to terminals T1 and T2, a current path through winding A will give an indication of continuity. If the ohmmeter is connected to terminals T2 and T3, a current path through winding B will indicate continuity. Although it is true that connection to terminals T1 and T3 indicate double the amount of resistance as connections from T1 and T2 or T2 and T3, large motors have such low resistance in their stator windings the difference may not be significant. As a general rule, the motor must be connected to power to determine if the winding is bad. An open winding causes the motor to single-phase and not start.

Testing for a Shorted Winding

Determining if a stator winding is shorted can be very difficult. Often, a winding will short internally and never make a connection to ground. It may, therefore, test good with a megohmmeter. Shorted windings can sometimes be determined with an ohmmeter if the motor is a small horsepower motor. Small horsepower motors contain enough resistance in their stator windings to make a meaningful comparison with an ohmmeter. If one winding is shorted, testing the resistance between all the T leads generally shows one reading to have a higher resistance than the other two. Assume that winding C in the wye-connected motor illustrated in *Figure 19-6* is shorted. The resistance between terminals T1 and T3 will be greater than the resistance between T2 and T1 or T2 and T3.

Large horsepower motors have such low resistance stator winding that they almost appear to be a short circuit under normal conditions. It is sometimes possible to determine that a winding is shorted with an instrument that measures inductance, such as an inductance bridge or a Z meter. A shorted winding has a lower inductance. It may be necessary to connect the motor to power and measure the current on each phase. A shorted winding causes a higher ampere reading than the other two phases. In the event the winding is severely shorted, it may cause a fuse to blow or a circuit breaker to trip when power is applied. Connecting the motor to power should be the last test procedure.

TESTING DUAL-VOLTAGE MOTORS

Dual-voltage motors are generally characterized as having more than three T leads. Nine-lead motors are the most common, but some motors have twelve leads. Nine-lead motors are connected either wye or delta *(Figure 19-7)*. It is possible to determine if the motor is connected wye or delta by checking the

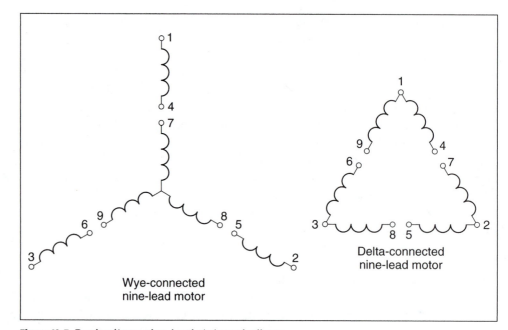

Figure 19-7 Dual voltage nine-lead stator windings.

continuity of the winding. With all stator leads disconnected and separated, a wye-connected motor will have continuity between T7, T8, and T9. In a delta-connected motor, there will be no continuity between T7, T8, and T9. It is important to know how the motor is connected in order to test the stator for grounds or shorts. Each set of windings is tested with a megohmmeter for grounds by connecting one lead to the case of the motor and the other lead to each individual set of winding. In a wye motor, for example, connecting one megohmmeter lead to the motor case and the other to T7, T8, or T9 tests that set of windings for grounds. Connecting one megohmmeter lead to T1 or T4 tests that winding.

Testing the windings for shorts is basically the same as testing a single voltage motor. When testing small motors, resistance values are measured for each set of windings and compared. When testing a wye-connected motor, the resistance reading between T7, T8, and T9 should be approximately equal. The resistance of winding T1 to T4 should be approximately equal to windings T2 to T5 and T3 to T6. If one winding exhibits a much lower value, it is probably shorted.

Delta-connected motors are tested in a similar manner. The resistance measurements should be approximately the same between each set of winding such as T1 to T4, T1 to T9, T2 to T5, T2 to T7, T3 to T6, and T3 to T8. As with single-voltage motors, large motors have such low resistance

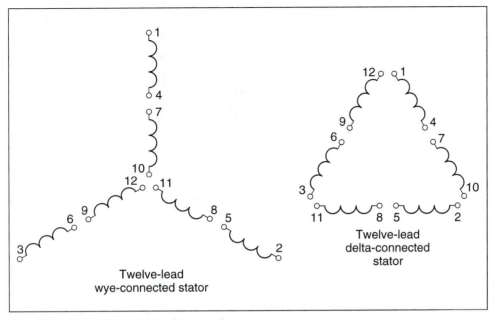

Figure 19-8 Twelve-lead dual voltage motors.

windings that they generally appear as a short circuit. To test for a shorted winding, it is generally necessary to use a device that measures the inductance of the winding.

Twelve-lead motors are basically the same as nine-lead motors except that there are no internal tie points *(Figure 19-8)*. Most twelve-lead motors are intended to be operated with their windings connected in delta. Generally, all twelve stator leads are brought out to permit wye-delta starting. Having access to all twelve leads permits the stator winding to be connected in wye during the starting period and then reconnected as a delta for normal operation. Wye-delta starting is used to reduce the motor inrush current during the starting period. The inrush current will be one-third the value if the stator winding is connected in wye than it will be if it is connected in delta.

Testing the Rotor

The rotor of a squirrel-cage motor is virtually impossible to test in the field. To test a squirrel-cage rotor, it is necessary to access the rotor bars concealed inside the rotor laminations.

Wound rotors with three slip rings are tested in the same basic ways as testing stator windings for grounds, opens, and shorts. The ends of the three-phase winding terminate at each of the slip rings *(Figure 19-9)*. The

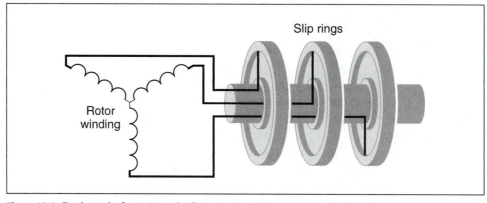

Figure 19-9 Each end of a rotor winding connects to a separate slip ring.

rotor can be tested for grounds by connecting one lead of a megohmmeter to the rotor shaft and the other lead to one of the slip rings. An ohmmeter is used to check for an open winding by testing for continuity between each of the slip rings. An inductance bridge or Z meter is used to test the inductance of the rotor windings. Often it may be necessary to check the current in each one of the M leads (rotor leads) while the motor is in operation. Each phase should indicate approximately the same amount of current flow.

Testing the Rotor of a Synchronous Motor

Synchronous motors fall into two basic categories: slip ring and brushless. Slip ring rotors contain two slip rings, which are used to connect direct current to the rotor winding while the motor is in operation. The rotor in this type of motor is tested in the same basic way as described for testing the rotor of a wound-rotor motor. The main difference between this type of rotor and the rotor of a wound-rotor motor is the synchronous motor has one large winding intended to produce electromagnetic poles in the rotor when direct current is applied *(Figure 19-10)*.

Brushless synchronous motors contain a separate three-phase winding mounted on the rotor shaft—this winding is used to supply direct current to the main rotor winding—as well as rectifiers and fuses. Refer to brushless exciters in Unit 16. A stationary direct current winding is used to provide the magnetic field for the small three-phase winding located on the rotor shaft. All three windings are tested for grounds, opens, and shorts with an ohmmeter and a megohmmeter. Fuses and diodes should also be tested. Diodes are tested with an ohmmeter. The ohmmeter should indicate continuity through the diode in only one direction.

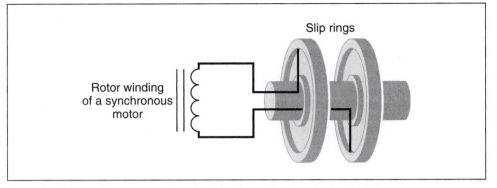

Figure 19-10 A synchronous motor contains two slip rings instead of three.

Identifying the Leads of a Three-Phase Wye-Connected Dual-Voltage Motor

The terminal markings of a three-phase motor are standardized and used to connect the motor for operation on 240 or 480 V. *Figure 19-11* shows these terminal markings and their relationship to the other motor windings. If the motor is connected to a 240-V line, the motor windings are connected parallel to each other as shown in *Figure 19-12*. If the motor is to be operated on a 480-V line, the motor windings are connected in series as shown in *Figure 19-13*.

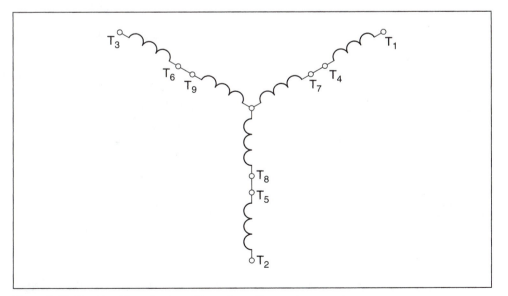

Figure 19-11 Standard terminal markings for a three-phase motor.

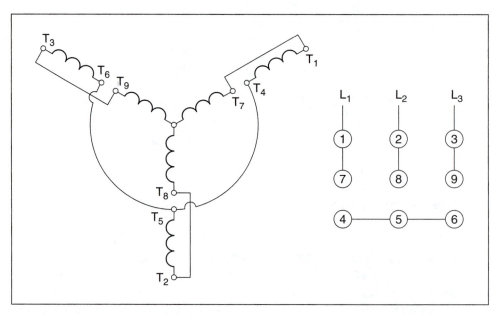

Figure 19-12 Low-voltage connection.

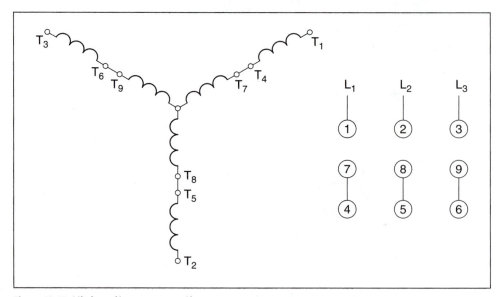

Figure 19-13 High-voltage connection.

As long as these motor windings remain marked with proper numbers, connecting the motor for operation on a 240- or 480-V power line is relatively simple. If these numbers are removed or damaged, however, the lead must be reidentified before the motor can be connected. The following procedure can be used to identify the proper relationship of the motor windings:

1. Using an ohmmeter, divide the motor windings into four separate circuits. One circuit will have continuity to three leads, and the other three circuits will have continuity between only two leads *(Figure 19-11)*.

    Caution: The circuits that exhibit continuity between two leads must be identified as pairs, but do not let the ends of the leads touch anything.

2. Mark the three leads that have continuity with each other as T7, T8, and T9. Connect these three leads to a 240-V three-phase power source *(Figure 19-14)*. (Note: Since these windings are rated at 240 V each, the motor can be safely operated on one set of windings as long as it is not connected to a load.)

3. With the power turned off, connect one end of one of the paired leads to the paired leads to the terminal marked T7. Turn the power on and using an AC voltmeter set for a range not less than 480 V, measure the voltage from the unconnected end of the paired lead to terminals T8 and T9 *(Figure 19-15)*. If the measured voltages are unequal, the wrong paired lead is connected to terminal T7. Turn the power off and connect another paired lead to T7. When the correct set of paired leads is connected to T7, the voltage readings to T8 and T9 are equal.

4. After finding the correct pair of leads, a decision must be made as to which lead should be labeled T4 and which should be labeled T1. Since an induction motor is basically a transformer, the phase windings

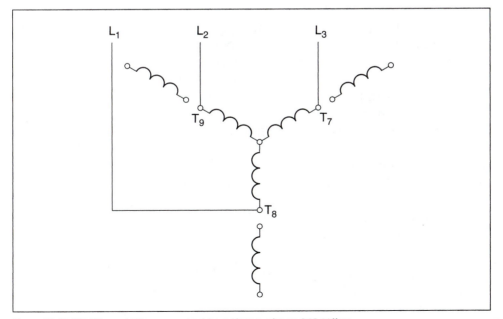

Figure 19-14 T7, T8, and T9 connected to a three-phase 240-V line.

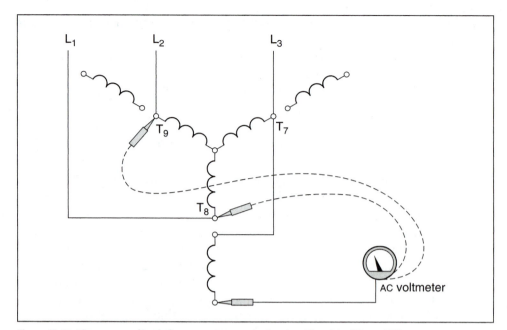

Figure 19-15 Measure voltage from unconnected paired lead to T8 and T9.

act very similar to a multiwinding autotransformer. If terminal T1 is connected to terminal T7, it will operate in a manner similar to that of a transformer with its windings connected to form subtractive polarity. If an AC voltmeter is connected to T4, a voltage of about 140 V should be seen between T4 and T8 or T4 and T9 *(Figure 19-16)*.

If terminal T4 is connected to T7, the winding will operate similar to a transformer with its windings connected for additive polarity. If an AC voltmeter is connected to T1, a voltage of about 360 V will be indicated when the other lead of the voltmeter is connected to T8 or T9 *(Figure 19-17)*. Label leads T1 and T4 using the preceding procedure to determine which lead is correct. Then disconnect and separate T1 and T4.

5. To identify the other leads, follow the same basic procedure. Connect one end of the remaining pairs to T8. Measure the voltage between the unconnected lead and T7 and T9 to determine if it is the correct lead pair for terminal T8. When the correct lead pair is connected to T8, the voltage between the unconnected terminal and T7 or T9 will be equal. Then determine which is T5 or T2 by measuring for a high or low voltage. When T5 is connected to T8, about 360 V can be measured between T2 and T7 or T2 and T9.

6. The remaining pair can be identified as T3 or T6. When T6 is connected to T9, voltage of about 360 V can be measured between T3 and T7 or T3 and T8.

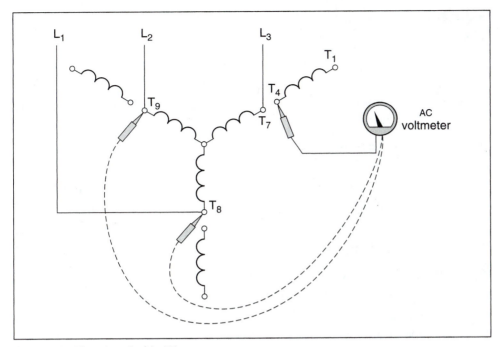

Figure 19-16 T1 connected to T7.

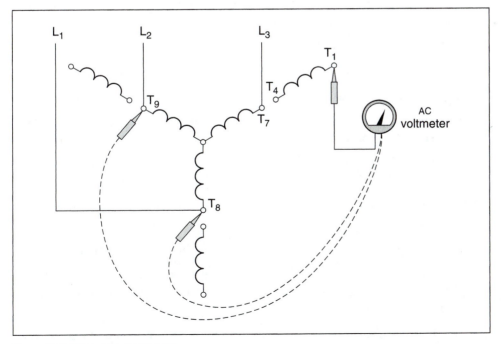

Figure 19-17 T4 connected to T7.

Summary

1. Electric motors generally require little maintenance as compared to internal combustion engines.

2. Direct current motors generally require more maintenance than alternating current motors because DC motors contain a wound armature, commutator, and brushes.

3. Maintenance records should be kept to help establish a periodic maintenance schedule.

4. Many companies elect to change motor bearings and other mechanical components at regular intervals during a shutdown rather than wait until the equipment fails while in use.

5. A growler is used to check the armature of direct current motors for shorts.

6. A direct current armature is tested for grounds with an ohmmeter or milliammeter and a source of direct current.

7. Shunt- and series-field windings are tested for opens with an ohmmeter or milliammeter and a source of direct current.

8. Megohmmeters are used to test windings for grounds.

9. When testing the insulation resistance with a megohmmeter, the insulation is tested at rated voltage or slightly above.

10. Brush tension is tested with a spring gage.

11. The stator windings of AC motors are tested for shorts, opens, and grounds.

12. A megohmmeter is used to test the insulation resistance of stator windings.

13. It is possible to determine if a nine-lead dual-voltage three-phase motor is connected wye or delta by checking for continuity between T7, T8, and T9.

14. Twelve-lead dual-voltage motors permit the motor to be started as a wye and changed to delta during the running period.

15. The rotor windings of a wound-rotor induction motor terminate at three slip rings located on the motor shaft.

16. Two basic types of synchronous motors are slip ring and brushless.

17. Synchronous motors with brushless exciters contain a separate three-phase winding, rectifiers, and fuses, as well as the main rotor winding on the same shaft.

18. The T leads of a wye-connected dual-voltage motor are identified by connecting power to T7, T8, and T9 and determining the polarity of the other windings with a voltmeter.

Review Questions

1. Why is it necessary to keep lubrication records for motors?

2. Why is it necessary to pick between the commutator bars of DC machines?

3. What are the effects of improper brush tension in a DC machine?

4. What instrument is used to test the insulation of motor windings?

5. How do you check for an open armature winding?

6. What instrument is generally used to test an armature for a shorted winding?

7. When testing the stator winding of a nine-lead dual-voltage three-phase motor, it is found that continuity exists between T7, T8, and T9. Is this motor connected internally as a wye or as a delta?

8. Assume that a motor contains twelve T leads. If one lead of an ohmmeter is connected to T10, connecting the other ohmmeter lead to what T lead should give an indication of continuity?

9. A nine-lead dual-voltage motor is being tested with an ohmmeter. Assume that one lead of the ohmmeter is connected to T7 and the other is connected to T8. Now assume that the ohmmeter indicates that there is no continuity between the two windings. Is this an indication that the stator winding is open?

10. Assume that a nine-lead dual-voltage motor is being tested with an ohmmeter. The meter indicates resistance values as follows: T7–T8 = 6Ω, T7–T9 = 6Ω, and T8–T9 = 10Ω. What would these readings indicate?

UNIT **20**

Motor Installation

Objectives

After studying this unit; the student will be able to:

- Determine the full load current rating of different types of motors using the *National Electrical Code (NEC)*®.

- Determine the conductor size for installing motors.

- Determine the overload size for different types of motors.

- Determine the size of the short circuit protective device for individual motors and multi-motor connections.

- Select the proper size starter for a particular motor.

DETERMINING MOTOR CURRENT

There are different types of motors, such as direct current, single-phase AC, two-phase AC, and three-phase AC. Different tables from the NEC are used to determine the running current for these different types of motors. Table 430.247 *(Figure 20-1)* is used to determine the full-load running current for a direct current motor. Table 430.248 *(Figure 20-2)* is used to determine the full-load running current for single-phase motors; Table 430.249 *(Figure 20-3)* is used to determine the running current for two-phase motors; and Table 430.250 *(Figure 20-4)* is used to determine the full-load running current for three-phase motors. Note that the tables list the amount of current that the motor is expected to draw under a full load condition. The motor will exhibit less current draw if it is not under full load. These tables list the ampere rating of the motors according to horsepower (hp) and connected voltage. It should also be noted that NEC Section 430.6(A) states

Table 430.247 Full-Load Current in Amperes, Direct-Current Motors
The following values of full-load currents* are for motors running at base speed.

| Horsepower | Armature Voltage Rating* | | | | | |
	90 Volts	120 Volts	180 Volts	240 Volts	500 Volts	550 Volts
¼	4.0	3.1	2.0	1.6	—	—
⅓	5.2	4.1	2.6	2.0	—	—
½	6.8	5.4	3.4	2.7	—	—
¾	9.6	7.6	4.8	3.8	—	—
1	12.2	9.5	6.1	4.7	—	—
1½	—	13.2	8.3	6.6	—	—
2	—	17	10.8	8.5	—	—
3	—	25	16	12.2	—	—
5	—	40	27	20	—	—
7½	—	58	—	29	13.6	12.2
10	—	76	—	38	18	16
15	—	—	—	55	27	24
20	—	—	—	72	34	31
25	—	—	—	89	43	38
30	—	—	—	106	51	46
40	—	—	—	140	67	61
50	—	—	—	173	83	75
60	—	—	—	206	99	90
75	—	—	—	255	123	111
100	—	—	—	341	164	148
125	—	—	—	425	205	185
150	—	—	—	506	246	222
200	—	—	—	675	330	294

*These are average dc quantities.

Figure 20-1 *NEC*® Table 430.247.

Table 430.248 Full-Load Currents in Amperes, Single-Phase Alternating-Current Motors

The following values of full-load currents are for motors running at usual speeds and motors with normal torque characteristics. The voltages listed are rated motor voltages. The currents listed shall be permitted for system voltage ranges of 110 to 120 and 220 to 240 volts.

Horsepower	115 Volts	200 Volts	208 Volts	230 Volts
⅙	4.4	2.5	2.4	2.2
¼	5.8	3.3	3.2	2.9
⅓	7.2	4.1	4.0	3.6
½	9.8	5.6	5.4	4.9
¾	13.8	7.9	7.6	6.9
1	16	9.2	8.8	8.0
1½	20	11.5	11.0	10
2	24	13.8	13.2	12
3	34	19.6	18.7	17
5	56	32.2	30.8	28
7½	80	46.0	44.0	40
10	100	57.5	55.0	50

Figure 20-2 *NEC*® Table 430.248.

Table 430.249 Full-Load Current, Two-Phase Alternating-Current Motors (4-Wire)

The following values of full-load current are for motors running at speeds usual for belted motors and motors with normal torque characteristics. Current in the common conductor of a 2-phase, 3-wire system will be 1.41 times the value given. The voltages listed are rated motor voltages. The currents listed shall be permitted for system voltage ranges of 110 to 120, 220 to 240, 440 to 480, and 550 to 600 volts.

| Horsepower | Induction-Type Squirrel Cage and Wound Rotor (Amperes) | | | | |
	115 Volts	230 Volts	460 Volts	575 Volts	2300 Volts
½	4.0	2.0	1.0	0.8	—
¾	4.8	2.4	1.2	1.0	—
1	6.4	3.2	1.6	1.3	—
1½	9.0	4.5	2.3	1.8	—
2	11.8	5.9	3.0	2.4	—
3	—	8.3	4.2	3.3	—
5	—	13.2	6.6	5.3	—
7½	—	19	9.0	8.0	—
10	—	24	12	10	—
15	—	36	18	14	—
20	—	47	23	19	—
25	—	59	29	24	—
30	—	69	35	28	—
40	—	90	45	36	—
50	—	113	56	45	—
60	—	133	67	53	14
75	—	166	83	66	18
100	—	218	109	87	23
125	—	270	135	108	28
150	—	312	156	125	32
200	—	416	208	167	43

Figure 20-3 *NEC*® Table 430.249.

to use these tables, instead of the nameplate rating of the motor, to determine conductor size, short circuit protection size, and ground fault protection size. The motor overload size, however, is to be determined by the nameplate rating of the motor.

Direct Current Motors

Figure 20-1 lists the full-load running currents for direct current motors. The horsepower rating of the motor is given in the far left-hand column. Rated voltages are listed across the top of the table. The table shows that a 1-hp motor will have a full-load current of 12.2 amperes when connected to 90 volts DC. If a 1-hp motor is designed to be connected to 240 volts, it will have a current draw of 4.7 amperes.

Table 430.250 Full-Load Current, Three-Phase Alternating-Current Motors
The following values of full-load currents are typical for motors running at speeds usual for belted motors and motors with normal torque characteristics.
    The voltages listed are rated motor voltages. The currents listed shall be permitted for system voltage ranges of 110 to 120, 220 to 240, 440 to 480, and 550 to 600 volts.

Horsepower	Induction-Type Squirrel Cage and Wound Rotor (Amperes)							Synchronous-Type Unity Power Factor* (Amperes)			
	115 Volts	200 Volts	208 Volts	230 Volts	460 Volts	575 Volts	2300 Volts	230 Volts	460 Volts	575 Volts	2300 Volts
½	4.4	2.5	2.4	2.2	1.1	0.9	—	—	—	—	—
¾	6.4	3.7	3.5	3.2	1.6	1.3	—	—	—	—	—
1	8.4	4.8	4.6	4.2	2.1	1.7	—	—	—	—	—
1½	12.0	6.9	6.6	6.0	3.0	2.4	—	—	—	—	—
2	13.6	7.8	7.5	6.8	3.4	2.7	—	—	—	—	—
3	—	11.0	10.6	9.6	4.8	3.9	—	—	—	—	—
5	—	17.5	16.7	15.2	7.6	6.1	—	—	—	—	—
7½	—	25.3	24.2	22	11	9	—	—	—	—	—
10	—	32.2	30.8	28	14	11	—	—	—	—	—
15	—	48.3	46.2	42	21	17	—	—	—	—	—
20	—	62.1	59.4	54	27	22	—	—	—	—	—
25	—	78.2	74.8	68	34	27	—	53	26	21	—
30	—	92	88	80	40	32	—	63	32	26	—
40	—	120	114	104	52	41	—	83	41	33	—
50	—	150	143	130	65	52	—	104	52	42	—
60	—	177	169	154	77	62	16	123	61	49	12
75	—	221	211	192	96	77	20	155	78	62	15
100	—	285	273	248	124	99	26	202	101	81	20
125	—	359	343	312	156	125	31	253	126	101	25
150	—	414	396	360	180	144	37	302	151	121	30
200		552	528	480	240	192	49	400	201	161	40
250	—	—	—	—	302	242	60	—	—	—	—
300	—	—	—	—	361	289	72	—	—	—	—
350	—	—	—	—	414	336	83	—	—	—	—
400	—	—	—	—	477	382	95	—	—	—	—
450	—	—	—	—	515	412	103	—	—	—	—
500	—	—	—	—	590	472	118	—	—	—	—

*For 90 and 80 percent power factor, the figures shall be multiplied by 1.1 and 1.25, respectively.

Figure 20-4 *NEC®* Table 430.250.

Single-Phase AC Motors

The current ratings for single-phase AC motors are given in *Figure 20-2*. Particular attention should be paid to the statement preceding the table. The table asserts that the values listed in this table are for motors that operate under normal speeds and torques. Motors especially designed for low speed and high torque, or multispeed motors, shall have their running current determined from the nameplate rating of the motor.

The voltages listed in the table are 115, 200, 208, and 230. The last sentence of the preceding statement states that the currents listed shall be permitted for voltages of 110 to 120 volts and 220 to 240 volts. This means that if the motor is connected to a 120-volt line, it is permissible to use the currents listed in the 115 volt column. If the motor is connected to a 220-volt line, the 230 volt column can be used.

Example 1

A 3-hp single-phase AC motor is connected to a 208-volt line. What will be the full-load running current of this motor?

Locate 3-hp in the far left-hand column. Follow across to the 208 volt column. The full-load current will be 18.7 amperes.

Two-Phase Motors

Although two-phase motors are seldom used, *Figure 20-3* lists the full-load running currents for these motors. Like single-phase motors, two-phase motors, which are especially designed for low speed, high torque applications and multispeed motors, used the nameplate rating instead of the values shown in the table. When using a two-phase, three-wire system, the size of the neutral conductor must be increased by the square root of 2 (1.41). The reason for this is that the voltages of a two-phase system are 90° out-of-phase with each other as shown in *Figure 20-5*. The principle of two-phase power generation is shown in *Figure 20-6*. In a two-phase alternator, the phase windings are arranged 90° apart. The magnet is the rotor of the alternator. When the rotor turns, it induces voltage into the phase windings, which are 90° apart. When one end of each phase winding is joined to form a common terminal, or neutral, the current in the neutral conductor will be greater than the current in either of the two phase conductors. An example of this is shown in *Figure 20-7*. In this example, a two-phase alternator is connected to a two-phase motor. The current draw on each of the phase windings is 10 amperes. The current flow in the neutral, however, is 1.41 times greater than the current flow in the phase windings, or 14.1 amperes.

Example 2

Compute the phase current and neutral current for a 60-hp, 460-volt two-phase motor.

The phase current can be taken from *Figure 20-3*.

Phase current = 67 amperes

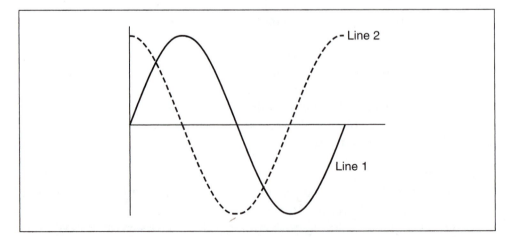

Figure 20-5 The voltages of a two-phase system are 90° out of phase with each other.

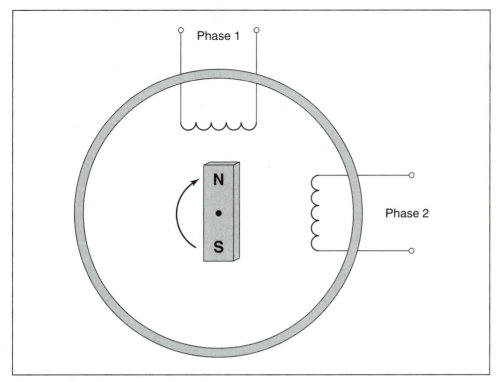

Figure 20-6 Two-phase alternator.

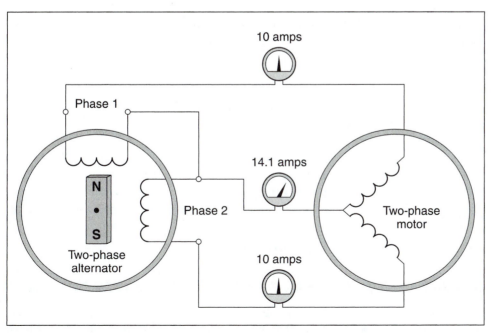

Figure 20-7 The neutral conductor must be larger than the phase conductors.

459

The neutral current will be 1.41 times higher than the phase current.

Neutral current = 67 x 1.41

Neutral current = 94.5 amperes

Three-Phase Motors

Figure 20-4 is used to determine the full-load current of three-phase motors. The notes at the top of the table are very similar to the notes of *Figures 20-2* and *20-3*. The full-load current of low speed, high torque, and multispeed motors is to be determined from the nameplate rating instead of from the values listed in the table. *Figure 20-4* has an extra note that deals with synchronous motors. Notice that the right side of *Figure 20-4* is devoted to the full-load currents of synchronous type motors. The currents listed are for synchronous type motors that are to be operated at unity or 100% power factor. Since synchronous motors are often made to have a leading power factor by overexcitation of the rotor current, the full-load current rating must be increased when this is done. If the motor is operated at 90% power factor, the rated full-load current in the table must be increased by 10%. If the motor is to be operated at 80% power factor, the full-load current is to be increased by 25%.

Example 3

A 150-hp, 460-volt synchronous motor is to be operated at 80% power factor. What will be the full-load current rating of the motor?

The table indicates a current value of 151 amperes for this motor. To determine the running current at 80% power factor, multiply this current by 125% or 1.25. (Multiplying by 1.25 results in the same answer that would be obtained by dividing by 0.80.)

151 x 1.25 = 188.75 or 189 amperes

Example 4

A 200-hp, 2300-volt synchronous motor is to be operated at 90% power factor. What will be the full-load current rating of this motor?

Locate 200 horsepower in the far left-hand column. Follow across to the 2300-volt column listed under synchronous type motors. Increase this value by 10%.

40 x 1.10 = 44 amperes

DETERMINING CONDUCTOR SIZE FOR A SINGLE MOTOR

NEC 430.6(A) states that the conductor for a motor connection shall be based on the values from Tables 430.247, 430.248, 430.249, and 430.250 instead of the motor nameplate current. NEC 430.22(A) states that conductors supplying a single motor shall have an ampacity of not less than 125% of the motor full load current. NEC Section 310 is used to select the conductor size after the ampacity has been determined. The exact table employed will be determined by the wiring conditions. Probably the most frequently used table is 310.16 *(Figure 20-8).*

Table 310.16 Allowable Ampacities of Insulated Conductors Rated 0 Through 2000 Volts, 60°C Through 90°C (140°F Through 194°F), Not More Than Three Current-Carrying Conductors in Raceway, Cable, or Earth (Directly Buried), Based on Ambient Temperature of 30°C (86°F)

	Temperature Rating of Conductor (See Table 310.13.)						
	60°C (140°F)	75°C (167°F)	90°C (194°F)	60°C (140°F)	75°C (167°F)	90°C (194°F)	
Size AWG or kcmil	Types TW, UF	Types RHW, THHW, THW, THWN, XHHW, USE, ZW	Types TBS, SA, SIS, FEP, FEPB, MI, RHH, RHW-2, THHN, THHW, THW-2, THWN-2, USE-2, XHH, XHHW, XHHW-2, ZW-2	Types TW, UF	Types RHW, THHW, THW, THWN, XHHW, USE	Types TBS, SA, SIS, THHN, THHW, THW-2, THWN-2, RHH, RHW-2, USE-2, XHH, XHHW, XHHW-2, ZW-2	Size AWG or kcmil
	COPPER			ALUMINUM OR COPPER-CLAD ALUMINUM			
18	—	—	14	—	—	—	—
16	—	—	18	—	—	—	—
14*	20	20	25	—	—	—	—
12*	25	25	30	20	20	25	12*
10*	30	35	40	25	30	35	10*
8	40	50	55	30	40	45	8
6	55	65	75	40	50	60	6
4	70	85	95	55	65	75	4
3	85	100	110	65	75	85	3
2	95	115	130	75	90	100	2
1	110	130	150	85	100	115	1
1/0	125	150	170	100	120	135	1/0
2/0	145	175	195	115	135	150	2/0
3/0	165	200	225	130	155	175	3/0
4/0	195	230	260	150	180	205	4/0
250	215	255	290	170	205	230	250
300	240	285	320	190	230	255	300
350	260	310	350	210	250	280	350
400	280	335	380	225	270	305	400
500	320	380	430	260	310	350	500
600	355	420	475	285	340	385	600
700	385	460	520	310	375	420	700
750	400	475	535	320	385	435	750
800	410	490	555	330	395	450	800
900	435	520	585	355	425	480	900
1000	455	545	615	375	445	500	1000
1250	495	590	665	405	485	545	1250
1500	520	625	705	435	520	585	1500
1750	545	650	735	455	545	615	1750
2000	560	665	750	470	560	630	2000

CORRECTION FACTORS

Ambient Temp. (°C)	For ambient temperatures other than 30°C (86°F), multiply the allowable ampacities shown above by the appropriate factor shown below.						Ambient Temp. (°F)
21–25	1.08	1.05	1.04	1.08	1.05	1.04	70–77
26–30	1.00	1.00	1.00	1.00	1.00	1.00	78–86
31–35	0.91	0.94	0.96	0.91	0.94	0.96	87–95
36–40	0.82	0.88	0.91	0.82	0.88	0.91	96–104
41–45	0.71	0.82	0.87	0.71	0.82	0.87	105–113
46–50	0.58	0.75	0.82	0.58	0.75	0.82	114–122
51–55	0.41	0.67	0.76	0.41	0.67	0.76	123–131
56–60	—	0.58	0.71	—	0.58	0.71	132–140
61–70	—	0.33	0.58	—	0.33	0.58	141–158
71–80	—	—	0.41	—	—	0.41	159–176

* See 240.4(D).

Figure 20-8 *NEC*® Table 310.16.

Termination Temperature

Another factor that must be taken into consideration when determining the conductor size is the temperature rating of the devices and terminals as specified in NEC 110.14(C). This section states that the conductor is to be selected and coordinated as to not exceed the lowest temperature rating of any connected termination, any connected conductor, or any connected device. This means that regardless of the temperature rating of the conductor, the ampacity must be selected from a column that does not exceed the temperature rating of the termination. The conductors listed in the first column of Table 310.16 have a temperature rating of 60°C, the conductors in the second column have a rating of 75°C, and the conductors in the third column have a rating of 90°C. The temperature ratings of devices such as circuit breakers, fuses, and terminals are found in the UL (Underwriters Laboratories) product directories. Occasionally, the temperature rating may be found on the piece of equipment, but this is the exception and not the rule. As a general rule the temperature rating of most devices will not exceed 75°C.

When the termination temperature rating is not listed or known, NEC 110.14(C)(1)(a) states that for circuits rated at 100 amperes or less, or for #14 AWG through #1 AWG conductors, the ampacity of the wire, regardless of the temperature rating, will be selected from the 60°C column. This does not mean that only those types of insulation listed in the 60°C column can be used, but that the *ampacities* listed in the 60°C column must be used to select the conductor size. For example, assume that a copper conductor with type XHHW insulation is to be connected to a 50-ampere circuit breaker that does not have a listed temperature rating. According to NEC Table 310.16, a #8 AWG copper conductor with XHHW insulation is rated to carry 55 amperes of current. Type XHHW insulation is located in the 90°C column, but the temperature rating of the circuit breaker is not known. Therefore, the wire size must be selected from the ampacity ratings in the 60°C column. A #6 AWG copper conductor with type XHHW insulation would be used.

NEC 110.14(C)(1)(a)(4) has a special provision for motors with marked NEMA (National Electrical Manufacturers Association) design codes B, C, D, or E. This section states that conductors rated at 75°C or higher may be selected from the 75°C column even if the ampacity is 100 amperes or less. This code will not apply to motors that do not have a NEMA design code marked on their nameplate. Most motors manufactured before 1996 will not have a NEMA design code. The NEMA design code letter should not be confused with the code letter that indicates the type of squirrel cage rotor used in the motor.

Example 5

A 30-hp three-phase squirrel cage induction motor is connected to a 480-volt line. The conductors are run in a conduit to the motor. The motor does not have a NEMA design code listed on the nameplate. The termination temperature rating of the devices is not known. Copper conductors with THWN insulation are to be used for this motor connection. What size conductors should be used?

The first step is to determine the full-load current of the motor. This is determined from Table 430.250. The table indicates a current of 40 amperes for this motor. The current must be increased by 25% according to NEC 430.22(A).

40 x 1.25 = 50 amperes

Table 310.16 is used to determine the conductor size. Locate the column that contains THWN insulation in the copper section of the table. THWN is located in the 75°C column. Since this circuit is rated less than 100 amperes and the termination temperature is not known, and the motor does not contain a NEMA design code letter, the conductor size must be selected from the ampacities listed in the 60°C column. A #6 AWG copper conductor with type THWN insulation will be used.

For circuits rated over 100 amperes, or for conductor sizes larger than #1 AWG, NEC 110.14(C)(1)(b) states that the ampacity ratings listed in the 75°C column may be used to select wire sizes unless conductors with a 60°C temperature rating have been selected for use. For example, types TW and UF insulation are listed in the 60°C column. If one of these two insulation types has been specified, the wire size must be chosen from the 60°C column regardless of the ampere rating of the circuit.

OVERLOAD SIZE

When determining the overload size for a motor, the *nameplate* current rating of the motor is used instead of the current values listed in the tables (NEC 430.6(A)). Other factors such as the service factor (SF) or temperature rise (°C) of the motor are also to be considered when determining the overload size for a motor. The temperature rise of the motor is an indication of the type of insulation used on the motor windings and should not be confused with termination temperature discussed in NEC 110.14(C). NEC 430.32 *(Figure 20-9)* is used to determine the overload size for motors of one horsepower or more. The overload size is based on a percentage of the full-load current of the motor listed on the motor nameplate.

430.32 Continuous-Duty Motors.

(A) More Than 1 Horsepower. Each motor used in a continuous duty application and rated more than 1 hp shall be protected against overload by one of the means in 430.32(A)(1) through (A)(4).

(1) Separate Overload Device. A separate overload device that is responsive to motor current. This device shall be selected to trip or shall be rated at no more than the following percent of the motor nameplate full-load current rating:

Motors with a marked service factor 1.15 or greater	125%
Motors with a marked temperature rise 40°C or less	125%
All other motors	115%

Figure 20-9 *NEC®* Section 430.32.

Example 6

A 25-hp three-phase induction motor has a nameplate rating of 32 amperes. The nameplate also shows a temperature rise of 30°C. Determine the ampere rating of the overload for this motor.

NEC 430.32(A)(1) indicates the overload size is 125% of the full-load current rating of the motor.

32 x 1.25 = 40 amperes

If for some reason this overload size does not permit the motor to start without tripping out, NEC 430.32(C) permits the overload size to be increased to a maximum of 140% for this motor. If this increase in overload size does not solve the starting problem, the overload may be shunted out of the circuit during the starting period in accordance with NEC 430.35(A)&(B).

Example 7

A 250-hp squirrel-cage induction motor is connected to a 560-volt line. The motor has a nameplate current of 232 amperes and a service factor of 1. Determine the overload size for this motor. NEC Table 430.32(A) indicates that the full-load current should be increased by a factor of 115%.

232 x 1.15 = 226.8 amperes

Figure 20-10 Current transformers are used to reduce the overload current.

An overload heater that could conduct 226.8 amperes would be extremely large. When an overload relay must protect a motor that requires a large current draw, current transformers are used to reduce the amount of current the overload heater must conduct. In this example, current transformers with a ratio of 300:5 are used *(Figure 20-10)*. The proper size for an overload heater can be computed using a ratio.

$$\frac{300}{5} = \frac{266.8}{X}$$

$$300 \, X = 1334 \, (266.8 \times 5)$$

$$X = 4.45 \text{ amperes } (1334 \, / \, 300)$$

The secondary side of the current transformer is connected to the overload heaters sized for a current of 4.45 amperes *(Figure 20-11)*. When a current of 266.8 amperes flows through the primary of the current transformer, a current of 4.45 amperes will flow through the overload heaters.

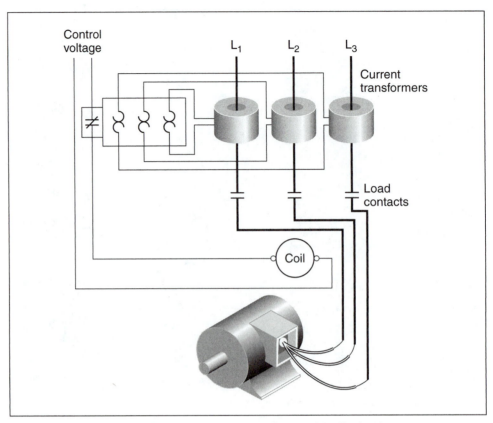

Figure 20-11 Current transformers reduce overload current to the heaters.

DETERMINING LOCKED-ROTOR CURRENT

There are two basic methods for determining the locked-rotor current (starting current) of a squirrel-cage induction motor, depending on the information available. If the motor nameplate lists code letters that range from A to V, they indicate the type of rotor bars used when the rotor was made. Different types of bars are used to make motors with different operating characteristics. The type of rotor bars largely determines the maximum starting current of the motor. NEC Table 430.7(B) *(Figure 20-12)* lists the different code letters and gives the locked-rotor kilovolt-amperes per horsepower. The starting current can be determined by multiplying the kVA rating by the horsepower rating and then dividing by the applied voltage.

Example 8

A 15-hp, three-phase squirrel-cage motor with a code letter of K is connected to a 240-volt line. Determine the locker-rotor current.

Table 430.7(B) Locked-Rotor Indicating Code Letters

Code Letter	Kilovolt-Amperes per Horsepower with Locked Rotor
A	0–3.14
B	3.15–3.54
C	3.55–3.99
D	4.0–4.49
E	4.5–4.99
F	5.0–5.59
G	5.6–6.29
H	6.3–7.09
J	7.1–7.99
K	8.0–8.99
L	9.0–9.99
M	10.0–11.19
N	11.2–12.49
P	12.5–13.99
R	14.0–15.99
S	16.0–17.99
T	18.0–19.99
U	20.0–22.39
V	22.4 and up

Figure 20-12 *NEC®* Table 430.7(B).

The table lists 8.0 to 8.99 kVA per horsepower for a motor with a code letter of K. An average value of 8.5 will be used.

$$8.5 \times 15 = 127.5 \text{ kVA or } 127,500 \text{ VA}$$

$$\frac{127,500}{240 \times \sqrt{3}} = 306.7 \text{ amperes}$$

The second method of determining locked rotor current is to use Tables 430.251(A) and (B) *(Figure 20-13)* if the motor nameplate contains NEMA design letters for the motor. Motors manufactured before 1996 do not generally contain a NEMA design letter. Table 430.251(A) lists the locked rotor currents for single-phase motors and Table 430.251(B) lists locked rotor currents for three-phase motors.

SHORT-CIRCUIT PROTECTION

The rating of the short-circuit protective device is determined by NEC Table 430.52 *(Figure 20-14)*. The far left-hand column lists the type of motor that is to be protected. To the right of this are four columns that list

Table 430.251(A) Conversion Table of Single-Phase Locked-Rotor Currents for Selection of Disconnecting Means and Controllers as Determined from Horsepower and Voltage Rating

For use only with 430.110, 440.12, 440.41, and 455.8(C).

Rated Horsepower	Maximum Locked-Rotor Current in Amperes, Single Phase		
	115 Volts	208 Volts	230 Volts
½	58.8	32.5	29.4
¾	82.8	45.8	41.4
1	96	53	48
1	120	66	60
2	144	80	72
3	204	113	102
5	336	186	168
7½	480	265	240
10	600	332	300

Table 430.251(B) Conversion Table of Polyphase Design B, C, and D Maximum Locked-Rotor Currents for Selection of Disconnecting Means and Controllers as Determined from Horsepower and Voltage Rating and Design Letter

For use only with 430.110, 440.12, 440.41 and 455.8(C).

Rated Horsepower	Maximum Motor Locked-Rotor Current in Amperes, Two- and Three-Phase, Design B, C, and D*					
	115 Volts	200 Volts	208 Volts	230 Volts	460 Volts	575 Volts
	B, C, D	B, C, D	B, C, D	B, C, D	B, C, D	B, C, D
½	40	23	22.1	20	10	8
¾	50	28.8	27.6	25	12.5	10
1	60	34.5	33	30	15	12
1½	80	46	44	40	20	16
2	100	57.5	55	50	25	20
3	—	73.6	71	64	32	25.6
5	—	105.8	102	92	46	36.8
7½	—	146	140	127	63.5	50.8
10	—	186.3	179	162	81	64.8
15	—	267	257	232	116	93
20	—	334	321	290	145	116
25	—	420	404	365	183	146
30	—	500	481	435	218	174
40	—	667	641	580	290	232
50	—	834	802	725	363	290
60	—	1001	962	870	435	348
75	—	1248	1200	1085	543	434
100	—	1668	1603	1450	725	580
125	—	2087	2007	1815	908	726
150	—	2496	2400	2170	1085	868
200	—	3335	3207	2900	1450	1160
250	—	—	—	—	1825	1460
300	—	—	—	—	2200	1760
350	—	—	—	—	2550	2040
400	—	—	—	—	2900	2320
450	—	—	—	—	3250	2600
500	—	—	—	—	3625	2900

*Design A motors are not limited to a maximum starting current or locked rotor current.

Figure 20-13 NEC® Tables 430.251(A) and (B).

different types of short-circuit protective devices: non-time delay fuses, dual-element time delay fuses, instantaneous-trip circuit breakers, and inverse-time circuit breakers. Although it is permissible to used non-time-delay fuses and instantaneous-trip circuit breakers, most motor circuits

Table 430.52 Maximum Rating or Setting of Motor Branch-Circuit Short-Circuit and Ground-Fault Protective Devices

	Percentage of Full-Load Current			
Type of Motor	Nontime Delay Fuse[1]	Dual Element (Time-Delay) Fuse[1]	Instantaneous Trip Breaker	Inverse Time Breaker[2]
Single-phase motors	300	175	800	250
AC polyphase motors other than wound-rotor				
Squirrel cage — other than Design B energy-efficient	300	175	800	250
Design B energy-efficient	300	175	1100	250
Synchronous[3]	300	175	800	250
Wound rotor	150	150	800	150
Direct current (constant voltage)	150	150	250	150

Note: For certain exceptions to the values specified, see 430.54.

[1]The values in the Nontime Delay Fuse column apply to Time-Delay Class CC fuses.

[2]The values given in the last column also cover the ratings of nonadjustable inverse time types of circuit breakers that may be modified as in 430.52(C), Exception No. 1 and No. 2.

[3]Synchronous motors of the low-torque, low-speed type (usually 450 rpm or lower), such as are used to drive reciprocating compressors, pumps, and so forth, that start unloaded, do not require a fuse rating or circuit-breaker setting in excess of 200 percent of full-load current.

Figure 20-14 *NEC*® Table 430.52.

are protected by dual-element time-delay fuses or inverse-time circuit breakers.

Each of these columns lists the percentage of motor current that is to be used in determining the ampere rating of the short-circuit protective device. The current listed in the appropriate motor table is to be used instead of the name plate current. NEC 430.52(C)(1) states that the protective device is to have a rating or setting not exceeding the value calculated in accord with Table 430.52. Exception No. 1 of this section, however, states that if the calculated value does not correspond to a standard size or rating of a fuse or circuit breaker that it shall be permissible to use the next higher standard size. The standard sizes of fuses and circuit breakers are listed in NEC 240.6 (*Figure 20-15*).

In 1996, Table 430.52 began listing squirrel-cage motor types by NEMA design letters instead of code letters. Section 430.7(A)(9) requires that motor

240.6 Standard Ampere Ratings.

(A) Fuses and Fixed-Trip Circuit Breakers. The standard ampere ratings for fuses and inverse time circuit breakers shall be considered 15, 20, 25, 30, 35, 40, 45, 50, 60, 70, 80, 90, 100, 110, 125, 150, 175, 200, 225, 250, 300, 350, 400, 450, 500, 600, 700, 800, 1000, 1200, 1600, 2000, 2500, 3000, 4000, 5000, and 6000 amperes. Additional standard ampere ratings for fuses shall be 1, 3, 6, 10, and 601. The use of fuses and inverse time circuit breakers with nonstandard ampere ratings shall be permitted.

Figure 20-15 *NEC*® Section 240.6.

nameplates be marked with design letters B, C, or D. Motors manufactured before this requirement, however, do not lists design letters on the nameplate. Most common squirrel-cage motors used in industry actually fall in the Design B classification, and for purposes of selecting the short-circuit protective device, are considered to be Design B unless otherwise listed.

Example 9

A 100-hp three-phase squirrel-cage induction motor is connected to a 240-volt line. The motor does not contain a NEMA design code. A dual-element time-delay fuse is to be used as the short-circuit protective device. Determine the size needed.

Table 430.250 lists a full-load current of 248 amperes for this motor. Table 430.52 indicates that the rating of a dual-element time-delay fuse is to be calculated at 175% of the full-load current rating for an AC polyphase (more than one phase) squirrel-cage motor. Since the motor does not list a NEMA Design code on the nameplate, it will be assumed that the motor is Design B.

$$248 \times 1.75 = 434 \text{ amperes}$$

The nearest standard fuse size above the computed value listed in NEC 240.6 is 450 amperes. 450-ampere fuses will be used to protect this motor.

If for some reason this fuse will not permit the motor to start without blowing, NEC 430.52(C)(1) Exception 2(b) states that the rating of a dual-element time-delay fuse may be increased to a maximum of 225% of the full load motor current.

STARTER SIZE

Another factor that must be considered when installing a motor is the size of the starter used to connect the motor to the line. Starter sizes are rated by motor type, horsepower, and connected voltage. The two most common

Motor Starter Sizes and Ratings

| NEMA Size | Load Volts | Maximum Horsepower Rating—Nonplugging and Nonjogging Duty | | NEMA Size | Load Volts | Maximum Horsepower Rating—Nonplugging and Nonjogging Duty | |
		Single Phase	Poly Phase			Single Phase	Poly Phase
00	115	$\frac{1}{2}$	. . .	3	115	$7\frac{1}{2}$	. . .
	200	. . .	$1\frac{1}{2}$		200	. . .	25
	230	1	$1\frac{1}{2}$		230	15	30
	380	. . .	$1\frac{1}{2}$		380	. . .	50
	460	. . .	2		460	. . .	50
	575	. . .	2		575	. . .	50
0	115	1	. . .	4	200	. . .	40
	200	. . .	3		230	. . .	50
	230	2	3		380	. . .	75
	380	. . .	5		460	. . .	100
	460	. . .	5		575	. . .	100
	575	. . .	5				
1	115	2	. . .	5	200	. . .	75
	200	. . .	$7\frac{1}{2}$		230	. . .	100
	230	3	$7\frac{1}{2}$		380	. . .	150
	380	. . .	10		460	. . .	200
	460	. . .	10		575	. . .	200
	575	. . .	10				
*1P	115	3	. . .	6	200	. . .	150
	230	5	. . .		230	. . .	200
					380	. . .	300
					460	. . .	400
					575	. . .	400
2	115	3	. . .	7	230	. . .	300
	200	. . .	10		460	. . .	600
	230	$7\frac{1}{2}$	15		575	. . .	600
	380	. . .	25	8	230	. . .	450
	460	. . .	25		460	. . .	900
	575	. . .	25		575	. . .	900

Tables are taken from NEMA Standards.
*$1\frac{3}{4}$, 10 hp is available.

Figure 20-16 NEMA motor starter sizes and rating.

ratings bodies are NEMA and IEC (International Electrotechnical Commission). A chart showing common NEMA size starters for alternating current motors is shown in *Figure 20-16*. A chart showing IEC starters for alternating current motors is shown in *Figure 20-17*. Each of these charts lists the minimum size starter designed to connect the listed motors to the line. It is not uncommon to employ larger size starters than those listed. This is especially true when using IEC-type starters because of their smaller load contact size.

IEC Motor Starters (60 Hz.)

Size	Max Amps	Motor Voltage	Maximum Horsepower 1∅	Maximum Horsepower 3∅	Size	Max Amps	Motor Voltage	Maximum Horsepower 1∅	Maximum Horsepower 3∅
A	7	115 200 230 460 575	1/4 1/2	 1 1/2 1 1/2 3 5	M	105	115 200 230 460 575	10 10	 30 40 75 100
B	10	115 200 230 460 575	1/2 1	 2 2 5 7 1/2	N	140	115 200 230 460 575	10 10	 40 50 100 125
C	12	115 200 230 460 575	1/2 2	 3 3 7 1/2 10	P	170	115 200 230 460 575		 50 60 125 125
D	18	115 200 230 460 575	1 3	 5 5 10 15	R	200	115 200 230 460 575		 60 75 150 150
E	25	115 200 230 460 575	2 3	 5 7 1/2 15 20	S	300	115 200 230 460 575		 75 100 200 200
F	32	115 200 230 460 575	2 5	 7 1/2 10 20 25	T	420	115 200 230 460 575		 125 125 250 250
G	37	115 200 230 460 575	3 5	 7 1/2 10 25 30	U	520	115 200 230 460 575		 150 150 350 250
H	44	115 200 230 460 575	3 7 1/2	 10 15 30 40	V	550	115 200 230 460 575		 150 200 400 400
J	60	115 200 230 460 575	5 10	 15 20 40 40	W	700	115 200 230 460 575		 200 250 500 500
K	73	115 200 230 460 575	5 10	 20 25 50 50	X	810	115 200 230 460 575		 250 300 600 600
L	85	115 200 230 460 575	7 1/2 10	 25 30 60 75	Z	1215	115 200 230 460 575		 450 450 900 900

Figure 20-17 IEC motor starters rated by size, horsepower, and voltage for 60 Hz. circuits.

Example 10

A 40-hp three-phase squirrel-cage motor is connected to a 208-volt line. What are the minimum size NEMA and IEC starters that should be used to connect this motor to the line?

NEMA: The 200 volt listing is used for motors rated at 208 volts. Locate the NEMA size starter that corresponds to 200 volts and 40 horsepower. Since the motor is three-phase, 40 hp will be in the polyphase column. A NEMA size 4 starter is the minimum size for this motor.

IEC: As with the NEMA chart, the IEC chart lists 200 volts instead of 208 volts. A size N starter lists 200 volts and 40 hp in the three-phase column.

Example Calculation 1

A 40-hp 240-volt DC motor has a nameplate current rating of 132 amperes. The conductors are to be copper with type TW insulation. The short-circuit protective device is to be an instantaneous-trip circuit breaker. The termination temperature rating of the connected devices is not known. Determine the conductor size, overload size, and circuit breaker size for this installation. Refer to *Figure 20-18*.

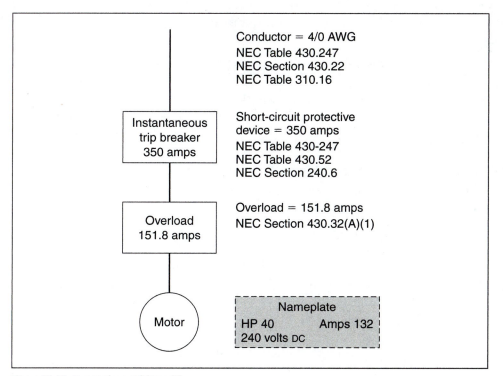

Figure 20-18 Example problem #1.

The conductor size must be determined from the current listed in Table 430.247. This value is to be increased by 25%. (NOTE: multiplying by 1.25 has the same effect as multiplying by 0.25 and then adding the product back to the original number (140 × 0.25 = 35) (35 + 140 = 175 amperes).

140 x 1.25 = 175 amperes

Table 310.16 is used to find the conductor size. Although section 110.14(C) states that for currents of 100 amperes or greater that the ampacity rating of the conductor is to be determined from the 75°C column, in this instance, the insulation type is located in the 60°C column. Therefore, the conductor size must be determined using the 60°C column instead of the 75°C column. A 4/0 AWG copper conductor with type TW insulation will be used.

The overload size is determined from NEC Section 430.32(A)(1). Since there is no service factor or temperature rise listed on the motor nameplate, the heading ALL OTHER MOTORS will be used. The motor nameplate current will be increased by 15%.

132 x 1.15 = 151.8 amperes

The circuit breaker size is determined from Table 430.52. The current value listed in Table 430.247 is used instead of the nameplate current. Under dc motors (constant voltage), the instantaneous trip circuit breaker rating is given at 250%.

140 x 2.50 = 350 amperes

Since 350 amperes is one of the standard sizes of circuit breakers listed in NEC 240.6, that size breaker will be employed as the short-circuit protective device.

Example Calculation 2

A 150-hp three-phase squirrel-cage induction motor is connected to a 440-volt line. The motor nameplate lists the following information:

Amps 175   SF 1.25   Code D   NEMA code B

The conductors are to be copper with type THHN insulation. The short-circuit protective device is to be an inverse-time circuit-breaker. The termination temperature rating is not known. Determine the conductor size, overload size, circuit breaker size, minimum NEMA starter size, and IEC starter size. Refer to *Figure 20-19*.

The conductor size is determined from the current listed in Table 430.250 and is then increased by 25%.

180 x 1.25 = 225 amperes

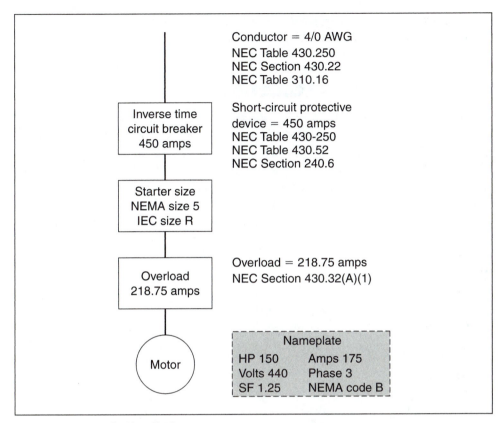

Figure 20-19 Example Circuit #2.

Table 310.16 is used to determine the conductor size. Type THHN insulation is located in the 90°C column. Since the motor nameplate lists NEMA Code B, and the amperage is over 100 amperes, the conductor will be selected from the 75°C column. The conductor size will be 4/0 AWG.

The overload size is determined from the nameplate current and NEC 430.32(A)(1). The motor has a marked service factor of 1.25. The motor nameplate current will be increased by 25%.

175 x 1.25 = 218.75 amps

The circuit breaker size is determined by Tables 430.250 and 430.52. Table 430.52 indicates a factor of 250% for squirrel-cage motors with NEMA design code B. The value listed in Table 430-250 will be increased by 250%.

180 x 2.50 = 450 amperes

One of the standard circuit breaker sizes listed in NEC Section 240.6 is 450 amperes. A 450 ampere inverse-time circuit breaker will be used as the short-circuit protective device.

The proper motor-starter sizes are selected from the NEMA and IEC charts shown in *Figures 20-16* and *20-17*. The minimum size NEMA starter is 5 and the minimum size IEC starter is R.

MULTIPLE MOTOR CALCULATION

The main feeder short-circuit protective devices and conductor sizes for multiple motor connections are set forth in NEC 430.62(A) and 430.24. In this example, three motors are connected to a common feeder. The feeder is 480 volts three-phase and the conductors are to be copper with type THHN insulation. Each motor is to be protected with dual-element time-delay fuses and a separate overload device. The main feeder is also protected by dual-element time-delay fuses. The termination temperature rating of the connected devices is not known. The motor nameplates state the following:

Motor #1

Phase: 3	HP: 20
SF: 1.25	NEMA code: C
Volts: 480	Amperes: 23
Type: Induction	

Motor #2

Phase: 3	HP: 60
Temp.: 40°C	Code: J
Volts: 480	Amperes: 72
Type: Induction	

Motor #3

Phase: 3	HP: 100
Code: A	Volts: 480
Amperes: 96	PF: 90%
Type: Synchronous	

Motor #1 Calculation

The first step is to calculate the values for motor amperage, conductor size, overload size, short-circuit protection size, and starter size for each motor. Both NEMA and IEC starter sizes will be determined. The values for motor #1 are shown in *Figure 20-20*.

The ampere rating from Table 430.250 is used to determine the conductor and fuse size. The amperage rating must be increased by 25% for the conductor size.

$$27 \times 1.25 = 33.75 \text{ amps}$$

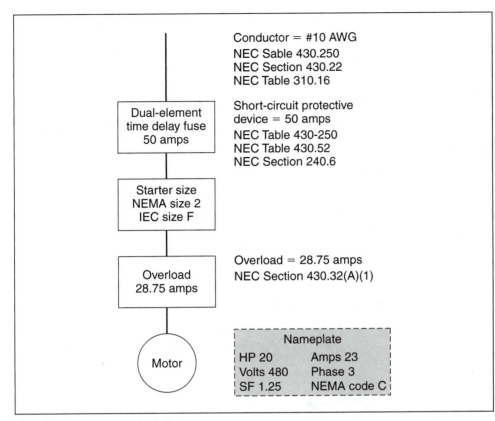

Conductor = #10 AWG
NEC Sable 430.250
NEC Section 430.22
NEC Table 310.16

Dual-element time delay fuse 50 amps

Short-circuit protective device = 50 amps
NEC Table 430-250
NEC Table 430.52
NEC Section 240.6

Starter size NEMA size 2 IEC size F

Overload 28.75 amps

Overload = 28.75 amps
NEC Section 430.32(A)(1)

Motor

Nameplate
HP 20   Amps 23
Volts 480   Phase 3
SF 1.25   NEMA code C

Figure 20-20 Motor #1 calculation.

The conductor size is chosen from Table 310.16. Although type THHN insulation is located in the 90°C column, the conductor size will be chosen from the 75°C column. Although the current is less than 100 amperes, NEC Section 110.14(C)(1)(d) permits the conductors to be chosen from the 75°C column if the motor has a NEMA design code.

33.75 amps = #10 AWG

The overload size is computed from the nameplate current. The demand factors in Section 430.32(A)(1) are used for the overload calculation.

23 x 1.25 = 28.75 amps

The fuse size is determined by using the motor current listed in Table 430.250 and the demand factor from Table 430.52. The percentage of full-load current for a dual-element time-delay fuse protecting a squirrel-cage motor listed as Design C is 175%. The current listed in Table 430.250 will be increased by 175%.

27 x 1.75 = 47.25 amperes

The nearest standard fuse size listed in Section 240.6 is 50 amperes. Fuses of 50 amperes will be used.

The starter sizes are determined from the NEMA and IEC charts shown in *Figures 20-16* and *20-17*. A 20 hp motor connected to 480 volts would require a NEMA size 2 starter and an IEC size F starter.

Motor #2 Calculation

Figure 20-21 shows an example for the calculation for motor #2. Table 430.250 lists a full-load current of 77 amperes for this motor. This value

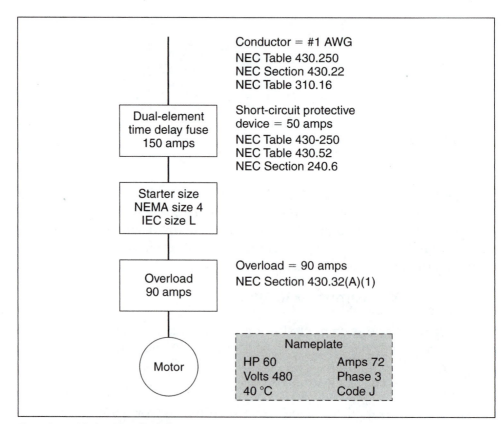

Figure 20-21 Motor #2 calculation.

of current is increased by 25% for the calculation of the conductor current.

77 X 1.25 = 96.25 amperes

Table 310.16 indicates a #1 AWG conductor should be used for this motor connection. The conductor size is chosen from the 60°C column because the circuit current is less than 100 amperes in accord with NEC 110.14(C), and the motor nameplate does not indicate a NEMA design code. (The code J indicates the type of bars used in the construction of the rotor.)

The overload size is determined from Section 430.32(A)(1). The motor nameplate lists a temperature rise of 40°C for this motor. The nameplate current will be increased by 25%.

72 X 1.25 = 90 amperes

The fuse size is determined from Table 430.52. The table current is increased by 175% for squirrel-cage motors other than Design E.

77 X 1.75 = 134.25 amperes

The nearest standard fuse size listed in Section 240.6 is 150 amperes. 150 ampere fuses will be used to protect this circuit.

The starter sizes are chosen from the NEMA and IEC starter charts. This motor would require a NEMA size 4 starter or a size L IEC starter.

Motor #3 Calculation

Motor #3 is a synchronous motor intended to operate with a 90% power factor. *Figure 20-22* shows an example of this calculation. The notes at the bottom of Table 430.250 indicate that the listed current is to be increased by 10% for synchronous motors with a listed power factor of 90%.

101 X 1.10 = 111 amperes

The conductor size is computed by using this current rating and increasing it by 25%.

111 X 1.25 = 138.75 amperes

Table 310.16 indicates that a #1/0 AWG conductor will be used for this circuit. Since the circuit current is over 100 amperes, the conductor size is chosen from the 75°C column.

This motor does not have a marked service factor or a marked temperature rise. The overload size will be calculated by increasing the nameplate current by 15% as indicated in Section 430.32(A)(1) under the heading "All other motors."

96 X 1.15 = 110.4 amperes

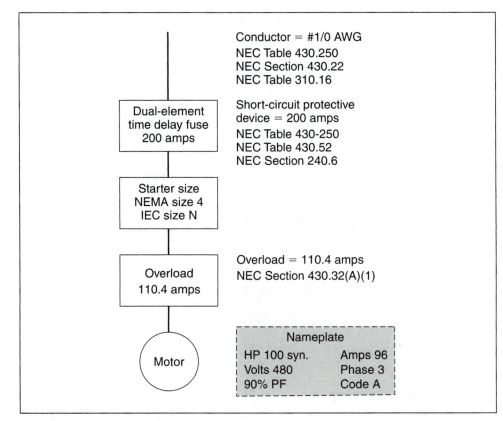

Figure 20-22 Motor #3 calculation.

The fuse size is determined from Table 430.52. The percent of full-load current for a synchronous motor is 175%.

$$111 \times 1.75 = 194.25 \text{ amperes}$$

The nearest standard size fuse listed in Section 240.6 is 200 amperes. 200-ampere fuses will be used to protect this circuit.

The NEMA and IEC starter sizes are chosen from the charts shown in *Figures 20-16* and *20-17*. The motor will require a NEMA size 4 starter and an IEC size N starter.

Main Feeder Calculation

An example of the main feeder connections is shown in *Figure 20-23*. The conductor size is computed in accord with NEC Section 430.24 by increasing the largest amperage rating of the motors connected to the

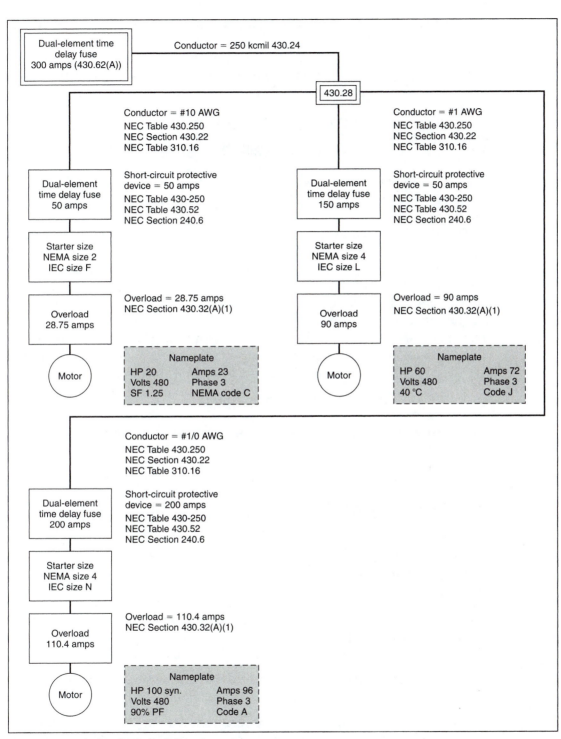

Figure 20-23 Main feeder calculation.

feeder by 25% and then adding the ampere rating of the other motors to this amount. In this example, the 100-hp synchronous motor has the largest running current. This current will be increased by 25% and then the running currents of the other motors as determined from Table 430.250 will be added.

$$111 \times 1.25 = 138.75 \text{ amperes}$$

$$138.75 + 77 + 27 = 242.75 \text{ amperes}$$

Table 310.16 states that 250-kcmil copper conductors are to be used as the main feeder conductors. The conductors were chosen from the 75°C column.

The size of the short-circuit protective device is determined by Section 430.62(A). The code states that the rating or setting of the short-circuit protective device *shall not be greater than* the largest rating or setting of the largest branch circuit short-circuit and ground-fault protective device for any motor supplied by the feeder plus the sum of the full-load running currents of the other motors connected to the feeder. The largest fuse size in this example is used in the 100-hp synchronous motor. The fuse calculation for this motor is 200 amperes. The running currents of the other two motors will be added to this value to determine the fuse size for the main feeder.

$$200 + 77 + 27 = 304 \text{ ampere}$$

The closest standard fuse size listed in Section 240.6 without going over 304 amperes is 300 amperes. Fuses of 300 amperes will be used to protect this circuit.

REVIEW QUESTIONS

1. A 20-hp DC motor is connected to a 500-volt DC line. What is the full-load running current of this motor?

2. What rating is used to find the full-load running current of a torque motor?

3. A $^3/_4$ hp, single-phase squirrel-cage motor is connected to a 240-volt AC line. What is the full-load current rating of this motor and what are the minimum size NEMA and IEC starters that should be used?

4. A 30-hp, two-phase motor is connected to a 230-volt AC line. What is the rated current of the phase conductors and the rated current of the neutral?

5. A 125-hp synchronous motor is connected to a 230-volt three-phase AC line. The motor is intended to operate at 80% power factor. What is

the full-load running current of this motor? What are the minimum size NEMA and IEC starters that should be used to connect this motor to the line?

6. What is the full-load running current of a three-phase, 50-hp motor connected to a 560-volt line? What are the minimum size NEMA and IEC starters that should be used to connect this motor to the line?

7. A 125-hp, three-phase squirrel-cage induction motor is connected to 560 volts. The nameplate current is 115 amperes. It has a marked temperature rise of 40°C and a code letter J. The conductors are to be type THHN copper and they are run in a conduit. The short-circuit protective device is dual-element time-delay fuses. Find the conductor sizes, overload size, fuse-size, minimum NEMA and IEC starter sizes, and the upper and lower range of starting current for this motor.

8. A 7.5 hp, single-phase squirrel-cage induction motor is connected to 120 VAC. The motor has a code letter of H. The nameplate current is 76 amperes. The conductors are copper with type TW insulation. The short-circuit protection device is a non-time-delay fuse. Find the conductor size, overload size, fuse size, minimum NEMA and IEC starter sizes, and upper and lower starting currents.

9. A 75-hp, three-phase synchronous motor is connected to a 230-volt line. The motor is to be operated at 80% power factor. The motor nameplate lists a full-load current of 185 amperes, a temperature rise of 40°C, and a code letter A. The conductors are to be made of copper and have type THHN insulation. The short-circuit protective device is to be an inverse-time circuit breaker. Determine the conductor size, overload size, circuit-breaker size, minimum size NEMA and IEC starters, and the upper and lower starting current.

10. Three motors are connected to a single branch circuit. The motors are connected to a 480-volt three-phase line. Motor #1 is a 50-hp induction motor with a NEMA code B. Motor #2 is 40 hp with a code letter of H, and motor #3 is 50 hp with a NEMA code C. Determine the conductor size needed for the branch circuit supplying these three motors. The conductors are copper with type THWN-2 insulation.

11. The short-circuit protective device supplying the motors in question #10 is an inverse-time circuit breaker. What size circuit breaker should be used?

12. Five 5-hp, three-phase motors with NEMA code B are connected to a 240-volt line. The conductors are copper with type THWN insulation. What size conductor should be used to supply all of these motors?

13. If dual-element time-delay fuses are to be used as the short circuit protective device, what size fuses should be used to protect the circuit in question #12?

14. A 75-hp, three-phase squirrel-cage induction motor is connected to 480 volts. The motor has a NEMA code D. What is the starting current for this motor?

15. A 20-hp, three-phase squirrel-cage induction motor has a NEMA code E. The motor is connected to 208 volts. What is the starting current for this motor?

Laboratory Experiments

These experiments are intended to provide the electrician with hands-on experience dealing with transformers and motors. The transformers used are standard control transformers with two high-voltage windings rated at 240 volts each. The transformers are generally used to provide primary voltage of 480/240, and they have one low-voltage winding rated at 120 volts. The transformers have a rating of 0.5 kVA. Loads are standard 100-watt lamps that may be connected in parallel or series. It is assumed that the power supply is 208/120-volt three-phase four-wire. It is also possibly used with a 240/120-volt three-phase high-leg system, provided adjustments are made in the calculations.

As in industry, these transformers will be operated with full voltage applied to the windings. The utmost caution must be exercised when dealing with these transformers. *These transformers can provide enough voltage and current to seriously injure or kill.* The power should be disconnected before attempting to make or change any connections.

Standard meters such as AC voltmeters, ohmmeters, and AC ammeters will be used to make circuit measurements.

Experiment 1

Transformer Basics

Objectives

After completing this experiment, you will be able to:

- Discuss the construction of an isolation transformer.
- Determine the winding configuration with an ohmmeter.
- Connect a transformer and make voltage measurements.
- Compute the turns ratio of the windings.

Materials needed:

480-240/120 volt, 0.5 kVA control transformer
Ohmmeter
AC voltmeter

The transformer used in this experiment contains two high-voltage windings and one low-voltage winding. The high-voltage windings are labeled H_1 - H_2 and H_3 - H_4. The low-voltage winding is labeled X_1 - X_2.

1. Set the ohmmeter to the Rx1 range and measure the resistance between the following terminals:

H_1 - H_2 _____ Ω
H_1 - H_3 _____ Ω
H_1 - H_4 _____ Ω
H_1 - X_1 _____ Ω

H_1 - X_2 _____ Ω
H_2 - H_3 _____ Ω
H_2 - H_4 _____ Ω
H_2 - X_1 _____ Ω
H_2 - X_2 _____ Ω
H_3 - H_4 _____ Ω
H_3 - X_1 _____ Ω
H_3 - X_2 _____ Ω
H_4 - X_1 _____ Ω
H_4 - X_2 _____ Ω
X_1 - X_2 _____ Ω

2. Using the information provided by the preceding measurements, which sets or terminals form complete circuits within the transformer?

_____

_____

_____

These circuits represent the connections to the three separate windings within the transformer.

3. Which of the windings exhibits the lowest resistance and why?

_____

_____

_____

_____

4. The H_1 - H_2 terminals are connected to one of the high-voltage windings, and the H_3 - H_4 terminals are connected to the second high-voltage winding. Each of these windings are rated at 240 volts. When this transformer is connected for 240-volt operation, the two high-voltage windings are connected in parallel to form one winding by connecting H_1 to H_3 and H_2 to H_4 *(Figure Exp. 1-1)*. This provides a 2:1 turns ratio with the low-voltage winding.

When this transformer is operated with 480 volts connected to the primary, the high-voltage windings are connected in series by connecting H_2 to H_3 and connecting power to H_1 and H_4 *(Figure Exp. 1-2)*. This effectively doubles the primary turns, providing a 4:1 turns ratio with the low-voltage winding.

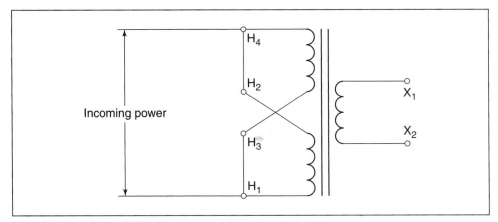

Figure Exp. 1-1 High-voltage windings connected in parallel.

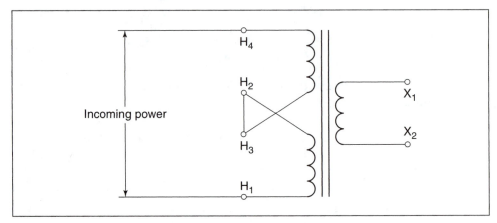

Figure Exp. 1-2 High-voltage windings connected in series.

5. Connect the two high-voltage windings for parallel operation as shown in *Figure Exp. 1-1*. Assuming a voltage of 208 volts is applied to the high-voltage windings, compute the voltage that should be present on the low-voltage winding between terminals X_1 and X_2.
   _____ volts

6. Apply a voltage of 208 volts to the transformer and measure the voltage across terminals X_1 and X_2.
   _____ volts

7. Notice that the measured voltage is slightly higher than the computed voltage. The rated voltage of a transformer is based on full load. It is

normal for the secondary voltage to be slightly higher than rated voltage with no load connected to the transformer. Transformers are generally wound with a few extra turns of wire in the winding that is intended to be used as the load side. This helps overcome the voltage drop when load is added. The slight change in turns ratio generally does not affect the operation of the transformer to a great extent.

8. Disconnect the power from the transformer.

9. Reconnect the high-voltage windings for a series connection as shown in *Figure Exp. 1-2.*

10. Assume that a voltage of 208 volts is applied to the high-voltage windings. Compute the voltage across the low-voltage winding.
    _____ volts

11. Apply a voltage of 208 volts across the high-voltage windings. Measure the voltage across terminals X_1 - X_2.
    _____ volts

12. Disconnect the power from the transformer.

13. Assume that the low-voltage winding is to be used as the primary and the high-voltage winding as the secondary. If the high-voltage windings are connected in series, the turns ratio will become 1:4, which means the secondary voltage will be 4 times greater than the primary voltage. Assume that a voltage of 120 volts is connected to terminals X_1 - X_2. If the high-voltage windings are connected in series, compute the voltage across terminals H_1 - H_4.
    _____ volts

14. Connect a 120-volt source to terminals X_1 - X_2. Make certain the AC voltmeter is set for a higher range than the computed value of voltage.
    Caution: The secondary voltage in this step will be 480 volts or higher. Use extreme caution when making this measurement.

15. Turn on the power supply and measure the voltage across terminals H_1 - H_4.
    _____ volts

Notice that the voltage is slightly lower than what was computed. Recall that previously it was stated that a few extra turns of wire are generally added to the winding that is intended to be used as the load side. This transformer is generally used with the low-voltage winding supplying power to the load. Since a few extra turns have been added, the actual turns ratio is probably 3.8:1 or 3.9:1 instead of 4:1.

16. Turn off the power supply.

17. Reconnect the high-voltage windings to form a parallel connection as shown in *Figure Exp. 1-1.*

18. Assume that a voltage of 120 volts is connected to the low-voltage winding. Compute the voltage across the high-voltage winding.
    _____ volts

19. Connect a 120-volt AC source across terminals X_1 - X_2. Turn on the power supply and measure the voltage across terminals H_1 - H_4.
    _____ volts

20. Turn off the power supply and disconnect the transformer. Return the components to their proper place.

Experiment 2

Single-Phase Transformer Calculations

Objectives

After completing this experiment, you will be able to:

- Discuss transformer excitation current.
- Compute values of primary current using the secondary current and the turns ratio.
- Compute the turns ratio of a transformer using measured values.
- Connect a step-down or step-up isolation transformer.

Materials needed:

480-240/120 volt, 0.5 kVA control transformer
AC voltmeter
AC ammeter

In this experiment the excitation current of an isolation transformer will be measured. The transformer will then be connected as both a step-down and a step-up transformer. The turns ratio will be determined from measured values and the primary current will be computed and then measured.

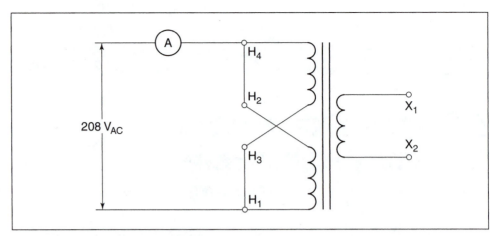

Figure Exp. 2-1 Connecting the high-voltage winding for parallel operation.

1. Connect the high-voltage windings of the transformer in parallel for 240-volt operation.

2. Connect the high-voltage winding to a 208-volt AC source with an AC ammeter connected in series with one of the lines *(Figure Exp. 2-1)*. If an inline ammeter is not available and a clamp-on type meter is to be used, form a 10:1 scale divider by wrapping 10 turns of wire around the jaw of the ammeter. This will produce a more accurate reading. When the current value is read, divide the value by 10 by moving the decimal point one place to the left.

3. Turn on the power source and measure the current. This is the *excitation* current of the transformer. The excitation current is the amount of current necessary to magnetize the iron in the transformer and will remain constant regardless of the load on the transformer.
   _____ amp(s)

4. Measure the voltage across the low-voltage winding at terminals X_1 - X_2.
   _____ volts

5. Compute the turns ratio by dividing the primary voltage by the secondary voltage. Since the primary has the highest voltage, the larger number will be placed on the left side of the ratio, such as 3:1 or 4:1.
   _____ ratio

6. Turn off the power supply.

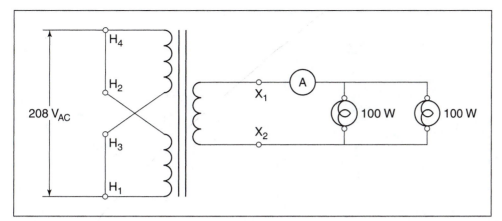

Figure Exp. 2-2 Two lamps are connected in parallel to the secondary winding.

7. Connect two 100-watt incandescent lamps in parallel with the low-voltage winding of the transformer. Connect an AC ammeter in series with one of the lines *(Figure Exp. 2-2)*.

8. Turn on the power and measure the current flow in the secondary circuit of the transformer.

   _____ amp(s)

9. Turn off the power supply.

10. Compute the amount of primary current by using the turns ratio. Since the primary voltage is higher, the amount of primary current will be less. Divide the secondary current by the turns ratio, and then add the excitation current to this value.

$$I_{(Primary)} = \frac{I_{(Secondary)}}{\text{Turns Ratio}} + \text{Excitation Current}$$

   _____ $I_{(Primary)}$

11. Reconnect the AC ammeter in one of the primary lines *(Figure Exp. 2-3)*.

12. Turn on the power supply and measure the primary current. Compare this value with the computed value.

   _____ $I_{(Primary)}$

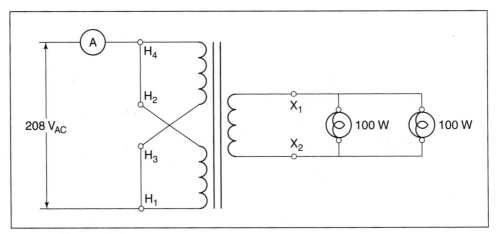

Figure Exp. 2-3 Measuring primary current.

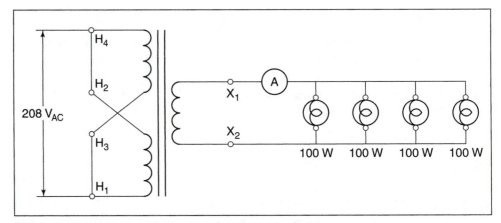

Figure Exp. 2-4 Adding load to the transformer secondary.

13. Reconnect the AC ammeter in the secondary circuit and add two 100-watt incandescent lamps in parallel with the transformer secondary *(Figure Exp. 2-4).*

14. Turn on the power and measure the secondary current.
    _____ amp(s)

15. Turn off the power.

16. Compute the amount of current flow that should be in the primary circuit using the turns ratio. Be sure to add the excitation current.
    _____ $I_{(Primary)}$

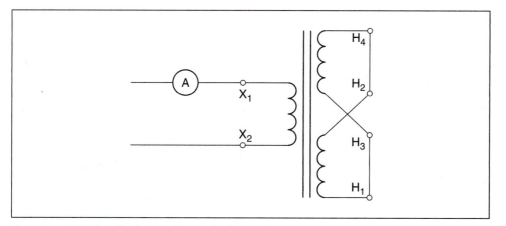

Figure Exp. 2-5 Using the low-voltage winding as the primary.

17. Reconnect the AC ammeter in series with one of the lines of the primary winding of the transformer.

18. Turn on the power and measure the current flow. Compare the measured value with the computed value.

    _____ $I_{(Primary)}$

19. Turn off the power supply.

20. Reconnect the transformer by connecting the low-voltage terminals, X_1 - X_2, to a 120-volt AC source. Connect an AC ammeter in series with one of the power lines *(Figure Exp. 2-5)*.

21. Turn on the power and measure the excitation current of the transformer.
    _____ amp(s)

22. Measure the secondary voltage with an AC voltmeter.
    _____ volts

23. Determine the turns ratio by dividing the secondary voltage by the primary voltage. Since the primary voltage is lower, the larger number will be placed on the right-hand side of the ratio, 1:3 or 1:4.
    _____ ratio

24. Turn off the power supply.

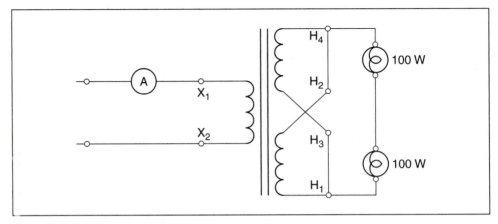

Figure Exp. 2-6 Connecting the load to the secondary.

5. Connect two 100-watt incandescent lamps in series. Connect these two lamps in parallel with the high-voltage winding. Connect an AC ammeter in series with one of the secondary leads *(Figure Exp. 2-6)*.

26. Turn on the power supply and measure the secondary current.
_____ amp(s)

27. Compute the primary current by using the turns ratio. Since the primary voltage is less than the secondary voltage, the primary current will be more than the secondary current. To determine the primary current, multiply the secondary current by the turns ratio and add the excitation current.

$$I_{(Primary)} = I_{(Secondary)} \times \textbf{Turns Ratio + Excitation Current}$$

_____ $I_{(Primary)}$

28. Turn off the power supply.

29. Reconnect the AC ammeter in series with the primary side of the transformer.

30. Turn on the power supply and measure the primary current. Compare this value with the computed value.
_____ $I_{(Primary)}$

31. Turn off the power supply.

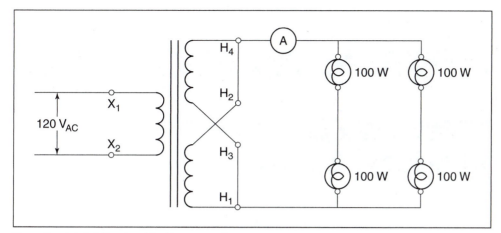

Figure Exp. 2-7 Adding load to the secondary.

32. Reconnect the AC ammeter in series with the secondary winding. Add two more 100-watt lamps that have been connected in series to the secondary circuit. These two lamps should be connected in parallel with the first two lamps *(Figure Exp. 2-7)*.

33. Turn on the power supply and measure the secondary current.
    _____ amp(s)

34. Turn off the power supply.

35. Compute the amount of current that should flow in the primary circuit.
    _____ $I_{(Primary)}$

36. Reconnect the AC ammeter in series with one of the primary lines.

37. Turn on the power supply and measure the primary current. Compare this value with the computed value.
    _____ $I_{(Primary)}$

38. Turn off the power supply. Disconnect the circuit and return the components to their proper places.

Experiment 3

Transformer Polarities

Objectives

After completing this experiment, you will be able to:

- Discuss buck and boost connections for a transformer.
- Connect a transformer for additive polarity.
- Connect a transformer for subtractive polarity.
- Determine the turns ratio and calculate current values using measured values.

Materials needed:

    480-240/120 volt, 0.5 kVA control transformer
    AC voltmeter
    AC ammeter
    Four 100-watt lamps

In this experiment a control transformer will be connected for both additive (boost) and subtractive (buck) polarity. Buck and boost connections are made by physically connecting the primary and secondary windings together. If they are connected such that the primary and secondary voltages add, the transformer is additive or boost. If the windings are connected such that the primary and secondary voltages subtract, they are connected subtractive or buck.

In this exercise only one of the high-voltage windings will be used. The other will not be connected.

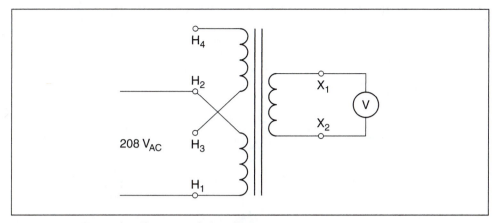

Figure Exp. 3-1 Measuring the secondary voltage.

1. Connect the circuit shown in *Figure Exp. 3-1*.

2. Turn on the power and measure the primary and secondary voltages.
   $E_{(Primary)}$ ——————— volts
   $E_{(Secondary)}$ ——————— volts

3. Turn off the power supply.

4. Determine the turns ratio of this transformer connection by dividing the higher voltage by the lower voltage. Recall that if the primary winding has the higher voltage, the larger number will be placed on the left and a 1 will be placed on the right. If the secondary winding has the higher voltage, a 1 will be placed on the left and the larger number placed on the right.

$$\text{Turns Ratio} = \frac{\text{Higher Voltage}}{\text{Lower Voltage}}$$

——————— ratio

5. Connect the circuit shown in *Figure Exp. 3-2* by connecting X_1 to H_1. Connect a voltmeter across terminals X_2 and H_2.

6. Turn on the power supply and measure the voltage across X_2 and H_2.
   ——————— volts

7. Turn off the power supply.

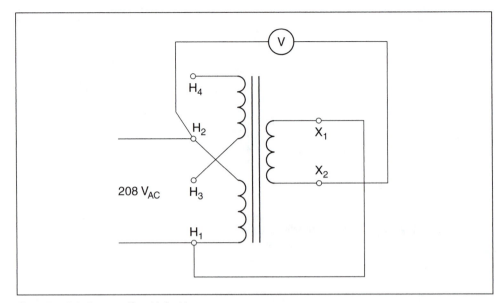

Figure Exp. 3-2 Connecting X_1 to H_1.

8. Determine the turns ratio of this transformer connection.

    _____ ratio

9. If the measured voltage is the difference between the applied voltage and the secondary voltage, the transformer is connected subtractive polarity or buck. If the measured voltage is the sum of the applied voltage and the secondary voltage, the transformer is connected additive or boost. Is the transformer connected buck or boost?

    _____

10. Connect an AC ammeter in series with one of the power supply lines.

11. Turn on the power supply and measure the excitation current of the transformer.

    $I_{(Exc)}$ _____ amp(s)

12. Turn off the power supply.

13. Reconnect the transformer as shown in *Figure Exp. 3-3* by connecting X_2 to H_1. Connect an AC voltmeter across terminals X_1 and H_2.

14. Turn on the power supply and measure the voltage across terminals X_1 and H_2.

    _____ volts

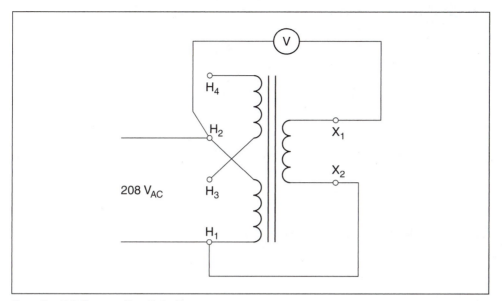

Figure Exp. 3-3 Connecting X_2 to H_1.

15. Turn off the power supply.

16. Is the transformer connected buck or boost?

17. Determine the turns ratio of this transformer connection.

    _____ ratio

18. Connect an AC ammeter in series with one of the primary leads.

19. Turn on the power supply and measure the excitation current of this
    connection.

    $I_{(Exc)}$ _____amp(s)

20. Compare the value of excitation current for the buck and boost connec-
    tions. Is there any difference between these two values?

    _____

21. Turn off the power supply.

22. *Figure Exp. 3-4* shows the proper location for the placement of polarity
    dots. Recall that polarity dots are used to indicate which windings of a
    transformer have the same polarity at the same time. To better understand

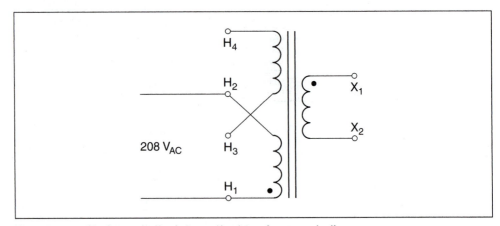

Figure Exp. 3-4 Placing polarity dots on the transformer windings.

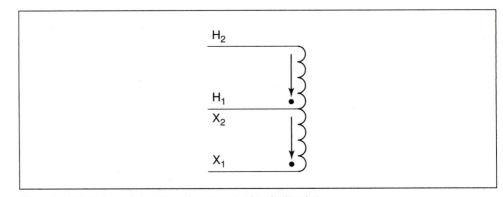

Figure Exp. 3-5 Determining the placement of polarity dots.

how the dots are placed, redraw the two transformer windings in a series connection as shown in *Figure Exp. 3-5*. Place a dot beside one of the high-voltage terminals. In this example, a dot has been placed beside the H_1 terminal. Next draw an arrow pointing to the dot. To place the second dot, draw an arrow in the same direction as the first arrow. This arrow should point to the dot that is to be placed beside the secondary terminal. Since terminal X_2 is connected to H_1, the arrow must point to terminal X_1.

23. Reconnect the transformer for subtractive polarity. If two ammeters are available, place one ammeter in series with one of the primary leads and the second ammeter in series with the secondary lead that is not connected to the H_1 terminal. Connect a 100-watt lamp in the secondary circuit and connect a voltmeter in parallel with the lamp *(Figure Exp. 3-6)*.

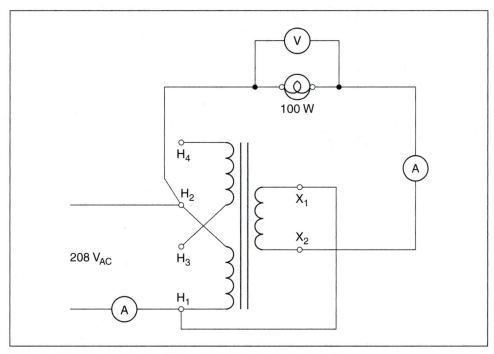

Figure Exp. 3-6 Connecting load to a subtractive polarity transformer.

24. Turn on the power supply and measure the secondary current.

    $I_{(Secondary)}$ _____ amp(s)

25. Measure the secondary voltage. Since the lamp is the only load connected to the secondary, the voltage drop across the lamp will be the secondary voltage.

    $E_{(Secondary)}$ _____ volts

26. Calculate the amount of primary current using the measured value of secondary current and the turns ratio. Be sure to use the turns ratio for this connection as determined in step 8. Since the primary voltage is greater than the secondary voltage, the primary current should be less. Therefore, divide the secondary current by the turns ratio and then add the excitation current measured in step 11.

$$I_{(Primary)} = \frac{I_{(Secondary)}}{\text{Turns Ratio}} + I_{(EXC)}$$

    $I_{(Primary)}$ _____ amp(s)

27. If necessary, turn off the power supply and connect an AC ammeter in series with one of the primary leads.

28. Turn on the power supply and measure the primary current. Compare this value with the calculated value.
    I$_{(Primary)}$ _____ amp(s)

29. Turn off the power supply.

30. Connect another 100-watt lamp in parallel with the first as shown in *Figure Exp. 3-7*. Reconnect the AC ammeter in series with the secondary winding if necessary.

31. Turn on the power supply and measure the amount of secondary current.
    I$_{(Secondary)}$ _____ amp(s)

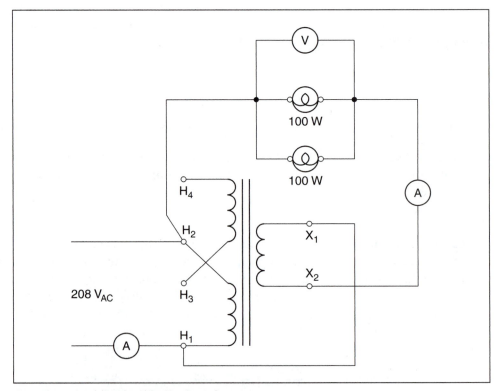

Figure Exp. 3-7 Adding load to the transformer.

32. Calculate the primary current.

   $I_{(Primary)}$ _____ amp(s)

33. If necessary, turn off the power supply and connect the AC ammeter in series with one of the primary leads.

34. Turn on the power supply and measure the primary current. Compare this value with the computed value.

   $I_{(Primary)}$ _____ amp(s)

35. Turn off the power supply.

36. Reconnect the transformer for the boost connection by connecting terminal X_2 to H_1. If two ammeters are available, connect one AC ammeter in series with one of the power supply leads and the second AC ammeter in series with the secondary. Connect four 100-watt lamps in series with terminals X_1 and H_2 as shown in *Figure Exp. 3-8*. Connect an AC voltmeter across terminals X_2 and H_1.

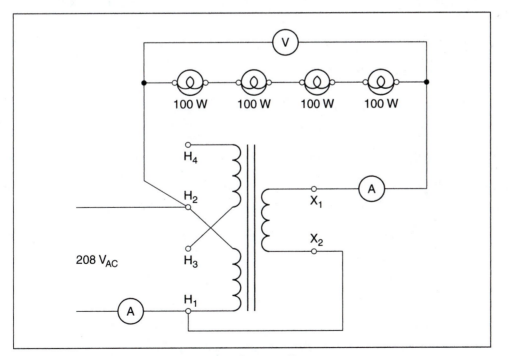

Figure Exp. 3-8 Connecting load to the boost connection.

37. Turn on the power and measure the secondary current.

    $I_{(Secondary)}$ —————————— amp(s)

38. Turn off the power supply.

39. Compute the primary current by using the turns ratio. Be sure to use the turns ratio for this connection as determined in step 17. Since the primary voltage in this connection is less than the secondary voltage, the primary current will be greater. To calculate the primary current, multiply the secondary current by the turns ratio and then add the excitation current.

$$I_{(Primary)} = (I_{(Secondary)} \times \text{Turns Ratio}) + I_{(Exc)}$$

   $I_{(Primary)}$ —————————— amp(s)

40. If necessary, connect the AC ammeter in series with one of the power supply leads.

41. Turn on the power supply.

42. Measure the primary current. Compare this value with the calculated value.

    $I_{(Primary)}$ —————————— amp(s)

43. Turn off the power supply.

44. Disconnect the circuit and return the components to their proper places.

Experiment 4
Autotransformers

Objectives

After completing this experiment, you will be able to:

- Discuss the operation of an autotransformer.
- Connect a control transformer as an autotransformer.
- Calculate the turns ratio from measured voltage values.
- Calculate primary current using the secondary current and the turns ratio.
- Connect an autotransformer as a step-down transformer.
- Connect an autotransformer as a step-up transformer.

Materials needed:
  480-240/120 volt, 0.5 kVA control transformer
  AC voltmeter
  AC ammeter
  Four 100-watt lamps

   In this experiment, the control transformer will be connected for operation as an autotransformer. The low-voltage winding will not be used in this experiment. The two high-voltage windings will be connected in series to form one continuous winding. The transformer will be connected as both a step-down and a step-up transformer.

1. Series connect the two high-voltage windings by connecting terminals H_2 and H_3 together. The H_1 and H_4 terminals will be connected to a

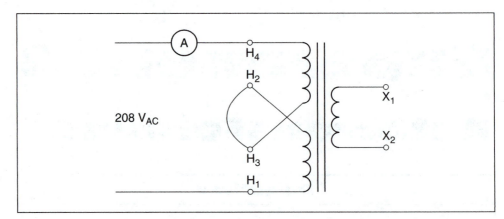

Figure Exp. 4-1 Connecting the high-voltage windings as an autotransformer.

source of 208 VAC. Connect an ammeter in series with one of the power supply lines *(Figure Exp. 4-1)*.

2. Turn on the power supply and measure the excitation current. (The current will be small, and it may be difficult to determine this current value.)
   $I_{(Ecx)}$ _____ amp(s)

3. Measure the primary voltage across terminals H_1 and H_4.
   $E_{(Primary)}$ _____ volts

4. Measure the secondary voltage across terminals H_1 and H_2. (Note: It is also possible to use terminals H_3 and H_4 as the secondary winding.)
   $E_{(Secondary)}$ _____ volts

5. Determine the turns ratio of this transformer connection.

$$\text{Turns Ratio} = \frac{\text{Higher Voltage}}{\text{Lower Voltage}}$$

   _____ ratio

6. Turn off the power supply.

7. Connect an AC ammeter in series with the H_2 terminal and a 100-watt lamp as shown in *Figure Exp. 4-2*. The secondary winding of the transformer will be between terminals H_2 and H_1.

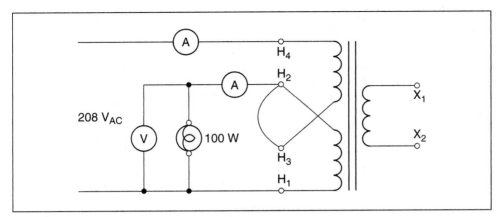

Figure Exp. 4-2 Connecting a load to the autotransformer.

8. Turn on the power supply and measure the amount of current flow in the secondary winding.

   $I_{(Secondary)}$ ⎯⎯⎯⎯⎯⎯ amp(s)

9. Measure the voltage drop across the secondary winding with an AC voltmeter.

   $E_{(Secondary)}$ ⎯⎯⎯⎯⎯⎯ volts

10. Turn off the power supply.

11. Calculate the primary current by using the turns ratio. Since the primary voltage is greater than the secondary voltage, the primary current will be less than the secondary current. To determine the primary current, divide the secondary current by the turns ratio and add the excitation current.

$$I_{(Primary)} = \frac{I_{(Secondary)}}{\text{Turns Ratio}} + I_{(EXC)}$$

   $I_{(Primary)}$ ⎯⎯⎯⎯⎯⎯ amp(s)

12. If necessary, reconnect the AC ammeter in series with one of the power supply leads.

13. Turn on the power and measure the primary current. Compare this value with the computed value.

   $I_{(Primary)}$ ⎯⎯⎯⎯⎯⎯ amp(s)

14. Turn off the power supply.

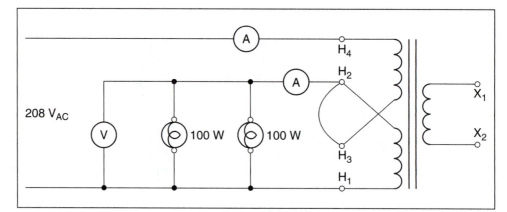

Figure Exp. 4-3 Adding load to the autotransformer.

15. Connect another 100-watt lamp in parallel with the existing lamp *(Figure Exp. 4-3)*.

16. If necessary, reconnect the AC ammeter in series with the secondary winding of the transformer.

17. Turn on the power supply and measure the secondary current.
    $I_{(Secondary)}$ _____ amp(s)

18. Calculate the primary current by using the turns ratio.
    $I_{(Primary)}$ _____ amp(s)

19. Turn off the power supply.

20. If necessary, reconnect the AC ammeter in series with one of the power supply leads.

21. Turn on the power supply and measure the primary current. Compare this value with the computed value.
    $I_{(Primary)}$ _____ amp(s)

22. Turn off the power supply.

23. Reconnect the circuit as shown in *Figure Exp. 4-4*. Terminals H_1 and H_2 will be connected to a source of 120 VAC. Connect an AC ammeter in series with terminal H_2. The entire winding between terminals H_1 and H_4 will be used as the secondary.

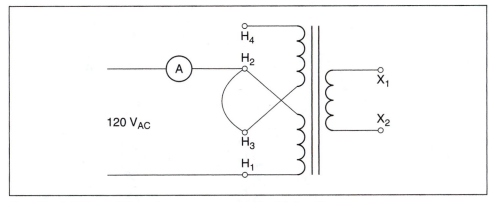

Figure Exp. 4-4 The autotransformer connected for high voltage.

24. Turn on the power and measure the excitation current of this transformer connection.

$I_{(Exc)}$ _____ amp(s)

25. Measure the voltage across terminals H_4 and H_1.

_____ volts

26. Compute the turns ratio of this transformer connection.

_____ ratio

27. Turn off the power supply.

28. Connect an AC ammeter and four 100-watt lamps in series with terminals H_4 and H_1 *(Figure Exp. 4-5)*.

29. Turn on the power and measure the secondary current.

$I_{(Secondary)}$ _____ amp(s)

30. Turn off the power supply.

31. If necessary, connect the AC ammeter in series with one of the primary leads.

32. Compute the value of primary current by using the turns ratio and the measured value of secondary current.

$I_{(Primary)}$ _____ amp(s)

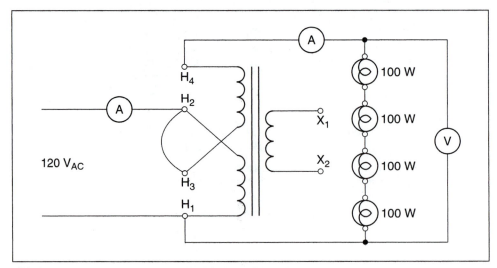

Figure Exp. 4-5 Adding load to the secondary winding.

33. Turn on the power and measure the primary current. Compare this value with the computed value.

    $I_{\text{(Primary)}}$ _____ amp(s)

34. Measure the voltage across terminals H_1 and H_4.

    $E_{\text{(Secondary)}}$ _____ volts

35. Turn off the power supply.

36. Disconnect the circuit and return the components to their proper place.

Experiment 5

Three-Phase Circuits

Objectives

After completing this experiment, you will be able to:

- Connect a wye-connected three-phase load.
- Calculate and measure voltage and current values for a wye-connected load.
- Connect a delta-connected load.
- Calculate and measure voltage and current values for a delta-connected load.

Materials needed:

> AC voltmeters
> AC ammeters
> Six 100-watt lamps

In this experiment, six 100-watt lamps will be connected to form different three-phase loads. Two lamps will be connected in series to form three separate loads. These loads will be connected to form wye or delta connections.

1. Connect two 100-watt lamps in series to form three separate load banks. Connect the load banks in wye by connecting one end of each bank together to form a center point *(Figure Exp. 5-1)*. It is assumed that this load is to be connected to a 208-VAC three-phase

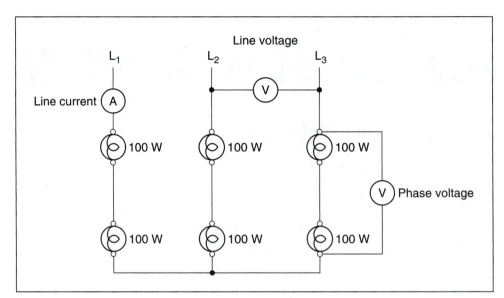

Figure Exp. 5-1 Measuring the line current and voltages of a wye connection.

line. Connect an AC ammeter in series with the line supplying power to the load.

2. Turn on the power and measure the line voltage supplied to the load.

   $E_{(Line)}$ _____ volts

3. Calculate the value of phase voltage for a wye-connected load.

$$E_{(Phase)} = \frac{E_{(Line)}}{\sqrt{3}}$$

   $E_{(Phase)}$ _____ volts

4. Measure the phase voltage and compare this value to the computed value.

   $E_{(Phase)}$ _____ volts

5. Measure the line current.

   $I_{(Line)}$ _____ amp(s)

6. Turn off the power supply.

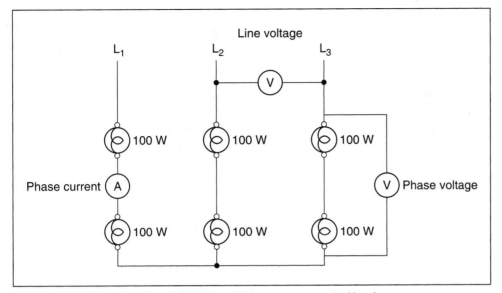

Figure Exp. 5-2 Measuring the phase current in a wye-connected load.

7. In a wye-connected system, the line current and phase current are the same. Reconnect the circuit as shown in *Figure Exp. 5-2.*

8. Turn on the power and measure the phase current.
$I_{(Phase)}$ _____ amp(s)

9. Turn off the power supply.

10. Reconnect the three banks of lamps to form a delta-connected load *(Figure Exp. 5-3).*

11. Turn on the power and measure the line voltage supplied to the load.
$E_{(Line)}$ _____ volts

12. Measure the phase value of voltage.
$E_{(Phase)}$ _____ volts

13. Are the line and phase voltage values the same or different?

_____

14. Measure the line current.
$I_{(Line)}$ _____ amp(s)

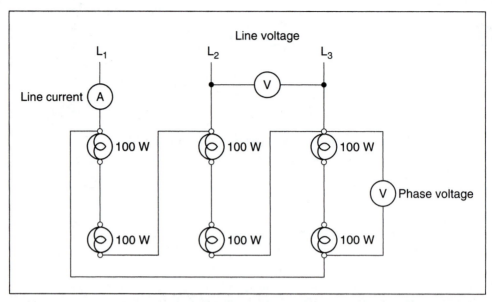

Figure Exp. 5-3 Measuring the voltage and line current values of a delta-connected load.

15. Turn off the power supply.

16. In a delta-connected system, the phase current will be less than the line current by a factor of 1.732. Calculate the phase current value for this connection.

$$I_{(Phase)} = \frac{I_{(Line)}}{1.732}$$

$I_{(Phase)}$ _____ amp(s)

17. Reconnect the circuit as shown in *Figure Exp. 5-4.*

18. Turn on the power supply and measure the phase current. Compare this value with the computed value.

$I_{(Phase)}$ _____ amp(s)

19. Turn off the power supply.

20. Disconnect the circuit and return the components to their proper places.

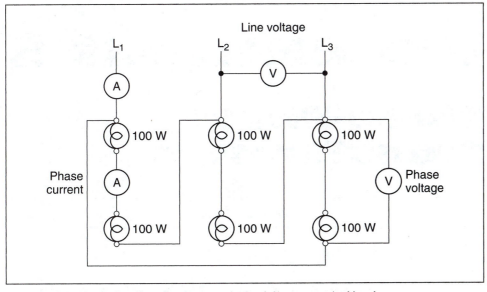

Figure Exp. 5-4 Measuring the phase current of a delta-connected load.

Experiment 6

Three-Phase Transformers

Objectives

After completing this experiment, you will be able to:

- Connect three single-phase transformers to form a three-phase bank.
- Connect transformer windings in a delta configuration.
- Connect transformer windings in a wye configuration.
- Compute values of voltage, current, and turns ratio for different three-phase connections.
- Compute the values for an open delta-connected transformer bank.

Materials needed:

Three 480-240/120 volt, 0.5 kVA, control transformers
AC voltmeter
AC ammeter
Six 100-watt lamps

In this experiment, three single-phase control transformers will be connected to form different three-phase transformer banks. Values of voltage, current, and turns ratios will be computed and then measured. The three transformers will be operated with their high-voltage windings connected in

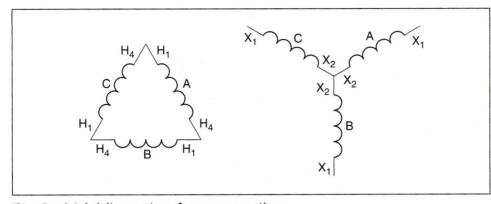

Figure Exp. 6-1 A delta-wye transformer connection.

parallel for low-voltage operation. The high-voltage windings are used as the primary for each connection.

Delta-Wye Connection

A delta-wye connected three-phase transformer bank has its primary windings connected in a delta configuration and its secondary windings connected in a wye configuration *(Figure Exp. 6-1)*. Notice that the three primary windings have been labeled A, B, and C. The H_1 terminal of transformer A is connected to the H_4 terminal of transformer C. The H_4 terminal of transformer A is connected to the H_1 terminal of transformer B, and the H_4 terminal of transformer B is connected to the H_1 terminal of transformer C. The secondary windings form a wye by connecting all the X_2 terminals together.

1. Connect the circuit shown in *Figure Exp. 6-2*. Notice that the three transformers have been labeled A, B, and C. The H_1 terminal of transformer A is connected to the H_4 terminal of transformer C. The H_4 terminal of transformer A is connected to the H_1 terminal of transformer B, and the H_4 terminal of transformer B is connected to the H_1 terminal of transformer C. This is the same connection shown in the schematic drawing of *Figure Exp. 6-1*. Also notice that the X_2 terminal of each transformer is connected together to form a wye-connected secondary. The 100-watt lamp loads form a wye connection also.

2. Turn on the power supply and measure the phase voltage of the secondary. Since the secondary windings of the three transformers form the

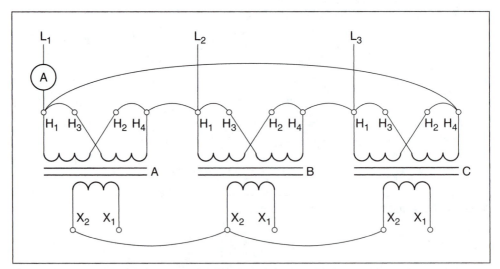

Figure Exp. 6-2 Transformer bank with a delta-connected primary and wye-connected secondary.

phases of the wye, the phase voltage can be measured across the X_1 - X_2 terminals of any transformer.

$E_{(Phase\ Secondary)}$ _____ volts

3. Calculate the line-to-line (line) voltage of the secondary.

$$E_{(Line)} = E_{(Phase)} \times \sqrt{3}$$

$E_{(Line\ Secondary)}$ _____ volts

4. Measure the line voltage of the secondary by connecting an AC voltmeter across any two of the X_1 terminals.

$E_{(Line\ Secondary)}$ _____ volts

5. Measure the excitation current flowing in the primary winding. The excitation current will remain constant as long as the primary windings remain connected in a delta configuration. Since this measurement indicates the *line* value of current for this delta connection, it will later be added to the *line* value of computed current.

$I_{(Exc)}$ _____ amp(s)

6. Turn off the power supply.

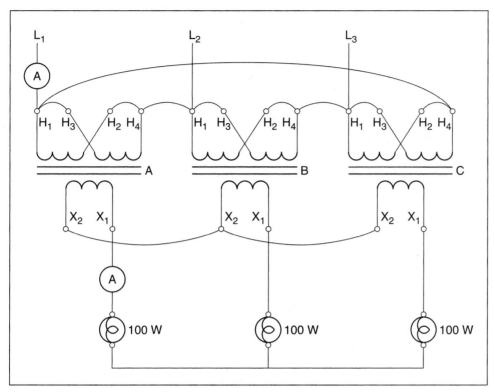

Figure Exp. 6-3 Adding load to the connection.

7. Connect three 100-watt lamps to the secondary of the transformer bank. These lamps will be connected in wye to form a three-phase load for the transformer. If available, connect a second AC ammeter in series with one of the secondary leads *(Figure Exp. 6-3)*.

8. Turn on the power supply and measure the secondary current.

   $I_{(Line\ Secondary)}$ _____ amp(s)

9. The measured value is the *line* current value. Since the secondary is connected in a wye configuration, the phase current value will be the same as the line current value. When calculating the primary current using secondary current and the turns ratio, *the phase current value must be used.* Calculate the phase current of the primary using the turns ratio. Since the phase voltage value of the primary is greater than the phase voltage value of the secondary, the phase current of the primary will be less than the phase current of the secondary. To determine the

primary phase current value, divide the phase current of the secondary by the turns ratio.

$$I_{(Primary)} = \frac{I_{(Secondary)}}{\text{Turns Ratio}}$$

$I_{(Phase\ Primary)}$ _____ amp(s)

10. Calculate the line current of the primary. Since the primary is connected as a delta, the line current will be greater than the phase current by a factor of 1.732. Be sure to add the line value of the excitation current in this calculation.

$$I_{(Line\ Primary)} = (I_{(Phase\ Primary)} \times 1.732) + I_{(Exc)}$$

$I_{(Line\ Primary)}$ _____ amp(s)

11. Measure the line current of the primary and compare this value with the calculated value.

$I_{(Line\ Primary)}$ _____ amp(s)

12. Turn off the power supply.

13. Add a 100-watt lamp in parallel with each of the three existing loads *(Figure Exp. 6-4)*.

14. Turn on the power supply and measure the line voltage of the secondary.

$E_{(Line\ Secondary)}$ _____ volts

15. Measure the line current of the secondary.

$I_{(Line\ Secondary)}$ _____ amp(s)

16. Calculate the phase current value of the primary using the phase current value of the secondary and the turns ratio.

$I_{(Phase\ Primary)}$ _____ amp(s)

17. Calculate the line current value of the primary.

$I_{(Line\ Primary)}$ _____ amp(s)

18. Measure the line current of the primary and compare this value with the computed value.

$I_{(Line\ Primary)}$ _____ amp(s)

19. Turn off the power supply.

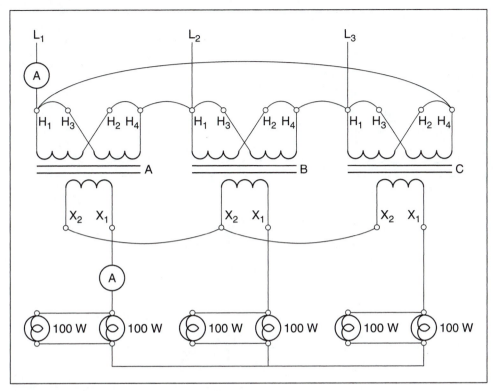

Figure Exp. 6-4 Adding load to the transformers.

Delta-Delta Connection

The three transformers will now be reconnected to form a delta-delta connection. The schematic diagram for a delta-delta connection is shown in *Figure Exp. 6-5.*

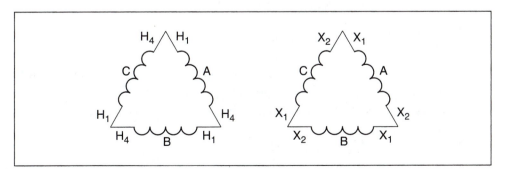

Figure Exp. 6-5 Delta-delta transformer connection.

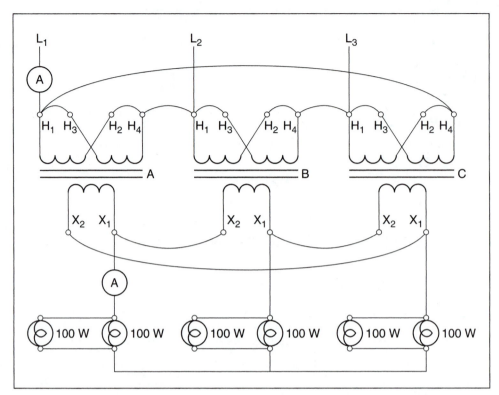

Figure Exp. 6-6 Transformers with a delta-connected primary, delta-connected secondary, and wye-connected load.

20. Reconnect the transformers as shown in *Figure Exp. 6-6*. In this connection, the primary windings remain connected in a delta configuration, but the secondary windings have been reconnected from a wye to a delta.

21. Turn on the power supply and measure the phase voltage of the secondary. The phase voltage can be measured across any set of X_1 - X_2 terminals.

    $E_{(Phase\ Secondary)}$ _____ volts

22. Since the secondary is connected as a delta, the line voltage value should be the same as the phase voltage value. Measure the line-to-line voltage of the secondary. The line voltage can be measured between any two X_1 terminals.

    $E_{(Line\ Secondary)}$ _____ volts

23. Measure the line current value of the secondary.

    $I_{(Line\ Secondary)}$ _____ amp(s)

24. Calculate the phase current value of the secondary.

$$I_{(Phase)} = \frac{I_{(Line)}}{1.732}$$

$I_{(Phase\ Secondary)}$ —————————— amp(s)

25. Using the phase current of the secondary and the turns ratio, calculate the phase current of the primary.
$I_{(Phase\ Primary)}$ —————————— amp(s)

26. Compute the line current of the primary.
$I_{(Line\ Primary)}$ —————————— amp(s)

27. Measure the line current of the primary and compare this value to the computed value.
$I_{(Line\ Primary)}$ —————————— amp(s)

28. Turn off the power supply.

29. Reconnect the lamps to form a delta-connected load instead of a wye-connected load. Each phase should have two 100-watt lamps connected in parallel as shown in *Figure Exp. 6-7*.

30. Turn on the power supply and measure the line voltage of the secondary.
$E_{(Line\ Secondary)}$ —————————— volts

31. Measure the line current of the secondary.
$I_{(Line\ Secondary)}$ —————————— amp(s)

32. Calculate the value of secondary phase current.
$I_{(Phase\ Seconary)}$ —————————— amp(s)

33. Calculate the phase current value of the primary by using the secondary phase current and the turns ratio.
$I_{(Phase\ Primary)}$ —————————— amp(s)

34. Calculate the line current value of the primary.
$I_{(Line\ Primary)}$ —————————— amps(s)

35. Measure the primary line current and compare this value with the computed value.
$I_{(Line\ Primary)}$ —————————— amps(s)

36. Turn off the power supply.

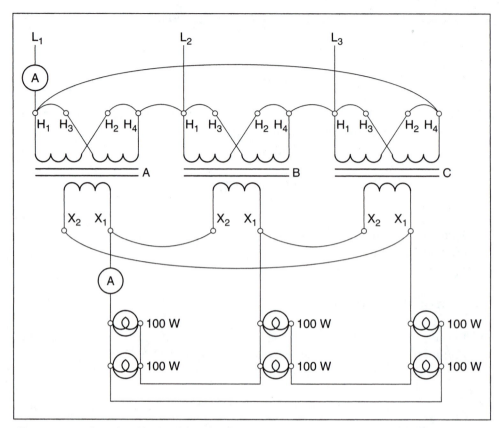

Figure Exp. 6-7 Changing the load from a wye connection to a delta connection.

Wye-Delta Connection

In the next part of the experiment, the three transformers will be reconnected to form a wye-delta transformer bank. The schematic drawing of the connection is shown in *Figure Exp. 6-8*. Notice that all of the H_4 terminals have been joined together to form the wye connection. Power will be applied to the H_1 terminals. The secondary winding will remain in a delta connection.

37. Reconnect the transformers as shown in *Figure Exp. 6-9*. For the first part of this experiment, be sure that no load is connected to the secondary.

38. Turn on the power supply and measure the phase voltage of the primary.
     $E_{(Phase\ Primary)}$ _____ volts

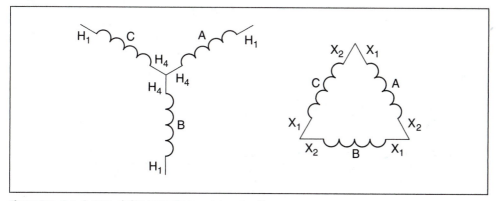

Figure Exp. 6-8 A wye-delta transformer connection.

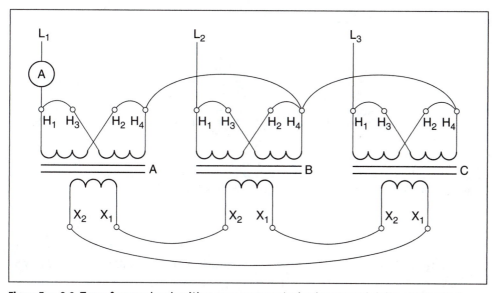

Figure Exp. 6-9 Transformer bank with a wye-connected primary and delta-connected secondary.

39. Measure the phase voltage of the secondary.

$E_{(Phase\ Secondary)}$ _____ volts

40. Compute the turns ratio of this transformer connection.

$$\text{Turns Ratio} = \frac{\text{Higher Voltage}}{\text{Lower Voltage}}$$

_____ ratio

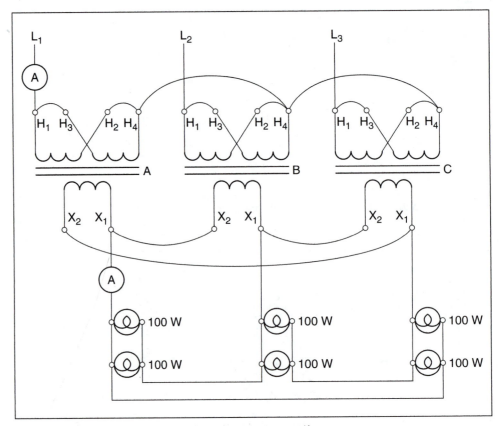

Figure Exp. 6-10 Adding load to the transformer connection.

41. Measure the excitation current of this connection. As long as the primary remains connected in a wye configuration, this excitation current will remain constant.

    $I_{(Exc)}$ _____ amps(s)

42. Turn off the power supply.

43. Reconnect the delta-connected lamp bank to the secondary of the transformer as shown in *Figure Exp. 6-10*.

44. Turn on the power supply and measure the line voltage of the secondary.

    $E_{(Line Secondary)}$ _____ volts

45. Measure the line current of the secondary.

    $I_{(Line Secondary)}$ _____ amps(s)

46. Calculate the phase current of the secondary.

$$I_{(Phase)} = \frac{I_{(Line)}}{1.732}$$

$I_{(Phase\ Secondary)}$ _____ amps(s)

47. Using the turns ratio and the secondary phase current, compute the phase current of the primary.

$I_{(Phase\ Primary)}$ _____ amps(s)

48. In a wye connection, the line current and the phase current are the same. To determine the line current for a transformer connection, however, the excitation current must be added to the line current value. Compute the value of the primary line current.

$I_{(Line\ Primary)}$ _____ amps(s)

49. Measure the primary line current and compare this value to the computed value.

$I_{(Line\ Primary)}$ _____ amps(s)

50. Turn off the power supply.

Wye-Wye Connection

The next section of this experiment deals with transformers connected in a wye-wye connection. The schematic diagram for this connection is shown in *Figure Exp. 6-11*. Notice that both the primary and secondary windings are connected to form a wye connection.

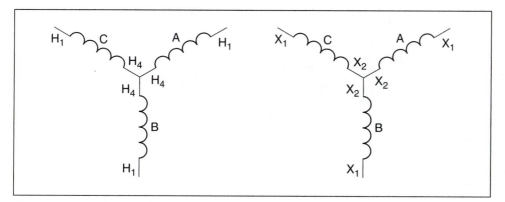

Figure Exp. 6-11 A wye-wye three-phase transformer connection.

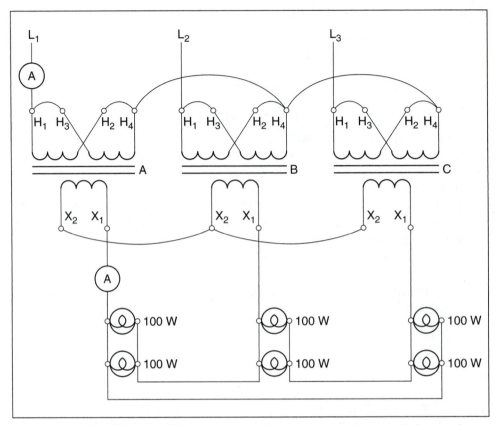

Figure Exp. 6-12 Transformers with a wye-connected primary, wye-connected secondary, and delta-connected load.

51. Connect the circuit shown in *Figure Exp. 6-12.*

52. Turn on the power supply and measure the phase voltage of the secondary.

    $E_{(Phase\ Secondary)}$ _____ volts

53. Calculate the line voltage value of the secondary.

    $$E_{(Line)} = E_{(Phase)} \times 1.732$$

    $E_{(Line\ Secondary)}$ _____ volts

54. Measure the line voltage of the secondary and compare this value to the computed value.

    $E_{(Line\ Secondary)}$ _____ volts

55. Measure the line current of the secondary.

$I_{(Line\ Secondary)}$ _____ amps(s)

56. Since the secondary is now connected in a wye configuration, the phase current will be the same as the line current. Compute the value of the primary phase current by using the secondary phase current and the turns ratio.

$I_{(Phase\ Primary)}$ _____ amps(s)

57. Since the primary is connected in a wye configuration also, the line current will be the same as the phase current plus the excitation current. Compute the total line current value for the primary.

$I_{(Line\ Primary)}$ _____ amps(s)

58. Measure the line current value and compare it to the computed value.

$I_{(Line\ Primary)}$ _____ amps(s)

59. Turn off the power supply.

Open-Delta Connection

The last connection to be made is the open delta. The open-delta connection requires the use of only two transformers to supply three-phase power to a load. The schematic diagram for an open-delta connection is shown in *Figure Exp. 6-13*. It should be noted that the open-delta connection can provide only about 87% of the combined kVA capacity of the two transformers.

60. Connect the circuit shown in *Figure Exp. 6-14*.

61. Turn on the power supply and measure the phase voltage of the primary.

$E_{(Phase\ Primary)}$ _____ volts

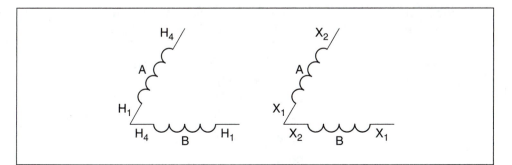

Figure Exp. 6-13 Open-delta connection.

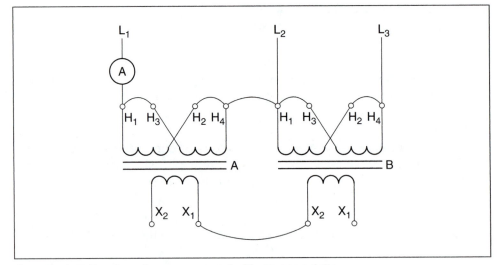

Figure Exp. 6-14 Two transformers connected in an open delta.

62. Measure the phase voltage of the secondary.

    $E_{(Phase\ Secondary)}$ _____ volts

63. Calculate the turns ratio of this transformer connection.

    Ratio _____

64. Measure the line-to-line voltage between all three of the secondary line terminals. Is there any variation in the voltages?

    _____

65. Turn off the power supply.

66. Connect three 100-watt lamps to form a delta connection. Connect these lamps to the line terminals of the secondary *(Figure Exp. 6-15)*.

67. Turn on the power supply and measure the line voltage between each of the three lines. Are the voltage values the same?

    _____

68. Measure the line current of the secondary.

    $I_{(Line\ Secondary)}$ _____ amps(s)

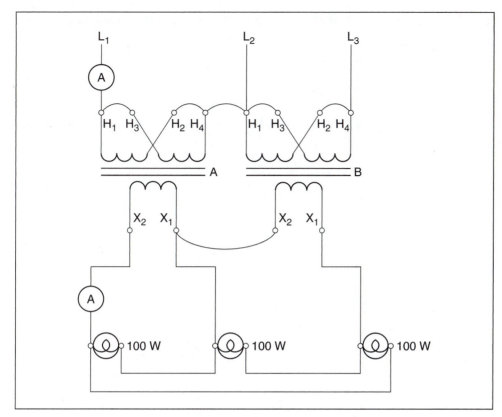

Figure Exp. 6-15 Connecting a three-phase load to the transformer bank.

69. The phase current value for an open delta is calculated in the same way as a closed delta. Calculate the phase current value for the secondary.

$$I_{(Phase)} = \frac{I_{(Line)}}{1.732}$$

$I_{(Phase\ Secondary)}$ ──────────── amps(s)

70. Using the secondary phase current and the turns ratio, calculate the phase current value for the primary.

$I_{(Phase\ Primary)}$ ──────────── amps(s)

71. Calculate the line current value.

$$I_{(Line)} = (I_{(Phase)} \times 1.732) + I(EXC)$$

$I_{(Line\ Primary)}$ ──────────── amps(s)

72. Measure the line current of the primary and compare this value to the computed value.

    $I_{(Line\ Primary)}$ _____ amps(s)

73. Turn off the power supply.

74. Disconnect the circuit and return the components to their proper places.

Experiment 7

Rotating Magnetic Field

Objectives

After completing this experiment, you should be able to:

- Discuss the operating principle of polyphase motors.
- Demonstrate a rotating magnetic field.
- Reverse the direction of rotation.

Materials needed:

   Three-phase stator winding
   Three-phase alternator or synchronous motor
   Direct current motor that can be coupled to the shaft of
      the alternator
   Method of coupling motor and alternator together
   Direct current power supply to operate DC motor
   Compass

A rotating magnetic field is the operating principle of all polyphase (more than one phase) motors such as two-phase or three-phase. A magnetic field is made to rotate because of three basic factors:
1. The voltages are out of phase with each other.
2. The voltages reverse polarity at regular intervals.
3. The manner in which the stator windings are arranged around the core material.

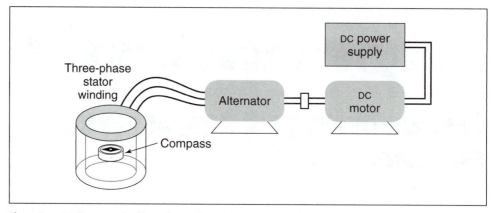

Figure Exp. 7-1 Demonstrating the action of a rotating magnetic field.

The following experiment demonstrates the operation of a magnetic field.

1. Connect the shaft of the alternator or synchronous motor to the shaft of the direct current machine *(Figure Exp. 7-1)*. The direct current motor should be connected to a variable voltage power supply so the motor speed can be controlled. The output of the alternator is connected to the stator winding of the three-phase motor. Place the compass inside the stator winding. It may be necessary to place the compass on some object so its height will be in about the middle of the stator winding.

2. Apply power to the DC motor so that it turns very slowly.

3. Excite the alternator. The output voltage and current will be very low, but it should generate enough current to make the compass needle turn in the direction of the rotating field inside the stator winding. If the motor speed is increased slightly, the speed of the compass will increase also. Note that the speed of the rotating magnetic field is proportional to the speed the alternator is turning. Recall that two factors determine the speed of the rotating magnetic field (synchronous speed):

   - The number of stator poles per phase
   - The frequency (The speed of the alternator determines the output frequency.)

4. Turn off the power and change any two of the leads from the alternator to the stator winding. Repeat the experiment and note if the compass needle turns in the same direction or the opposite direction.

5. Disconnect the components and return them to their proper place.

Experiment 8

Three-Phase Dual-Voltage Motors

Objectives

After completing this experiment, you should be able to:

- Determine if a three-phase dual-voltage motor is connected wye or delta.
- Connect a dual-voltage motor for low-voltage operation.
- Connect a dual-voltage motor for high-voltage operation.
- Reverse the direction of rotation of a three-phase motor.

Materials needed:

Three-phase nine-lead dual-voltage motor
Megohmmeter
Ohmmeter or continuity tester
AC ammeter

Many three-phase motors are designed to be operated on two different voltages. The most common voltage ranges for these motors are 240 volts and 480 volts. If the motor is to be operated on 240 volts, the windings will be connected in parallel. If the motor is to be operated on 480 volts, the windings will be connected in series.

Dual-voltage motors generally have nine T leads in the terminal connection box. Some motors have twelve, but these are the exception instead of

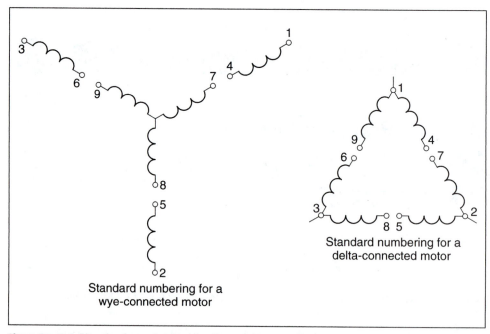

Figure Exp. 8-1 Standard numbering for three-phase motors.

the rule. Twelve-lead motors are generally large horsepower intended for wye-delta starting. Dual-voltage motors can be connected wye or delta *(Figure Exp. 8-1)*.

1. Remove the cover plate of the terminal connection box. Separate the nine T leads.

2. Determine if the motor is connected wye or delta. Using an ohmmeter or continuity tester, determine if there is continuity between T7, T8, and T9.

   _____

3. Determine if there is continuity between T1 and T9.

   _____

4. If there is continuity between T7, T8, and T9, the motor is wye connected. If there is continuity between T1 and T9, the motor is delta connected. If the motor is wye connected, perform steps 5 through 9. If the motor is delta connected, begin at step 10.

5. Test each T lead for grounds by connecting one probe of a megohmmeter to the case of the motor. The megohmmeter should be set for a

voltage of at least 500 volts. Connect the other megohmmeter lead to each of the motors T leads and measure the resistance to ground.

T1 _____ Ω   T2 _____ Ω

T3 _____ Ω   T4 _____ Ω

T5 _____ Ω   T6 _____ Ω

T7 _____ Ω   T8 _____ Ω

T9 _____ Ω

Each of the tests should indicate a resistance of several million ohms. If they do not, inform your instructor before connecting power to this motor.

6. If the motor is wye-connected, perform the test procedures described in steps 7 through 9. If the motor is delta-connected, perform the procedure listed in steps 10 through 12.

7. Test the remaining three windings for continuity, opens, and shorts. Check for continuity between the following terminals:

T1 and T4 _____   T1 and T2 _____

T1 and T5 _____   T1 and T3 _____

T1 and T6 _____   T1 and T7 _____

T1 and T8 _____   T1 and T9 _____

T4 and T2 _____   T4 and T5 _____

T4 and T3 _____   T4 and T6 _____

T4 and T7 _____   T4 and T8 _____

T4 and T9 _____

In the preceding test, the only set of T leads that should indicate continuity are T1 and T4. If there is no continuity between T1 and T4, the winding is open. If there is continuity to any other winding, they are shorted.

8. Test for continuity between the following T leads:

T2 and T3 _____   T2 and T4 _____

T2 and T5 _____   T2 and T6 _____

T2 and T7 _____   T2 and T8 _____

T2 and T9 _____   T5 and T6 _____

T5 and T7 _____   T5 and T8 _____

T5 and T9 _____

In the preceding test, continuity should be shown between T2 and T5. If there is no continuity, the winding is open. If there is continuity between either of these T leads and any others, the windings are shorted.

9. Test for continuity between the following T leads:

T3 to T6 _____   T3 to T7 _____

T3 to T8 _____   T3 to T9 _____

The test should reveal continuity between T3 and T6. If there is no continuity, the winding is open. If there is continuity between T3 or T6 and any other T leads, the windings are shorted.

10. If the motor is delta-connected, test the windings for shorts and open by testing for continuity between the following terminals:

T1 to T4 _____   T1 to T2 _____

T1 to T5 _____   T1 to T3 _____

T1 to T6 _____   T1 to T7 _____

T1 to T8 _____   T1 to T9 _____

T4 to T2 _____   T4 to T5 _____

T4 to T3 _____   T4 to T6 _____

T4 to T7 _____   T4 to T8 _____

T4 to T9 _____

There should be an indication of continuity between T1 and T4, T1 and T9, and T4 and T9. If there is no continuity, the winding is open. If there is continuity to any other T lead, the windings are shorted.

11. Test for continuity between the following T leads:

T2 and T3 _____   T2 and T4 _____

T2 and T5 _____   T2 and T6 _____

T2 and T7 _____   T2 and T8 _____

T2 and T9 _____   T5 and T6 _____

T5 and T7 _____   T5 and T8 _____

T5 and T9 _____

This test should reveal continuity between T2 and T5, T2 and T7, and T5 and T7. If there is no continuity between these T leads, the winding is open. If there is continuity between these T leads and any others, the windings are shorted.

12. Test the following T leads for continuity:

T3 to T6 _____   T3 to T7 _____

T3 to T8 _____   T3 to T9 _____

T6 to T7 _____   T6 to T8 _____

T6 to T9 _____

The test should indicate continuity between T3 and T6, T3 and T8, and T6 and T8. If there is no continuity between these leads, the winding is open. If there is continuity between these leads and any others, the windings are shorted.

13. The motor will now be connected to the three-phase power line. Determine if the motor is to be connected for low-voltage operation (208, 220, 230, or 240), or high-voltage operation (440, 460, or 480).

14. Connect the nine T leads for high- or low-voltage operation. The connection diagram for a wye-connected motor operating on low voltage is shown in *Figure Exp. 8-2*. The connection for a wye-connected motor operating on high voltage is shown in *Figure Exp. 8-3*. The connection for a delta-connected motor operating on low voltage is shown in *Figure Exp. 8-4*, and the connection for a delta-connected motor operating on high voltage is shown in *Figure Exp. 8-5*.

15. Connect the motor to the power line. Ask the instructor to check the connections before turning on the power.

16. Turn on the three-phase power and observe the direction of rotation of the motor.

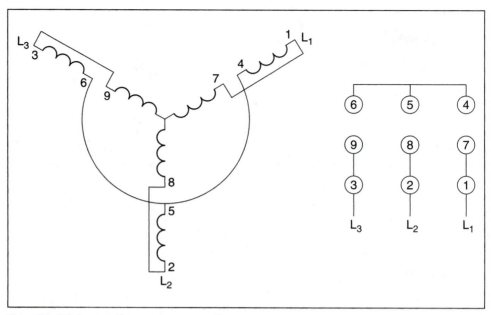

Figure Exp. 8-2 Low-voltage wye connection.

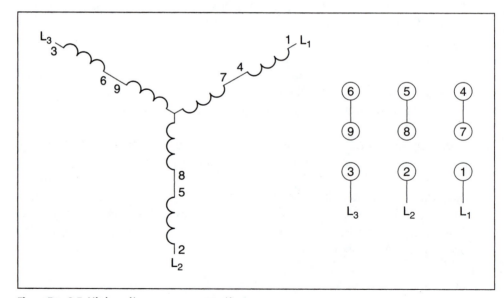

Figure Exp. 8-3 High-voltage wye connection.

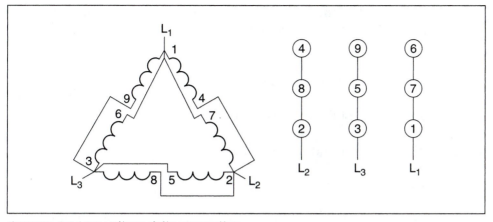

Figure Exp. 8-4 Low-voltage delta connection.

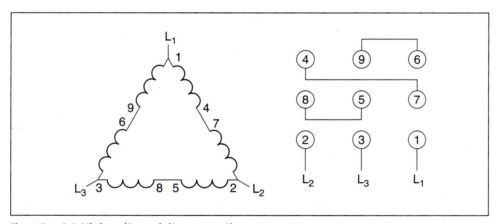

Figure Exp. 8-5 High-voltage delta connection.

17. Measure the current on each of the three-phase lines.

    L1 _____ amps

    L2 _____ amps

    L3 _____ amps

18. The current on each line should be approximately the same. It should also be much lower than the current listed on the motor nameplate, because the motor is operating without a load. The nameplate lists the

full-load current of the motor. If the currents are not approximately the same, it could indicate a winding that has shorted turns. Inform the instructor.

19. Turn off the three-phase power.

20. Reverse any two of the line leads to the motor.

21. Turn on the power and observe the direction of rotation. Did the motor reverse direction?

    _____

22. Turn off the power and disconnect the motor. Return the components to their proper places.

Experiment 9

Single-Phase Dual-Voltage Motors

Objectives

After completing this experiment, you should be able to:

- Test the winding of a single-phase dual-voltage motor.
- Connect a dual-voltage motor for 240-volt operation.
- Connect a dual-voltage motor for 120-volt operation.
- Reverse the direction of rotation of a dual-voltage motor.

Materials needed:

    Single-phase dual-voltage motor with external
      start winding leads
    Megohmmeter
    Ohmmeter
    AC ammeter

Single-phase motors are often designed to permit connection to 240 or 120-volt power sources. Single-phase motors can be wound in a variety of ways. Some dual-voltage motors will contain two run windings and two start windings *(Figure Exp. 9-1)*. One of the run windings will be labeled T1 and T2, and the other run winding will be labeled T3 and T4. One start winding will be labeled T5 and T6, and the other will be labeled T7 and T8.

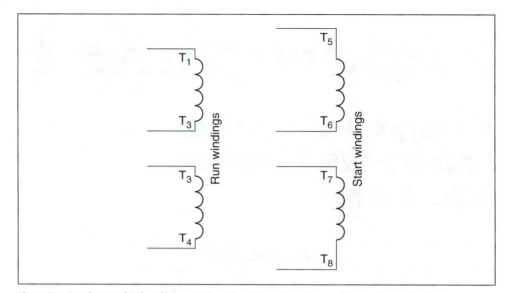

Figure Exp. 9-1 Some dual-voltage motors contain two run and two start windings.

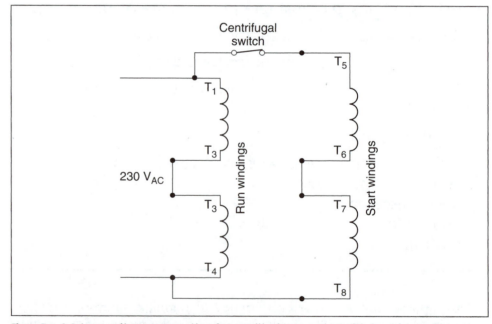

Figure Exp. 9-2 Low-voltage connection for a split-phase motor with two run and two start windings.

To connect this motor for high-voltage operation, series the run winding together by connecting T2 and T3. Series the start windings by connecting T6 and T7 *(Figure Exp. 9-2)*. The start windings are then connected in parallel with the run windings.

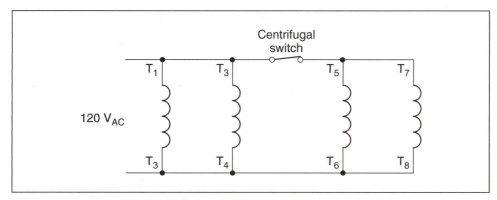

Figure Exp. 9-3 High-voltage connection for a split-phase motor with two run and two start windings.

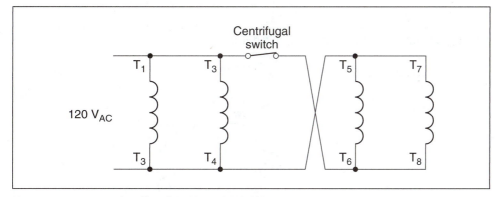

Figure Exp. 9-4 Reversing the direction of rotation.

If a dual-voltage motor with two run and two start windings is to be connected for low-voltage operation, the run and start windings are connected in parallel as shown in *Figure Exp. 9-3*. The direction of rotation can be reversed by changing the start windings in relation to the run winding. However, T5 and T7 should remain connected together, and T6 and T8 should remain connected (*Figure Exp. 9-4*).

Many dual-voltage single-phase motors are wound with only one start winding instead of two. The run winding labels will be the same, T1, T2, and T3, T4, but the start winding may be labeled T5, T6, or T5, T8 depending on the manufacturer (*Figure Exp 9-5*). If the motor is to be connected for 240-volt operation, the start winding will be connected in parallel with only one start winding (*Figure Exp. 9-6*). If the motor is connected for low-voltage operation, the start and run windings will be connected in parallel as shown in *Figure Exp. 9-7*. Direction of rotation can be changed by reversing the start winding in relation to the run winding.

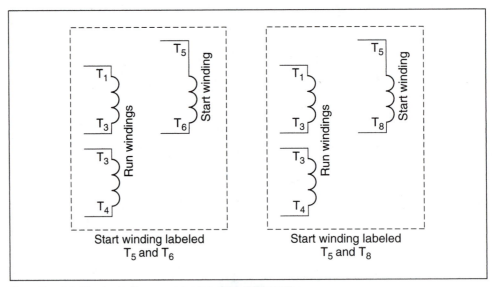

Figure Exp. 9-5 Start windings can be labeled differently.

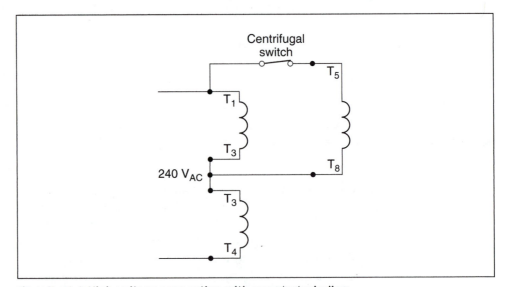

Figure Exp. 9-6 High-voltage connection with one start winding.

1. Separate the motor's T leads at the terminal connection box.

2. Determine if the motor is a capacitor-start or resistance-start motor. A capacitor-start motor will have a starting capacitor mounted on the motor.

   capacitor-start _____   resistance-start _____

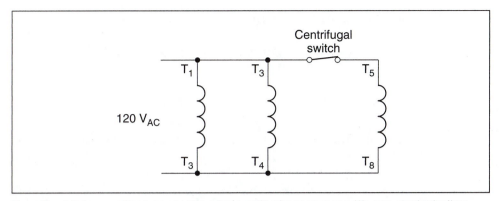

Figure Exp. 9-7 Low-voltage connection for a split-phase motor with one start winding.

3. Using an ohmmeter, measure the resistance of the windings. Note that the motor may have one start winding or two start windings. If the motor is a capacitor-start motor, it may be necessary to disconnect one side of the start lead from the capacitor to conduct this test.

   T1 to T2 _____ Ω   T3 to T4 _____ Ω

   T5 to T6(8) _____ Ω T7 to T8 _____ Ω

4. Compare the resistance of the run windings to each other. They should be relatively close. The start winding or windings should have a higher resistance than the run windings. If the ohmmeter indicates infinity between any of the test points, the winding is open.

5. Using a megohmmeter, measure the resistance of each set of windings to the case of the motor. Connect one lead of the megohmmeter to the motor frame and the other lead to each of the windings. Note: If the motor is a capacitor-start motor, it may be necessary to disconnect one end of the start winding from the capacitor to conduct this test.

   T1, T2 to GND _____ Ω   T3, T4 to GND _____ Ω

   T5, T6(8) to GND _____ Ω T7, T8 to GND _____ Ω

6. Referring to the motor nameplate, connect the motor for low-voltage operation. Ask the instructor to check the connection before applying power to the motor.

7. Connect the motor to a 120-volt AC power line.

8. Turn on the power and observe the direction of rotation.

9. Measure the current of the motor with an ammeter.
   _____ amps

10. Turn off the power.

11 Change the position of T5 and T6 (8). This should reverse the direction of rotation.

12. Turn on the power and observe the direction of rotation. Did the motor rotate in the opposite direction?
    _____

13. Turn off the power.

14. Using the motor nameplate, reconnect the motor for high-voltage operation. Ask the instructor to check the connection before connecting the motor to power.

15. Connect the motor to a 240-volt AC power line.

16. Turn on the power and observe the direction of rotation. Measure the motor current with an ammeter.
    _____ amps.

17. Is the current more than or less than the value recorded in step 9?
    _____

18. Turn off the power.

19. Reverse the connection of the start winding leads.

20. Turn on the power and observe the direction of rotation. Did the motor change direction?
    _____

21. Turn off the power.

22. Disconnect the motor and return all components to their proper location.

Appendix A
Greek Alphabet

Name of Letter	Upper Case	Lower Case	Designates
Alpha	A	α	Angles, coefficients, attenuation constant, absorption factor, area
Beta	B	β	Angles, coefficients, phase constant
Gamma	Γ	γ	Specific quantity, angles, electrical conductivity, propagation constant, complex propagation constant (cap.)
Delta	Δ	δ	Density, angles, increment or decrement, determinant, permittivity, change of quantity
Epsilon	E	ε	Dielectric constant, permittivity, base of natural logarithms, electrical intensity
Zeta	Z	ζ	Coordinates, coefficients, impedance
Eta	H	η	Hysteresis coefficient, efficiency, intrinsic impedance, surface charge density, coordinates
Theta	Θ	θ	Phase angle, time constant, reluctance
Iota	I	ι	Unity vector
Kappa	K	κ	Dielectric constant, coefficient of coupling, susceptibility
Lambda	Λ	λ	Wavelength, attenuation constant permeance (cap.)
Mu	M	μ	Permeability, prefix indicating micro, amplification factor
Nu	N	ν	Change of quantity, reluctivity, frequency
Xi	Ξ	ξ	Coordinates
Omicron	O	o	—
Pi	Π	π	3.1416 (circumference divided by diameter)
Rho	P	ρ	Resistivity, volume charge density, coordinates

Name of Letter	Upper Case	Lower Case	Designates
Sigma	Σ	σ	Surface charge density, complex propagation constant, electrical conductivity, leakage coefficient, sign of summation (cap.)
Tau	T	τ	Time constant, time phase displacement, volume resistivity, transmission factor, density
Upsilon	Y	υ	—
Phi	Φ	φ	Angles, magnetic flux, scalar potential
Chi	X	χ	Electric susceptibility, angles
Psi	Ψ	ψ	Dielectric flux, phase difference, angles
Omega	Ω	ω	Resistance in ohms (cap.), angular velocity (2πf), solid angles

Appendix B

Metals

Metal	Symbol	Specific Gravity	Melt Point °C	Melt Point °F	Elec. Cond. % Copper	Lb/in³
Aluminum	Al	2.710	660	1220	64.9	0.0978
Antimony	Sb	6.620	630	1167	4.42	0.2390
Arsenic	As	5.730	1280	2336	4.90	0.2070
Beryllium	Be	1.830	271	512	9.32	0.0660
Bismuth	Bi	9.800	900	1652	1.50	0.3540
Brass (70–30)		8.510	1000	1382	28.0	0.3070
Bronze (5% Sn)		8.870	321	610	18.0	0.3200
Cadmium	Cd	8.650	850	1562	22.7	0.3120
Calcium	Ca	1.550	1495	2723	50.1	0.0560
Cobalt	Co	8.900	1083	1981	17.8	0.3210
Copper	Cu	8.900	1063	1945	100.	0.3210
Gold	Au	19.30	3500	6332	71.2	0.6970
Graphite		2.250	156	311	0.001	0.0812
Indium	In	7.300	2450	4442	20.6	0.2640
Iridium	Ir	22.40	1400 to 1500	2552 to 2732	32.5	0.8090
Iron	Fe	7.200	1600 to 1500	2912 to 2732	17.6	0.2600
Malleable		7.200	1600	2912	10.0	0.2600
Wrought		7.700	327	621	10.0	0.2780
Lead	Pb	11.40	651	1204	8.35	0.4120
Magnesium	Mg	1.740	1245	2273	38.7	0.0628
Manganese	Mn	7.200	-38.9	-37.7	0.90	0.2600
Mercury	Hg	13.65	2620	4748	1.80	0.4930
Molybdenum	Mo	10.20	1300	2372	36.1	0.3680
Monel (63–37)		8.870	1452	2646	3.00	0.3200
Nickel	Ni	8.90			25.0	0.3210

Metal	Symbol	Specific gravity	Melt Point		Elec. Cond. % copper	Lb/in³
			°C	°F		
Phosphorus	P	1.82	44.1	111.4	10^{-17}	0.0657
Platinum	Pt	21.46	1773	3221	17.5	0.0657
Potassium	K	0.860	62.3	144.1	28.0	0.0310
Selenium	Se	4.81	220	428	14.4	0.1740
Silicon	Si	2.40	1420	2588	10^{-5}	0.0866
Silver	Ag	10.50	960	1760	106	0.3790
Steel (carbon)		7.84	1330 to 1380	2436 to 2516	10.0	0.2830
Stainless						
(18–8)		7.92	1500	2732	2.50	0.2860
(13–Cr)		7.78	1520	2768	3.50	0.2810
(18–Cr)		7.73	1500	2732	3.00	0.2790
Tantalum	Ta	16.6	2900	5414	13.9	0.5990
Tellurium	Te	6.20	450	846	10^{-5}	0.2240
Thorium	Th	11.7	1845	3353	9.10	0.4220
Tin	Sn	7.30	232	449	15.0	0.2640
Titanium	Ti	4.50	1800	3272	2.10	0.1620
Tungsten	W	19.3	3410	6170	31.5	0.6970
Uranium	U	18.7	1130	2066	2.80	0.6750
Vanadium	V	5.96	1710	3110	6.63	0.2150
Zinc	Zn	7.14	419	786	29.1	0.2580
Zirconium	Zr	6.40	1700	3092	4.20	0.2310

Appendix C

Full-Load Current Direct Current Motors

hp	90 Volts	120 Volts	180 Volts	240 Volts	500 Volts	550 Volts
¼	4.0	3.1	2.0	1.6	X	X
⅓	5.2	4.1	2.6	2.0	X	X
½	6.8	5.4	3.4	2.7	X	X
¾	9.6	7.6	4.8	3.8	X	X
1	12.2	9.5	6.1	4.7	X	X
1½	X	13.2	8.3	6.6	X	X
2	X	17	10.8	8.5	X	X
3	X	25	16	12.2	X	X
5	X	40	27	20	X	X
7½	X	58	X	29	13.6	12.2
10	X	76	X	38	18	16
15	X	X	X	55	27	24
20	X	X	X	72	34	31
25	X	X	X	89	43	38
30	X	X	X	106	51	46
40	X	X	X	140	67	61
50	X	X	X	173	83	75
60	X	X	X	206	99	90
75	X	X	X	255	123	111
100	X	X	X	341	164	148
125	X	X	X	425	205	185
150	X	X	X	506	246	222
200	X	X	X	675	330	294

Appendix B

Full-Load Current Alternating Current Single-Phase Motors

hp	115 Volts	200 Volts	208 Volts	230 Volts
⅙	4.4	2.5	2.4	2.2
¼	5.8	3.3	3.2	2.9
⅓	7.2	4.1	4.0	3.6
½	9.8	5.6	5.4	4.9
¾	13.8	7.9	7.6	6.9
1	16	9.2	8.8	8
1½	20	11.5	11	10
2	24	13.8	13.2	12
3	34	19.6	18.7	17
5	56	32.2	30.8	28
7½	80	46	44	40
10	100	57.5	55	50

Appendix E

Full-Load Current Alternating Current Three-Phase Motors

	Squirrel-Cage and Wound-Rotor Induction Type							Synchronous Type Unity Power Factor			
hp	115 Volts	200 Volts	208 Volts	230 Volts	460 Volts	575 Volts	2300 Volts	230 Volts	460 Volts	575 Volts	2300 Volts
½	4.4	2.5	2.4	2.2	1.1	0.9					
¾	6.4	3.7	3.5	3.2	1.6	1.3					
1	8.4	4.8	4.6	4.2	2.1	1.7					
1½	12	6.9	6.6	6	3	2.4					
2	13.6	7.8	7.5	6.8	3.4	2.7					
3		11	10.6	9.6	4.8	3.9					
5		17.5	16.7	15.2	7.6	6.1					
7½		25.3	24.2	22	11	9					
10		32.2	30.8	28	14	11					
15		48.3	48.2	42	21	17					
20		62.1	59.4	54	27	22					
25		78.2	74.8	68	34	27		53	26	21	
30		92	88	80	40	32		63	32	26	
40		120	114	104	52	41		83	41	33	
50		150	143	130	65	52		104	52	42	
60		177	169	154	77	61	16	123	61	49	12
75		221	211	192	96	77	20	155	78	62	15
100		285	273	248	124	99	26	202	101	81	20

For synchronous motors with 90% power factor, multiply above listed current by 1.1.
For synchronous motors with 80% power factor, multiply above listed current by 1.25.

hp	Squirrel-Cage and Wound Rotor Induction Type							Synchronous Type Unity Power Factor			
	115 Volts	200 Volts	208 Volts	230 Volts	460 Volts	575 Volts	2300 Volts	230 Volts	460 Volts	575 Volts	2300 Volts
125		359	343	312	156	125	31	253	126	101	25
150		414	396	360	180	144	37	302	151	121	30
200		552	528	480	240	192	49	400	201	161	40
250					302	242	60				
300					361	289	72				
350					414	336	83				
400					477	382	95				
450					515	412	103				
500					590	472	118				

For synchronous motors with 90% power factor, multiply above listed current by 1.1.
For synchronous motors with 80% power factor, multiply above listed current by 1.25.

Appendix F

NEMA Design Codes for Three-Phase Squirrel-Cage Motors

The National Electrical Manufacturers Association (NEMA) has established design codes for three-phase squirrel-cage induction motors. The codes include B, C, D, and E. Each of these design codes denotes motors that have different characteristics. The *National Electrical Code*®, beginning in 1996, uses these design codes to determine the fuse/circuit breaker size for motor circuits. These codes are also used in determining the locked-rotor (maximum inrush) current during starting.

Design Code B

Motors with a NEMA design code B are the most common motors found in industry. NEMA design code B motors are designed to have a 2% to 5% slip when loaded. Starting torque is approximately 150% of the full-load running torque. Locked-rotor current at full voltage will be approximately 600% to 725% of the full-load running current. Their rotor characteristics are most similar to a motor with code B. Most motors in industry that do not have a NEMA design code are generally considered to be design code B.

Design Code C

Design code C motors have approximately the same slip as design code B motors, 2% to 5%. Design code C motors are intended for use with loads that may require a greater starting torque than with generally encountered, such as is compressors. Design code C motors develop a starting torque of approximately 225% of the full-load running torque. Locked-rotor current will be approximately 600% to 650% of full-load running current.

Design Code D

Motors with a design code D are for loads that are difficult to start such as compressors, elevators, and high inertia loads such as flywheels. Design code D motors develop a starting torque that is approximately 275% of the full-load running torque. Their inrush current is relatively low when compared to other motors, 525% to 625% of full-load current. Design code D motors have relatively poor speed regulation. Their percent slip is from 5% to 13%. These motors are often referred to as high-slip motors. These motors have excellent starting torque and relatively low starting current, but poor speed regulation.

Design Code E

Design code E motors are considered to be high efficiency motors. They obtain a higher efficiency by reducing I^2R losses, mechanical and windage losses, and core losses. Core losses include eddy current and hysteresis losses. The starting torque and percent slip is determined by the type of rotor. As a general rule, design code E motors can have very high starting currents. Some design code E motors have starting currents of over 1500% of the full-load running current.

Answers to Practice Problems

UNIT 3 INDUCTANCE IN ALTERNATING-CURRENT CIRCUITS

Inductance (H)	Frequency (Hz)	Induct. Rec. (Ω)
1.2	60	452.4
0.085	400	213.628
0.75	1000	4712.389
0.65	600	2450.442
3.6	30	678.584
2.65	25	411.459
0.5	60	188.5
0.85	1200	6408.849
1.6	20	201.062
0.45	400	1130.973
4.8	80	2412.743
0.0065	1000	40.841

UNIT 4 SINGLE-PHASE ISOLATION TRANSFORMERS

1. E_P 120 E_s 24
    I_P 1.6 I_s 8
    N_P 300 N_s 60
    Ratio 5:1 $Z = 3 \ \Omega$

2. E_P 240 E_s 320
    I_P 0.853 I_s 0.643
    N_P 210 N_s 280
    Ratio 1:1.333 $Z = 500 \ \Omega$

3. E_p 64 E_s 160   4. E_p 48 E_s 240
     I_p 33.333 I_s 13.333 I_p 3.333 I_s 0.667
     N_P 32 N_s 80 N_P 220 N_s 1,100
     Ratio 1:2.5 $Z = 12 \ \Omega$ Ratio 1:5 $Z = 360 \ \Omega$

5. E_p 35.848 E_s 182   6. E_p 480 E_s 916.346
     I_p 16.5 I_s 3.25 I_p 1.458 I_s 0.764
     N_P 87 N_s 450 N_P 275 N_s 525
     Ratio 1:5.077 $Z = 56 \ \Omega$ Ratio 1:1.909 $Z = 1.2 \ k\Omega$

7. $E_p = 208$ $E_{s1} = 320$ $E_{s2} = 120$ $E_{s3} = 24$
     $I_p = 11.93$ $I_{s1} = 0.0267$ $I_{s2} = 20$ $I_{s3} = 3$
     $N_p = 800$ $N_{s1} = 1231$ $N_{s2} = 462$ $N_{s3} = 92$
          Ratio1 1:1.54   Ratio2 1.73:1   Ratio3 1:8.67
          $R_1 = 12 \ k\Omega$ $R_2 = 6 \ \Omega$ $R_3 = 8 \ \Omega$

8. $E_p = 277$ $E_{s1} = 480$ $E_{s2} = 208$ $E_{s3} = 120$
     $I_p = 8.93$ $I_{s1} = 2.4$ $I_{s2} = 3.47$ $I_{s3} = 5$
     $N_p = 350$ $N_{s1} = 606$ $N_{s2} = 263$ $N_{s3} = 152$
          Ratio1 1:1.73   Ratio2 1.33:1   Ratio3 2.31:1
          $R_1 = 200 \ \Omega$ $R_2 = 60 \ \Omega$ $R_3 = 24 \ \Omega$

UNIT 5 AUTOTRANSFORMERS

1.   A-B 89.1   A-C 148.5   A-D 252.5   A-E 297.1   B-C 59.4   B-D 163.4
     B-E 208   C-D 104   C-E 148.6   D-E 44.6

2. $I_p = 3.1$ amperes; $I_s = 4.3$ amperes

3.   280/60 = 4.67:1

4.   168 volts (480/400 = 1.2 volts per turn) (1.2 × 140 turns = 168 volts)

5.   1.35 amperes (325/240 = 1.35)

UNIT 7 THREE-PHASE CIRCUITS

1. $EP_{(A)}$ 138.57 $EP_{(L)}$ 240
     $IP_{(A)}$ 34.64 $IP_{(L)}$ 20
     $E_{L(A)}$ 240 $E_{L(L)}$ 240
     $I_{L(A)}$ 34.64 $I_{L(L)}$ 34.64
     P 14,399.16 $Z_{(Phase)}$ 12 Ω

2. $EP_{(A)}$ 4,160 $EP_{(L)}$ 2401.85
   $IP_{(A)}$ 23.11 $IP_{(L)}$ 40.03
   $E_{L(A)}$ 4,160 $E_{L(L)}$ 4,160
   $I_{L(A)}$ 40.03 $I_{L(L)}$ 40.03
   P 288,420.95 $Z_{(Phase)}$ 60 Ω

3. $EP_{(A)}$ 323.33 $E_{P(L1)}$ 323.33 $E_{P(L2)}$ 560
   $IP_{(A)}$ 185.91 $I_{P(L1)}$ 64.67 $I_{P(L2)}$ 70
   $E_{L(A)}$ 560 $E_{P(L1)}$ 560 $E_{L(L2)}$ 560
   $I_{L(A)}$ 185.91 $I_{L(L1)}$ 64.67 $I_{L(L2)}$ 121.24
   P 180,317.83 $Z_{(Phase)}$ 5 Ω $Z_{(Phase)}$ 8 Ω

4. $EP_{(A)}$ 277.14 $E_{P(L1)}$ 277.14 $E_{P_{(L2)}}$ 480 $E_{P(L3)}$ 277.14
   $IP_{(A)}$ 33.49 $I_{P(L1)}$ 23.1 $I_{P(L2)}$ 30 $I_{P(L3)}$ 27.71
   $E_{L(A)}$ 480 $E_{L(L1)}$ 480 $E_{L(L2)}$ 480 $E_{L(L3)}$ 480
   $I_{L(A)}$ 33.49 $I_{L(L1)}$ 23.1 $I_{L(L2)}$ 51.96 $I_{L(L3)}$ 27.71
   VA 27,843.39 $R_{(Phase)}$ 12 Ω $X_{L(Phase)}$ 16 Ω $X_{C(Phase)}$ 10 Ω
                       P 19,204.42 $VARS_L$ 43,197.47 $VARS_C$ 23,037

UNIT 8 THREE-PHASE TRANSFORMERS

1. E_P 2401.8 E_P 440 E_P 254.04
   I_P 7.67 I_P 41.9 I_P 72.58
   E_L 4160 E_L 440 E_L 440
   I_L 7.67 I_L 72.58 I_L 72.58
   Ratio 5.46:1   Z = 3.5 Ω

2. E_P 4157.04 E_P 240 E_P 138.57
   I_P 1.15 I_P 20 I_P 34.64
   E_L 7200 E_L 240 E_L 240
   I_L 1.15 I_L 34.64 I_L 34.64
   Ratio 17.32:1   Z = 4 Ω

3. E_P 13,800 E_P 277 E_P 480
   I_P 6.68 I_P 332.54 I_P 192
   E_L 13,800 E_L 480 E_L 480
   I_L 11.57 I_L 332.54 I_L 332.54
   Ratio 49.76:1   Z = 2.5 Ω

4. E_P 23,000 E_P 120 E_P 208
   I_P 0.626 I_P 120.08 I_P 69.33
   E_L 23,000 E_L 208 E_L 208
   I_L 1.08 I_L 120.08 I_L 120.08
   Ratio 191.66:1 Z = 3 Ω

GLOSSARY

Alternator a machine used to generate alternating current by rotating conductors through a magnetic field.

Amortisseur winding a squirrel-cage winding on the rotor of a synchronous motor used for starting purposes only.

Ampacity the maximum current-carrying capacity of a wire or device.

Ampere-turns determined by multiplying the number of turns of wire by the current flow.

Armature the rotating member of a motor or generator. The armature generally contains windings and a commutator.

Armature reaction the twisting or bending of the main magnetic field of a motor or generator. Armature reaction is proportional to armature current.

Askarel special type of dielectric oil used to cool electrical equipment such as transformers. Some types of Askarel contain polychlorinated biphenyl (PCB).

Back-EMF see Counter-EMF.

Bifilar has two windings that are wound together.

Brushes sliding contacts, generally made of carbon, used to provide connection to rotating parts of machines.

Brushless DC motors have armatures or rotors (rotating members) containing permanent magnets and are surrounded by fixed stator windings.

Brushless exciter an assembly where the armature, rectifier and rotor winding are connected to the the main rotor shaft without brushes or slip rings.

Centrifugal switch disconnects the start windings from the circuit in motors that are not hermetically sealed.

Commutator strips or bars of metal insulated from each other and arranged around one end of an armature. They provide connection between the armature windings and the brushes. The commutator is used to ensure proper direction of current flow through the armature windings.

Compensating windings the winding embedded in the main field poles of a DC machine. The compensating winding is used to help overcome armature reaction.

Compound generators contain both series and shunt fields.

Compounding the relationship of the strengths of the two fields in a generator.

Compound motor uses both a series field and a shunt field.

Conductive compensation compensating winding of a universal motor connected in series with the armature.

Consequent pole squirrel-cage motors permit the synchronous speed to be changed by changing the number of stator poles.

Consequent pole motor a multispeed single-phase motor that is used where high running torque must be maintained at different speeds.

Constant speed motors see Shunt motor.

Control transformer used to reduce the line voltage to the value needed to operate control circuits.

Cooling lessening the amount of heat generated by an alternator, usually by providing passage of air to the alternator.

Counter-EMF (CEMF) the voltage induced in the armature of a DC motor that opposes the applied voltage and limits armature current.

Countertorque the magnetic force developed in the armature of a generator that makes the shaft hard to turn. Countertorque is proportional to armature current and is a measure of the electrical energy produced by the generator.

Cumulative compound shunt and series fields connected so that they aid each other in the production of magnetism when current flows through them.

Cumulative-compounded motors see Cumulative compound.

Current relay a relay that is operated by a predetermined amount of current flow. Current relays are often used as one type of starting relay for air conditioning and refrigeration equipment.

Decimal impedance represents the percent impedance in decimal form.

Deenergized The condition of a line or transformer through which no current is flowing.

Delta connection a circuit formed by connecting three electrical devices in series to form a closed loop. It is used most often in three-phase connections.

Delta-wye transformer has its primary winding connected in a delta and its secondary connected in a wye.

Diametrical connection is a three-phase to six-phase connection requiring only one low-voltage winding on each transformer.

Dielectric oil provides electrical insulation between the windings and the case. It also helps to provide cooling and to prevent the formation of moisture.

Differential compound shunt and series fields connected so that they oppose each other in the production of magnetism when current flows through them.

Differential-compounded motors see Differential compound. Generally, a connection to be avoided.

Differential selsyn is used to produce the algebraic sum of the rotation of two other selsyn units.

Direction of rotation the direction of a motor's rotating magnetic field can be tested by momentarily applying power to the motor before it is coupled to the load.

Distribution transformer a very common type of isolation transformer used to supply power to most homes and many businesses.

Double-delta a three- to six-phase connection with transformers using two secondary windings that provide equal voltage and form two delta connections.

Double-wye a three- to six-phase connection with transformers using two secondary windings that provide equal voltage with the windings connected in two different wyes.

Dynamic braking is a type of braking action that causes the armature to decrease in speed.

Excitation current the direct current used to produce electromagnetism in the fields of a DC motor or generator, or in the rotor of an alternator or synchronous motor.

Field discharge resistor prevents induced voltage from becoming excessive.

Field excitation current the DC current that flows through the shunt field winding.

Field loss relay (FLR) a current relay connected in series with the shunt field of a direct current motor. The relay causes power to be disconnected from the armature in the event that field current should drop below a certain level.

Flat compounding is accomplished by permitting the series field to increase the output voltage by an amount that is equal to the losses of the generator.

Flux density measures magnetic strength in lines per square inch.

Frequency the number of complete cycles of AC voltage that occur in 1 s.

Frogleg wound armatures used in machines designed for use with moderate current and moderate voltage.

Gauss represents a magnetic force of one maxwell per square centimeter.

Generator a device used to convert mechanical energy into electrical energy.

Gilbert measurement used to represent magnetomotive force.

Growler a device constructed by wrapping a coil of wire around a set of v-shaped laminated cores, that is often used to test an armature.

Harmonics voltages or currents that operate at a frequency that is a multiple of the fundamental power frequency.

Henry (H) measures inductance.

High leg a line of voltage that is higher than the voltage between the neutral and either of the other two conductors.

Holtz motor uses a rotor that is cut with six slots that form six salient poles.

Hydrogen a gas that is used to cool large-capacity alternators because it has the ability to absorb and remove heat in an alternator.

Impedance the total current-limiting effect of the inductor, a combination of the inductive reactance and resistance.

Inductive compensation compensating winding of a universal motor connected by shorting its leads together.

Inductive reactance (XL) reactance that is caused by inductance.

Interconnected-wye see Zig-zag.

Interpoles small pole pieces inserted between the main field poles in correcting armature reaction.

Interrupting rating the ability of the fuse or circuit breaker to open under the maximum fault current.

Isolation transformer a transformer whose secondary winding is electrically isolated from its primary winding.

Lap-wound armatures used in machines designed for low voltage and high current.

Left-hand generator rule used to determine the relationship of the motion of the conductor in a magnetic field to the direction of the induced current.

Line voltage the voltage measured between the lines; also known as line-to-line voltage.

Long shunt the more common way of connecting the series and shunt fields in compound generators, by connecting the shunt field in parallel with both the armature and series field.

Magnetic domains are tiny magnetic structures contained in magnetic materials which can be affected by outside sources of magnetism.

Magneto a DC generator that uses permanent magnets as its field.

Main transformer in a T-connection, contains a center or 50% tap for both the primary and secondary windings.

Maxwell one magnetic line of force.

Motor a device used to convert electrical energy into mechanical energy.

Multispeed motors motors that can be operated at more than one speed.

Nameplate a type of current rating used to determine the overload size for a motor.

Neutral conductor a device or material that permits neutral current to flow through it easily.

Neutral plane position where the brushes should be set in a universal motor.

Open-delta transformer connections made with two transformers. Often used when the amount of three-phase power needed is not excessive, such as in a small business.

One-line diagram illustrates a delta-wye connection generally to show the main power distribution system of a large industrial plant.

Orange wire a tag used to identify the point where a high-leg line enters an enclosure with the neutral conductor.

Overcompounding characterized by the fact that the output voltage at full load will be greater than the output voltage at no load.

Parallel alternators are connected in parallel with each other.

Percent impedance an empirical value that can be used to predict transformer performance.

Percent slip measures the speed performance of an induction motor by subtracting the synchronous speed from the speed of the rotor.

Permanent magnet motors contain a wound armature and brushes with permanent magnets as pole pieces.

Permeability a measure of a material's willingness to become magnetized.

Phase rotation the direction of magnetic field rotation.

Phase rotation meter a six-terminal lead meter that compares the phase rotation of two different three-phase connections.

Phase voltage the voltage measured across a single winding.

Pole pieces provide the magnetic field necessary for the operation of a machine.

Primary winding winding of a transformer to which power is applied.

Printed circuit motor see ServoDisc® motor.

Pull-out torque the amount of torque necessary to pull a rotor out of sync with the rotating magnetic field.

Pulse-width modulation means of regulating the amount of voltage supplied to the armature.

Reactance property of the inductor that limits the flow of current through the circuit.

Regenerative braking see Dynamic braking.

Reluctance resistance to magnetism.

Repulsion motor an induction motor that works on the principle that like magnetic poles repel each other.

Residual magnetism the amount of magnetism left in a material after the magnetizing force has stopped.

Retentivity a material's ability to retain magnetism.

Revolving armature alternator that uses an armature with loops of wire connected to slip rings.

Revolving field alternator that uses a stationary armature (stator) and a rotating magnetic field.

Rotating magnetic field the principle of operation for all three-phase motors.

Rotating transformers AC induction motors.

Rotor the rotating member of an alternating current machine.

Rotor frequency of the current decreases as the rotor approaches synchronous speed and the rotor bars become less inductive.

Run winding made of relatively large wire placed near the bottom of the stator core of a split-phase motor.

Saturation occurs when all the molecules of the magnetic material are lined up.

Scott connection used to convert three-phase power into two-phase power using two single-phase transformers.

Secondary winding winding of a transformer to which the load is connected.

Selsyn motors are used to provide position control and angular feedback information in industrial applications.

Series field diverter a low-value variable resistor; also known as the series field shunt rheostat.

Series field winding a winding of large wire and few turns designed to be connected in series with the armature of a DC machine.

Series generator contains only a series field connected in series with the armature.

Series motor has only a series field connected in series with the armature.

ServoDisc® motor uses permanent magnets to provide a constant magnetic field. The armature design is completely different from conventional servomotors.

Servomotors permanent magnet motors with small, lightweight armatures that contain very little inertia.

Shaded-pole induction motor an AC induction motor that develops a rotating magnetic field by shading part of the stator windings with a shading coil.

Shading coil a large copper wire or band connected around part of a magnetic pole piece to oppose a change of magnetic flux.

Short shunt one way of connecting the series and shunt fields in compound generators, by connecting the short shunt in parallel with the armature.

Shunt field winding a coil wound with small wire and having many turns designed to be connected in parallel with the armature of a DC machine.

Shunt generators contain only a shunt field winding connected in parallel with the armature.

Shunt motor has the shunt field connected in parallel with the armature.

Single-phase loads a circuit energized by a single alternating current.

Single-phasing when one of the three lines supplying power to a three-phase motor opens, the motor will be connected to single-phase power.

Slip rings circular bands of metal placed on the rotating part of a machine. Carbon brushes riding in contact with the slip rings provide connection to the external circuit.

Speed regulation the amount by which the speed decreases as load is added.

Split-phase motor a type of single-phase motor that uses resistance or capacitance to cause a shift in the phase of the current in the run winding and the current in the start winding. The three primary types of split-phase motors are resistance-start induction-run, capacitor-start induction-run, and capacitor-start capacitor-run motors.

Squirrel-cage motor uses a rotor made by connecting bars attached to two end rings.

Star connection another name for wye connection.

Start winding made of small wire placed near the top of the stator core of a split-phase motor.

Stator the stationary winding of an AC motor.

Stepping motors are devices that convert electrical impulses into mechanical movement.

Step-down transformer a transformer that produces a lower voltage at its secondary than is applied to its primary.

Step-up transformer a transformer that produces a higher voltage at its secondary than is applied to its primary.

Synchronous condenser refers to a motor operated at no load and used for power factor correction only.

Synchronous motor operates on the principle of a rotating magnetic field. After DC excitation current is applied to the rotor, the motor will operate at the speed of the rotating magnetic field.

Synchronous speed the speed of the rotating magnetic field of an AC induction motor.

Synchroscope an instrument used to determine the phase angle difference between the voltages of two alternators.

T-connection uses two transformers to supply three-phase power. One transformer is generally referred to as the the main transformer and the other the teaser.

Tagging see Orange wire.

Three-phase bank formed by three single-phase transformers with their primary and secondary windings connected in a wye or delta connection.

Three-phase VARs computed in a similar fashion to three-phase watts, except that voltage and current values of pure reactive load are used.

Three-phase watts computed by multiplying the apparent power by the power factor.

Torque the turning force developed by a motor.

Transformer an electrical device that changes one value of AC voltage into another value of AC voltage.

Triplens a particular set of harmonics with a zero sequence. They are the odd multiples of the third harmonic (3rd, 9th, 21st, etc.).

Two-phase system produced by an alternator with two sets of coils wound 90° apart.

Turns ratio the ratio of the number of primary turns of wire as compared to the number of secondary turns.

Undercompounded characterized by the fact that the output voltage will be less at full load than it is at no load.

Universal motor similar to a DC motor, containing a wound armature, brushes, and a compensating winding; often referred to as an AC series motor.

Volts-per-turn ratio the ratio of the number of turns of wire in the secondary windings to those in the primary.

Warren motor constructed with laminated stator and a single coil.

Wave wound armatures used in machines designed for high voltage and low current.

Weber (Wb) in magnetic measurement, 100,000,000 lines of flux.

Wound-rotor induction motors with high starting torque and low starting current.

Wye connection a connection of three components made in such a manner that one end of each component is connected. This connection is generally used to connect devices to a three-phase power system.

Wye-delta transformer with its primary winding connected in a wye and its secondary winding connected in a delta.

Zig-zag transformer primarily used for grounding purposes to establish a neutral point for the grounding of fault currents.

INDEX